# 52 Structure and Bonding

# Structures versus Special Properties

With Contributions by
L. Banci A. Bencini C. Benelli
R. Bohra J.-M. Dance D. Gatteschi
V. K. Jain R. C. Mehrotra A. Tressaud
R. G. Woolley C. Zanchini

With 90 Figures and 19 Tables

Springer-Verlag
Berlin Heidelberg New York 1982

ISBN 3-540-11781-4 Springer-Verlag Berlin Heidelberg New York
ISBN 0-387-11781-4 Springer Verlag New York Heidelberg Berlin

Library of Congress Catalog Card Number 67-11280



Typesetting and printing: Schwetzinger Verlagsdruckerei. Bookbinding: J. Schäffer, Grünstadt.
2152/3140-543210

# Table of Contents

# Natural Optical Activity and the Molecular Hypothesis

**R. Guy Woolley***

Cavendish Laboratory, Madingley Road, Cambridge, CB3 OHE, UK**

An account of the quantum theory of the molecular hypothesis for chemical substances is presented, and is used as a basis for a critical discussion of the theory of natural optical activity. Atoms and molecules are characterized as composite elementary excitations (or quasi-particles) of the macroscopic quantum-mechanical system we call matter, and the spontaneously broken space inversion symmetry revealed by the existence of optical isomerism is studied in this context. Special attention is paid to the representation of the space-inversion operator $\hat{P}$. Finally a variety of microscopic quantum theories of natural optical activity are critically reviewed.



* SRC Advanced Fellow
** Present address: Trent Polytechnic, Burton Street, Nottingham NG 1 4 BU, United Kingdom

# 1 Introduction

Natural optical activity has always held a special fascination for chemists not least because of the central role of chiral substances in the chemistry of life. In the last few years there seems to have been a renewed interest in theoretical aspects of optical activity[1–16)]. Perhaps this is due to a greater awareness of the possibilities and relevance of quantum mechanics in chemistry. As is well-known the elucidation of the phenomenon of optical activity was a decisive factor in the elaboration of the classical molecular structure hypothesis, and greatly strengthened the acceptance of the molecular nature of chemical substances. The classical theory of optical activity was given in an essentially complete form by Le Bel[17)] and van't Hoff[18)], following earlier ideas of Pasteur[19)]. I should like to reconsider the conventional discussion of chiral substances. My aim is to examine the relationship between the classical discussion and a modern quantum mechanical description: this involves both a clear understanding of the notion of a "molecule" and a carefully formulated quantum theory of the experiment. It will become evident that the Hilbert space structure of quantum mechanics leads to restrictions on the conclusions that one may draw.

The classical theory of optical activity was proposed at a time when the molecular hypothesis was still highly controversial, and it is noteworthy that even van't Hoff and Le Bel had entirely different conceptions of the molecular hypothesis. For van't Hoff[18)], molecules were *objects* in ordinary 3-dimensional space which we could not perceive simply because of the limitations inherent in our senses, and thus molecules could be taken to be endowed with properties identical to those of other material bodies. The geometrical properties of molecules were entirely real to him, and were meant to be part of a logical argument to give a proof of their physical reality. By contrast, Le Bel[17)], who admittedly made use of the same geometrical relationships, interpreted chemical formulae as a *coded or symbolic* representation of the properties of substances. In order to correlate these images with optical rotatory power, it was *not* necessary for him to interpret them as concrete images of microscopic material objects, as van't Hoff had done; rather, the crucial point focussed on by Le Bel was the correlation between the symmetries of molecules and the optical activity of substances, an early and striking example of an association between a group theoretical argument and a physical property. In order to explain the rotatory power of organic substances it is sufficient but not *necessary* to assume a tetrahedral arrangement of the four affinities of the carbon atom, for this is but a particular realization of the underlying fundamental symmetry group associated with the substance.

While van't Hoff's view has become widely accepted in chemistry (and other sciences based on a molecular conception of matter, for example, molecular biology), Le Bel's abstract conception is closer to a quantum mechanical description of chemical phenomena in which molecular structure is understood as a *powerful and illuminating metaphor*[1] that provides a route to the construction of the quantum states (wavefunctions) that actually describe chemical phenomena[1–3, 8)]. Although the difference between these interpretations of chemical formulae (in a sense "what chemistry is about") has not had much effect on the actual practice of experimental chemistry, the evolution of ideas

1 i.e. molecular structure is an idea that is not to be understood in its literal (classical) sense

in natural science is important since, in Heisenberg's words[20)], "science progresses not only because it helps us to explain newly discovered facts, but also because it teaches us over and over again what the word "understanding" may mean". It is possible that for example the structure/function paradox in molecular biology might be resolved if a slavish adherence to a view of atoms as objects of *classical* physics, as so many bricks or grains of sand, were given up[20)]. I hope to show, using the example of natural optical activity, that the quantum theory of matter implies a considerable departure from the conventional understanding of the notion "molecule".

One can characterize the theory of matter according to classical chemistry very simply: the fundamental postulate is that "atoms" are the "building-blocks" of matter, and molecules are built up using atoms like the letters of an alphabet. When we speak of a molecule in classical chemistry we mean a semi-rigid collection of atoms held together by chemical bonds. The "laws" that govern the relative dispositions of the atoms in ordinary 3-dimensional space are the classical valency rules which therefore provide the syntax of chemical structural formulae. In classical chemistry molecular structure is an intrinsic molecular property that is not to be derived from any more basic set of principles than these: one can therefore say that the molecular structure hypothesis is the central dogma of classical chemistry. Indeed, the majority of chemists would never give (have never given?) the idea a second thought, it being so deeply ingrained in our thinking. Nevertheless it is equally the case that if we believe in quantum mechanics, consistency requires us to think critically about these old ideas in chemistry. In order to avoid gross misunderstanding, let me caution the reader at once that I am not about to suggest that classical molecular structure simply be thrown out – that would be silly; however, I am unaware of any convincing reason why we should not *try* to think accurately about the ideas we use in chemistry in terms of modern, rather than 19th century physics, even if this leads us into difficult and unfamiliar analysis. The above quotation from Heisenberg[20)] reminds us that whatever the conclusions we arrive at in a given conceptual framework, they are liable to have a provisional character because of the refinements and modifications in our understanding of the physics of matter that continually occur. An alternative position involves an explicit rejection of the relevance of quantum mechanics to chemical enquiry, asserting in effect that a theory of chemistry can exist autonomously from physics. There is a long tradition of support for such a position (which has obvious attractions), for the classical "atoms and bonds" model of molecules provides a tremendously useful conceptual framework for rationalizing chemical reactions which, after all, are the stuff of chemistry. There are however many physical measurements, including optical activity, for which such an approach is too limited.

One can take an apparently more "fundamental" approach by making a simple extension of the classical theory: instead of identifying atoms as the building blocks of matter we postulate that *electrons and nuclei*[2] are the elementary constituents of atoms. One can then start a discussion of quantum chemistry with some such definition as: a molecule consists of a bound system of electrons and nuclei that interact according to the laws of

2 The italics here are intended to emphasize that while the structure and properties of electrons and nuclei are the proper concerns of elementary particle and nuclear physics, they can be disregarded for all ordinary chemical discussions for which electrons and nuclei can be characterized as the structurally stable particles or quanta of quantized fields carrying appropriate electric charge, mass and spin, cf. C. K. Jørgensen, 1980, Structure and Bonding, *43*

quantum mechanics[1–3]. Such a definition leads at once to the question; Is the "atoms and bonds" model always compatible with the "electrons and nuclei" model of a molecule? This is a non-trivial question, for whereas classical objects have intrinsic properties irrespective of measurements, in quantum theory we have to base our description on the notion of *states* that are linked inevitably to the kinds of experiments we do. A paradigmatic example of such a discussion is the theory of position and momentum measurements on an electron using a "$\gamma$-ray microscope"[2]. So our question ought to be reformulated in terms such as; can a "molecular structure" be associated consistently with all the observable quantum states of a molecule, as seems to be implied by the central dogma, or only with a particular subset of the experimentally accessible states? It is an obvious statement that the exact energy eigenstates of a molecule have no structural interpretation: it is almost equally obvious that such states are only useful for the interpretation of certain kinds of experiments that lie outwith the scope of *classical* chemistry, for example, high-resolution spectroscopy of molecular beams of substances with low molecular weights[1–3]. For the phenomena of classical chemistry one needs what may be called "molecular structure states"[1–3], and it is for this reason of course that most discussions of quantum chemistry appeal to the molecular structure hypothesis at the earliest opportunity. The conventional wisdom however gives no insight into *how* molecular structure states arise, and declares in effect (as in classical chemistry) that the structure hypothesis is required by the facts, and does not require discussion within microscopic theory.

I have struggled with this approach to a quantum theory of chemistry for some years[1–3, 8], but now believe that it is based on an unsatisfactory set of initial premises, due in part to an uncritical acceptance of the legacy of the classical molecular model of matter. This legacy has lead to a widely held conviction that molecules are "real" microscopic material objects (as in classical physics), and that ordinary matter can be made up by taking a sufficiently large number (e.g. Avogadro's Number) of such objects: a corollary of this view is that it is meaningful to discuss the properties of a single molecule in, say, intergalactic space as an example of an "isolated molecule". Such an approach leads to grave difficulties when one comes to consider chiral molecules[8]. This paradox is resolved by the observation that one must not adopt a dogmatic attitude towards the universal applicability of any given set of ideas in physical science (the atoms and bonds model of a molecule, quantum chemistry, or whatever) since the overriding aim of a physical theory of matter is to make sense of matter as it *actually is*, not to elaborate on how it *might* be. Theories based on building the universe out of elementary constituents, whether they be quarks or molecules (the classical "building-block" approach) can very easily fall into the second category.

The approach to be used here is, to be sure, well known in parts of theoretical physics, but is novel as far as chemistry is concerned. It is based on the view that macroscopic matter is to be described by a suitably generalized formulation of quantum mechanics, namely Quantum Field Theory: *the traditional postulate that matter is made up or composed of microscopic elementary constituents (in the classical building-block sense) is given up,* and instead the fundamental postulate of the quantum theory of matter is, to paraphrase Gertrude Stein, *Matter is Matter is Matter.* Then if our interest is chemistry we have of course to confront the obvious question as to how we may construct the particles we call "atoms" and "molecules" i.e. we must establish how the notions of *atom* and *molecule* emerge from quantum theory construed in a general and modern way as the theory of matter. This is the subject matter of the next section of the review

(Sect. 2), and must be disposed of before a theory of optical activity can be considered. It will appear that exactly analogous discussions apply to other areas of physics – nuclei, elementary particles and so on. I propose to reinterpret the classical "building-blocks" of matter as the "quasi-particles" or "elementary excitations" of the quantum (field) theory of matter.

It is surely agreed that the empirical fact of optical activity requires the state of an optically active medium to be lacking in space-inversion symmetry. The conclusion of van't Hoff[18], like Pasteur[19], was simply that this was because the macroscopic optically active medium is composed of a very large number of "elementary objects" (molecules) which must themselves carry the space-inversion dissymmetry. However, in view of the discussion in the previous paragraphs the question then arises as to how we are to understand van't Hoff's claim in terms consistent with the quantum theory. I shall argue in this paper that the quantum mechanics of the loss of space-inversion symmetry is most naturally given in terms of a quantum mechanical description of the macroscopic system as a whole based on Quantum Field Theory. I emphasize again, this does *not* mean that the idea of a molecule has to be abandoned, or is not useful. It does mean that one should understand that the quantum-mechanical meaning of the notion of "a single molecule" is as an "elementary excitation" (= "quasi-particle" or "structurally stable excitation") in the macroscopic quantum mechanical system we call matter (Sect. 2). With this understanding the macroscopic nature of the quantum system may no longer be of interest, and one can relate the results of experiments directly to the elementary excitations: the macroscopic nature of the quantum system *is* of interest in optical activity precisely because of the spontaneously broken space inversion symmetry revealed by the experiment[1–3]. I am not the only person to have given such a description of optical activity, see for example the footnote in Bouchiat[21], but since my previous brief references to optical activity in terms of spontaneously broken space-inversion symmetry seem to have caused consternation among some chemists[5, 7], it may be helpful to describe in some detail what it entails. In so doing I hope at the same time to clarify the quantum mechanical interpretation of the classical molecular structure hypothesis.

The molecular interpretation of the properties of chemical substances is generally correct when understood in terms of elementary excitations; it can be argued however that in the context of optical activity an exclusive focus on "chiral molecules"[5, 7] leads to an incomplete understanding of the phenomenon. Consider the conventional wisdom as to the meaning of molecular structure in quantum theory. This says that all that matters with respect to the "low-symmetry" of molecules is the relation between the timescale of the experiment and the energy separation of the eigenstates of the symmetry operators (rotation, space-inversion, permutation of identical particles) that commute with the molecular Hamiltonian $\hat{H}$, Eq. (2.14),[5, 22]. If, for example, the system is in a (non-stationary) state of mixed-parity (e.g. a resolved enantiomer,[5]) which has a certain lifetime, then only that non-stationary state can be observed in any experiment done on a shorter timescale than the lifetime of the non-stationary state. This is obviously so; depending on one's view of what constitutes a satisfactory theory of chemistry it either explains everything or nothing. It offers *no insight* into the dynamical factors involved in the *origin* of the non-stationary state and as such is no advance on van't Hoff's original hypothesis. It is a phenomenology in which the introduction of a discussion of timescales is necessary for the consistency of the description, but it is hardly a substitute for a quantum *theory* of molecular structure[1–3]. Even though "isolated molecule states" are seldom realized experimentally, one's intuition is that the true stationary states of the Hamiltonian for an isolated molecule (Sect. 2) should provide a valid reference point from which useful, physical nonstationary states ("molecular structure states") for chemical substances can be constructed by regular perturbation theory[1-3]. This intuition fails in the case of natural optical activity which is a genuine spontaneously broken symmetry phenomenon[1, 8].

If we could produce an exact solution to quantum field theory one would be able to answer *ab initio* all questions about the nature of the wavefunctions of optically active isomers, and the representations of operators that they induce. As it is, *ab initio* solutions are out of the question and I shall follow a heuristic approach to describe the quantum mechanical structure that appears to be required for a description of optical activity. This semi-empirical structure is suggestive of the kind of microscopic theory that is required. I shall take the view that in quantum mechanics (T = 0) the description of optical activity requires a Hamiltonian possessing a doubly degenerate *ground state,* such that the components of the doublet (a) belong to incoherent Hilbert spaces and (b) are interconverted by the space-inversion, or parity, operator $\hat{P}$. Racemization is a phenomenon I associate with quantum statistical mechanics (T $\neq$ 0) and requires an understanding of long-lived metastable states[23].

The next section (Sect. 2) is devoted to a lengthy discussion of the molecular hypothesis from the point of view of quantum field theory, and this provides the basis for the subsequent discussion of optical activity. Having used linear response theory to establish the equations for optical activity (Sect. 3), we pause to discuss the properties of the wavefunctions of optically active isomers in relation to the space inversion operator (Sect. 4), before indicating how the general optical activity equations can be related to the usual Rosenfeld equation for the optical rotation in a chiral molecule. Finally (Sect. 5), there are critical remarks about what can currently be said in the microscopic quantum-mechanical theory of optical activity based on some approximate models of the field theory.

## 2 Quantum Field Theory and the Molecular Hypothesis for Chemical Substances

In Sect. 1 I remarked that the quantum mechanical aspects of systems such as chemical substances[3] must be described by Quantum Field Theory (QFT). The argument that leads from this starting point to the familiar description in terms of "molecules" is not mathematically rigorous, but is given here to show that the quantum mechanical considerations that apply to all other forms of matter apply just as much to chemical substances – all that really distinguishes the different regimes of physics (elementary particle physics, nuclear, atomic, chemical physics, etc.) is the *energy* at which one is working. When the true quantum nature of chemical substances is understood, the field theoretical context may be put into the background, to be used if and when convenient, or, as in the case of natural optical activity, necessary.

---

3 The logical distinction between "substance" and "molecule" is obvious, but easily overlooked

The basic building blocks of the theory are Heisenberg operators $\psi_n^+(x)$, $\psi_n(x)$ which create and destroy respectively, particles of type $n$ at the space-time point $x = \mathbf{x}, t_{\mathbf{x}}$. For the purposes of chemistry we can take the index $n$ as $e$ for electrons and $\alpha$ for nuclei only. Of course when energies are much larger than chemical energies, nuclei appear to be "composite" particles, and we must then introduce fields for their "constituents" (quarks, rishons). We shall not make any explicit reference to the spins carried by these fields beyond noting that odd-integral spins require fermi statistics, so that for fermi fields we have *canonical anticommutation relations* (CARS)

$$\begin{aligned} [\psi_n^+(x), \psi_{n'}(x')]_+ &= \delta_{nn'}\, \delta(x - x') \\ [\psi_n^+(x), \psi_{n'}^+(x')]_+ &= [\psi_n(x), \psi_{n'}(x')]_+ = 0 \end{aligned} \tag{2.1}$$

whereas integer spin fields require bose statistics and hence the field operators satisfy *canonical commutation relations* (CCRS),

$$\begin{aligned} [\psi_n^+(x), \psi_{n'}(x')] &= \delta_{nn'}\, \delta(x - x') \\ [\psi_n^+(x), \psi_{n'}^+(x')] &= [\psi_n(x), \psi_{n'}(x')] = 0 \end{aligned} \tag{2.2}$$

The Hamiltonian for the coupled electron and nuclear fields can then be written in the usual way as a sum of kinetic ($\hat{\mathcal{T}}$) and potential ($\hat{\mathcal{V}}$) energy terms,

$$\mathcal{H} = \hat{\mathcal{T}}_e + \sum_\alpha \hat{\mathcal{T}}_\alpha + \hat{\mathcal{V}}_{ee} + \sum_\alpha \hat{\mathcal{V}}_{e\alpha} + \sum_{\alpha \neq \beta} \hat{\mathcal{V}}_{\alpha\beta} \tag{2.3}$$

where for example,

$$\hat{\mathcal{T}}_n = \frac{1}{2\,m_n} \int d^3\mathbf{x}\, \nabla\psi_n^+(\mathbf{x}) \cdot \nabla\psi_n(\mathbf{x}) \tag{2.4}$$

is the kinetic energy for field $n$, and

$$\hat{\mathcal{V}}_{nn'} = \frac{1}{2} \int d^3\mathbf{x} \int d^3\mathbf{x}'\, \psi_n^+(\mathbf{x})\, \psi_{n'}^+(\mathbf{x}')\, V_{nn'}(\mathbf{x} - \mathbf{x}')\, \psi_n(\mathbf{x}')\, \psi_{n'}(\mathbf{x}) \tag{2.5}$$

describes the potential energy of interaction between fields $n$ and $n'$. We take the potential $V_{nn'}(\mathbf{x} - \mathbf{x}')$ to be a Coulomb potential,

$$V_{nn'}(\mathbf{x} - \mathbf{x}') = \frac{Z_n Z_{n'}}{|\mathbf{x} - \mathbf{x}'|} \tag{2.6}$$

This Hamiltonian is Galilean invariant, as is appropriate for chemical phenomena, rather than Lorentz invariant.

Quantum field theory justifies the assumption that to a first approximation the *structure* of the energy level spectrum obeys the same principle as that of the energy levels of an ideal gas. In other words, any energy level can be obtained as the sum of energies of a certain number of "quasi-particles" or "elementary excitations", with momenta $\mathbf{p}$ and energy $\varepsilon = \hbar\omega(\mathbf{p})$, moving in the volume occupied by the system: this does *not* require that $\varepsilon = \varepsilon_0^{(n)}(\mathbf{p}) \equiv p^2/2m_n$, the energy of free particles with mass $m_n$, which is the Ideal Gas limit[4]. Each different kind of elementary excitation is associated with a quantum field, described by annihilation and creation operators $\hat{\chi}_n(\mathbf{p})$, $\hat{\chi}_n^+(\mathbf{p})$ which may satisfy either boson or fermion statistics, and are such that their equations of motion are free (non-interacting) field equations,

$$[\hat{\mathcal{H}}, \hat{\chi}_n^+(\mathbf{p})] = i\hbar\omega(\mathbf{p})\, \hat{\chi}_n^+(\mathbf{p}) \tag{2.7}$$

$$[\hat{\mathcal{H}}, \hat{\chi}_n(\mathbf{p})] = -\, i\hbar\omega(\mathbf{p})\, \hat{\chi}_n(\mathbf{p}) \tag{2.8}$$

and

$$\hat{\chi}_n(\mathbf{p})|0\rangle = 0 \tag{2.9}$$

if $|0\rangle$ is the ground-state wavefunctional. Perhaps the best-known example of this description is a many-body system of electrons that is overall electrically neutral because of a compensating positive background: this is a model of ordinary matter that arises if the nuclei are either treated in the Born-Oppenheimer approximation or smeared out into a uniform background field, as for the electron gas. The mutual Coulomb interactions of the electrons lead to elementary excitations of two types: there are single-particle-like excitations which we call "quasi-electrons" (or just "electrons") and which obey fermi statistics like the original "bare" electron field, and also collective excitations called "plasmons" which describe a collective oscillation of the bare electrons, and are boson particles. In the simplest approximations, the quasi-electrons are described by the one-electron Hartree or Hartree-Fock self-consistent-field equations, and the plasmons are described as the quanta of a set of harmonic oscillator Hamiltonians with characteristic plasmon frequencies $\omega(\mathbf{p})$.

From the above example, it should be clear that all the elementary excitations are the result of the collective interactions of the bare fields in the system, and therefore pertain to the system as a whole[24)]. Elementary excitations, *which will be identified with the "physical particles"* we observe, correspond to superpositions of large numbers of exact stationary states of the field Hamiltonian $\hat{\mathcal{H}}$, Eq. (2.3), with a narrow spread in energy i.e. they are wave-packets. An equivalent way of saying this is that the elementary excitations interact with one another, and so have finite lifetimes: their interactions may lead to reactive, inelastic or elastic scattering processes.

---

4 According to quantum mechanics every plane wave state corresponds to a set of "particles" with momentum determined by the wave-vector $\mathbf{p}$ and energy determined by the frequency $\omega(\mathbf{p})$ moving in the volume of the system. Of course the particles in such a wave are identical copies of each other. We choose to introduce elementary excitations in terms of the particle picture simply because we shall identify atoms and molecules as quasi-particles in chemical substances. Hence all atoms or molecules of a given type have identical states by definition. Note that in principle it must be possible to describe the states of a chemical substance in the wave picture; although this is unfamiliar it should be recognized that the wave and particle descriptions are completely equivalent and our choice is dictated by convenience (and conventionality), cf. lattice waves and phonons, light waves and photons, *matterwaves and atoms etc.*

A fundamental difference between quantum field theory and the ordinary quantum mechanics of point-particle systems with finitely many degrees of freedom, is that in the latter, the Hamiltonian $\mathscr{H}$ can be defined unambiguously (in the sense of unitary equivalence) on a Hilbert space $\mathfrak{H}$. In quantum field theory the existence of inequivalent representations of the CCRS (CARS) means that we must specify the representation of Hilbert space we are going to work in. A natural, physical procedure is to formulate the theory in the Fock space of the elementary excitations only, since it is these that we can relate to our experiments[25)]. We only introduce the Heisenberg operators $\hat{\psi}_n^+$, $\hat{\psi}_n$ so as to be able to state the fundamental Hamiltonian $\hat{\mathscr{H}}$, and hence the equations of motion of the bare fields. The bare operators[5] $\{\hat{\psi}_n^+, \hat{\psi}_n\}$, and the Hamiltonian, $\hat{\mathscr{H}}$, are then to be interpreted as operators in the Fock space of the physical elementary excitations: we can set up an equivalence between the matrix elements of the bare operators in this Hilbert space, and the matrix elements of the quasi-particle field operators. Such relations may conveniently be written as "weak" equalities between bare and quasi-particle operators with the symbol $\approx$, and with the bra and ket omitted. If we write the bare fields $\{\hat{\psi}_n^+, \hat{\psi}_n\}$ and the quasi-particle fields $\{\hat{\chi}_n^+, \hat{\chi}_n\}$ collectively as $\hat{\Psi}$ and $\hat{X}$ respectively (e.g. as column vectors), then one seeks only those solutions of the Heisenberg equations of motion for $\hat{\Psi}$ that can be expressed in terms of a complete set (an irreducible operator ring[26)]) of free quasi-particle fields, $\hat{\Psi} \approx \hat{\Psi}(\hat{X})$.

It is vital to note that we do not know *a priori* how many different elementary excitations constitute such a complete set. One way of dealing with this problem is to set up a self-consistent quantum-field theory[27)] in which one is guided by physical considerations in the initial choice of elementary excitations in order to begin a self-consistent treatment. Of course there are many other approaches to the transformation of quantum field theory to a quasi-particle theory[28, 29)]. The self-consistent procedure is particularly convenient[25)] in situations where "composite" quasi-particles are of interest and so is appropriate for a scheme leading to atoms and molecules. In the first step of the calculation we choose the "incoming particles" (in scattering theory language) associated with the Heisenberg fields as the quasi-particles, that is, we use the weak limit,

$$\lim_{t \to -\infty} \hat{\Psi}(x) \approx a\hat{X}(x) \tag{2.10}$$

to define $\hat{X}(x)$. We then expand $\hat{\Psi}(x)$ in terms of *normal products* of the quasi-particle operators $\hat{X}$,

$$\hat{\Psi}(x) \approx a\hat{X}(x) + \int d^4y_1 \int d^4y_2\, g(x - y_1, x - y_2) : \hat{X}(y_1)\, \hat{X}(y_2) + \ldots \tag{2.11}$$

where the notation: $\hat{X}(y_1)\,\hat{X}(y_2)$ means the normal product in which all the annihilation operators stand to the *right* of all the creation operators. The + ... stands for higher-order normal products with more factors. The time integrations in this equation extend

5 The reader should not be deceived by the (conventional) reference to "creation" and "annihilation" of particles: the fields $(\psi_n^+, \psi_n)$ carry electric charge which is absolutely conserved. Physically we can only *move* charge from one space-time point to another. Charge conservation appears in the theory through the requirement that physically significant quantities be gauge-invariant: the bare operators $\psi_n^+$, $\psi_n$ are not gauge-invariant and should not be interpreted physically[68)]

from $-\infty$ to $t_x$ in accordance with the choice of (retarded) incoming fields as the quasi-particles. The normal product expansion (2.11) is substituted in the equation of motion for $\hat{\Psi}$ obtained from $\hat{\mathcal{H}}$, and one determines the coefficients g such that the quasi-particle operators $\hat{X}$ satisfy free-field Eqs. (2.7, 2.9): in the course of this process the energy spectra of the quasi-particles are also determined. When the equations for these coefficients have no solutions, we modify the initial set of free fields and repeat the computations: such a modification frequently involves the introduction of additional quasi-particle fields, for example, for *composite particles*. This process is continued until a self-consistent set is determined. One can then write the Hamiltonian $\hat{\mathcal{H}}$, Eq. (2.3), in terms of the quasi-particle fields (in the sense of a weak equality) as a sum of free (non-interacting) Hamiltonians: of course this can only be done to a certain accuracy, and as mentioned above, the residual terms in the Hamiltonian are responsible for decay processes of the quasi-particles[27)].

In terms of Green's functions, what we have just described can be expressed by saying that, for example, by studying the equation of motion of the single-particle Green function,

$$G^{(1)}(\mathbf{x}, \tau) = -\,i\langle 0|T\{\hat{\psi}(\mathbf{x}, \tau)\,\hat{\psi}^{+}(0)\}|0\rangle \tag{2.12}$$

(where T is the Dyson chronological operator), under the many-body Hamiltonian $\hat{\mathcal{H}}$, one can show that $G^{(1)}$ has an approximate representation, after Fourier transformation, in the form,

$$G^{(1)}(\mathbf{p}, \varepsilon) = \frac{Z_{\mathbf{p}}}{\varepsilon - \varepsilon(\mathbf{p}) + i\Gamma(\mathbf{p})} + \text{small terms} \tag{2.13}$$

One then seens that for times $\tau \lesssim \Gamma^{-1}(\mathbf{p})$ it is useful to speak of a single-particle like object with energy $\varepsilon(\mathbf{p})$ in the many-body system: this is the one-particle elementary excitation associated with the field theory. If we express the field operator $\hat{\psi}(x)$ in terms of a complete set of single-particle states,

$$\hat{\psi}(x) = \sum_{n} \hat{C}_n\, \varphi_n(x)$$

where $\hat{C}_n$ is an annihilation operator for particles in state $n$, then the $\{\varphi_n\}$ associated with $G^{(1)}(\mathbf{x}, \tau)$ are, in the simplest approximation, the solutions of the Hartree or Hartree-Fock equations.

Composite quasi-particles arise in the same way from higher-order Green functions, $G^{(n)}$, involving $n > 1$ pairs of annihilation and creation operators for the various fields involved in the composite (the $G^{(n)}$ are also known as $n$-point correlation functions). At finite temperatures the expectation value $\langle 0|\ldots|0\rangle$ must be replaced by a trace over the statistical operator $\zeta_T$ at temperature T. The new feature of $n$-particle Green functions ($n > 1$) as compared with $G^{(1)}(\mathbf{x}, \tau)$, is that their energy spectra contain "internal" excitations as well as the momentum $\mathbf{p}$ ("centre-of-mass" motion). As with the single-particle Green functions one can define a set of states for the composite quasi-particle as the solutions of some Schrödinger-like equation[30)]. As a simple example, consider the two-particle Green function which, in a certain approximation can be shown to satisfy the Bethe-Salpeter integral equation[30, 31)]. The inhomogeneous form of this integral equation describes scattering solutions, and is related to the response functions of the many-body system (dielectric susceptibility, Sect. 3); on the other hand the homogeneous integral equation describes bound states of the two-particle system (in the approximation considered). If one takes the $\hat{\psi}_n$ operators to refer to electron and proton fields then the two-particle Green function, obtained from the Bethe-Salpeter equation can be used to describe hydrogen atoms; by induction one can extend the argument to any number of particles and the $n$-particle Green function $\mathrm{G}^{(n)}$ satisfies a well-defined equation that is determined purely by the field Hamiltonian $\hat{\mathcal{H}}$. An equivalent statement is that the bound or resonant states of a composite system can be associated with the singularities (poles) in the scattering (S-) matrix obtained from the Hamiltonian $\hat{\mathcal{H}}$. The important point is that quantum field theory leads one to expect the existence of long-lived composite quasi-particles possessing internal excitations as well as centre-of-mass motion that can be described in terms of the quantum states of a system with a *finite* number of degrees of freedom; these states can be used to build up the Fock space of the elementary excitations as a tensor product. There is thus a natural association between the composite $\mathcal{N}$-particle elementary excitations and a finite-dimensional Hamiltonian operator $\hat{\mathrm{H}}$ on the Hilbert space $\mathfrak{H} = \mathcal{L}^2(\mathfrak{R}^{3\mathcal{N}})$. For a system of $n$ electrons and $\bar{n}$ nuclei, this fundamental molecular Hamiltonian is,

$$\begin{aligned}\hat{\mathrm{H}} &= \hat{\mathrm{T}}_e + \sum_{\alpha} \hat{\mathrm{T}}_{\alpha} + \hat{\mathrm{V}}_{ee} + \sum_{\alpha} \hat{\mathrm{V}}_{e\alpha} + \sum_{\alpha \neq \beta} \hat{\mathrm{V}}_{\alpha\beta} \\ &= \frac{1}{2\mathrm{m}_e} \sum_{i=1}^{n} \hat{\mathrm{p}}_i^2 + \frac{1}{2} \sum_{\alpha=1}^{\bar{n}} \frac{\mathrm{p}_\alpha^2}{\mathrm{m}_\alpha} + \sum_{i<j}^{n} \frac{e^2}{|\hat{\mathbf{q}}_i - \hat{\mathbf{q}}_j|} + \sum_{i,\alpha}^{n,\bar{n}} \frac{e Z_\alpha}{|\hat{\mathbf{q}}_i - \mathbf{q}_\alpha|} \\ &\quad + \sum_{\alpha<\beta}^{\bar{n}} \frac{Z_\alpha Z_\beta}{|\hat{\mathbf{q}}_\alpha - \hat{\mathbf{q}}_\beta|}\end{aligned} \tag{2.14}$$

In some circumstances the quantum states of these elementary excitations can be identified with sufficient accuracy with the stationary states (eigenstates) supported by $\hat{\mathrm{H}}$: this is true for example in high-resolution spectroscopy of dilute gases of low molecular weight[1–3, 32]. Moreover, at finite temperatures in such systems one can choose the Gibbs state $\hat{\xi} = \exp(-\hat{\mathrm{H}}/kT)$ as the statistical operator for calculating expectation values (canonical ensemble). Generally, however, one can only observe superpositions of the eigenstates (so-called metastable states[23]) because the atomic or molecular quasi-particles cannot be completely decoupled into free-fields. It must be recognized, however, that the quantum states of composite elementary excitations are determined by the dynamics of the macroscopic system, so that an *ab initio* theory of molecular states requires the full solution of the quantum field theory for a macroscopic system of given density and temperature. This is simply not practical! I have proposed elsewhere that this dilemma is "resolved" by the *molecular structure hypothesis*[1–3, 33] which suggests that we should describe the situation in terms of *a model time-independent Schrödinger equation for an individual molecule with a definite "equilibrium" structure.* This statement discloses the meaning of the molecular structure hypothesis in quantum mechanical terms. We make a classical mechanical model of a molecule as a quasi-rigid material object with some equilibrium structure, i.e. we make use of the legacy of van't Hoff[18]: this leads to a classical Hamiltonian which can be quantized, and we then calculate the eigenstates of this *model* Hamiltonian. The approach is useful when the resulting eigenvalues and eigenfunctions give an accurate description of the observed phenomena to within the resolution of the experiment performed: naturally one only aims to represent a limited portion of the excitation spectrum of the molecule with any one such model, for example, one associates different equilibrium structures with different electronic states of a molecule. Again, at finite temperatures one uses the Gibbs state for the *model* Hamiltonian. This approach is exemplified by the classical derivation of the Eckart Hamiltonian for a "vibrating, rotating" molecule; although one can carry through directly a quantum-mechanical transformation of the fundamental molecular Hamiltonian, $\hat{\mathsf{H}}$, Eq. (2.14), to the Eckart form, and so apparently avoid the classical model, it is important to understand that such a transformation is *not unitary*[34–36]. One then must ask what determines this non-unitary transformation? As I see it, the answer is given by recourse to the semi-empirical argument just outlined, which leads one *to put molecular structure into the theory as a short-cut that dodges the quantum field theory*. Claims in the literature that "molecular structure" can be deduced *ab initio* from quantum mechanics, if evaluated in the light of the foregoing arguments, will be seen to be vacuous.

If we disregard electronic phenomena, ignore the residual coupling terms between "atomic" quasi-particles, and specialize to the case of a single kind of composite quasi-particle, i.e. a pure chemical substance, at low density, the dynamical map $\hat{\Psi} \approx \hat{\Psi}(\hat{X})$ allows us to write the Hamiltonian $\mathscr{H}^\circ$ in the form,

$$\mathscr{H}^\circ \approx \frac{1}{2M} \int d^3\mathbf{x} \nabla\hat{\chi}^+(\mathbf{x}) \cdot \nabla\hat{\chi}(\mathbf{x}) \tag{2.15}$$

where the composite particle (atom or molecule) has mass M, and field operators $\hat{\chi}^+$, $\chi$. It is then a straightforward matter in statistical mechanics to show that this Hamiltonian describes the (trivial) kinetic and thermodynamic properties of the Ideal Gas. When the residual couplings are retained there is an intermolecular interaction term of the form,

$$\begin{aligned} \mathscr{V} &\approx \frac{1}{2} \int d^3\mathbf{x} \int d^3\mathbf{x}' \int \hat{\chi}^+(\mathbf{x})\, \hat{\chi}^+(\mathbf{x}')\, \mathrm{U}(\mathbf{x}-\mathbf{x}')\, \hat{\chi}(\mathbf{x})\, \hat{\chi}(\mathbf{x}') \\ \mathscr{H} &\approx \mathscr{H}^\circ + \mathscr{V} \end{aligned} \tag{2.16}$$

where of course U is of much shorter range than the original Coulomb interaction (2.6), if the composite particles are overall neutral $\left(\sum_i^n e_i + \sum_\alpha^n Z_\alpha = 0\right)$. This Hamiltonian $\mathscr{H}$ can be used to give a quantum field theory description of molecular fluids, and of perfect crystals and their defect structures, mechanical properties etc., by transformation to a set of field operators for new elementary excitations ("quasi-atoms or molecules", "phonons") that once again approximately diagonalize $\mathscr{H}$[37]. It should be clearly understood that *quantum field theory is the general and fundamental approach to the description of the properties of matter, whether one is interested in solids or molecules, nuclei or elementary particles.*

Finally it is appropriate to make some remarks about the symmetry aspects of quantum field theories, and to review the ideas implied by the expression "spontaneously broken symmetry". The density operator for the field $\hat{\psi}$ can be defined as

$$\hat{\mathsf{n}}(\mathbf{x}) = \hat{\psi}^+(\mathbf{x})\, \hat{\psi}(\mathbf{x}) \tag{2.17}$$

and it has a ground-state expectation value,

$$\mathsf{n}(\mathbf{x}) = \langle 0|\hat{\psi}^{+}(\mathbf{x})\,\hat{\psi}(\mathbf{x})|0\rangle \tag{2.18}$$

that may be studied in appropriate experiments. The fundamental Hamiltonian $\mathcal{H}$, (2.3), for a system of coupled electron and nuclear fields is invariant under the extended Galilean group[1–3], and hence one expects initially that these symmetries will still be manifest when one calculates the density $\mathsf{n}(\mathbf{x})$ in terms of the quasi-particle fields $\hat{X}$ using the dynamical map $\hat{\Psi} \approx \hat{\Psi}(\hat{X})$. Ordinary fluids are of course invariant under rotations, translations, space and time inversions. However we do commonly observe states of matter that have lower symmetry than the corresponding Hamiltonian. In such cases we speak of "spontaneously broken symmetries" or "dynamically rearranged symmetry"; for example the ground state density $\mathsf{n}(\mathbf{x})$ for a perfect crystal associated with an atomic or molecular field $\hat{\chi}$ is

$$\mathsf{n}(\mathbf{x}) = \langle 0|\hat{\chi}^{+}(\mathbf{x})\,\hat{\chi}(\mathbf{x})|0\rangle \tag{2.19}$$

and has the discrete symmetry property,

$$\mathsf{n}(\mathbf{x} + \mathbf{a}_{\iota}) = \mathsf{n}(\mathbf{x}) \qquad \iota = 1, 2, 3 \tag{2.20}$$

where $\{\mathbf{a}_{\iota}\}$ are the lattice vectors defining the crystal structure, whereas the associated Hamiltonian (2.16) is invariant under *arbitrary* translations $\mathbf{a}$. The point here is that in addition to the "quasi-atoms" there are other elementary excitations in the crystal, namely the lattice vibrations or *phonons*, and these play the role of recovering the translational symmetry broken by the lattice[37]. The "loss" of translational symmetry occurs because we choose to focus attention on one partner of the pair {lattice of quasi-atoms; phonons}. This is a general phenomenon and there is indeed a theorem due to Goldstone that says that in a field theory based on a Hamiltonian $\mathcal{H}$ that is invariant under a *continuous* symmetry group $\mathcal{G}$, the occurrence of a ground state $|0\rangle$ which is not invariant under $\mathcal{G}$ requires the existence of the appropriate number of mass-zero boson elementary excitations ("Goldstone bosons") that serve to recover the spontaneously broken or dynamically rearranged symmetry[38, 39].

In Sect. 1 I remarked that natural optical activity should be understood as a manifestation of spontaneously broken space-inversion symmetry. Space-inversion is a discrete symmetry, and therefore one does not expect Goldstone bosons to arise when it is broken. However, the phenomenon of chirality is best appreciated as involving the response of a (chiral) material medium to polarized light, and conversely circularly polarized light only occurs through the intervention of a chiral material medium. One can regard the photon, which is a mass-zero boson, as a Goldstone boson connected with spontaneously broken gauge invariance; that is, one supposes that there is some more primitive field theory involving bare operators $\hat{\Psi}$ which are invariant under gauge transformations, and which under the dynamical map $\hat{\Psi} \approx \hat{\Psi}(\hat{X})$ can be reformulated in terms of elementary excitations corresponding to ordinary matter and the radiation field (photons). As is well-known in quantum electrodynamics the usual canonical variables for a matter system and the radiation field are not separately gauge-invariant, and there is therefore a deep sense in which they have to be considered together as two different aspects of the same phenomenon just as one does with a crystal lattice and its associated phonons. This brings me to my earlier discussion of optical activity[1)]. One can argue that parity is conserved in natural optical activity in the sense that if one inverts the system – light beam + active medium – one obtains a new experiment which is also realized in nature[40)]. One can make a simple analogy of this situation; if we represent the complete system as a nut on a bolt, inversion of the coordinates leaves us with another possible nut and bolt. The "problem" is to explain how one can have nuts and bolts separately (instead of say rings and pins) since these are objects that have a definite handedness and therefore obviously cannot model systems in parity eigenstates[1)]. Thus in the wider context of quantum electrodynamics, one can argue that in optical activity, space inversion symmetry is rearranged, so that the material medium appears to be chiral, and circularly polarized photons of appropriate sense serve to restore the spontaneously broken space-inversion symmetry because of gauge invariance, the symmetry that says electric charge is conserved.

We shall recall in Sect. 4 that a system with a finite number of degrees of freedom such as a molecule cannot exhibit spontaneously broken symmetry in the absence of coupling to a quantized field. Hence a possible procedure to follow in setting up a theory of optical activity is to study the interaction of the molecular Hamiltonian with other quasi-particles in an electrodynamical system (Sect. 5): this at least begins to approach the idea of "molecules" as *elementary excitations* in the many-body system, including the electromagnetic field. It is not meaningful to make a sharp distinction between a "single molecule" and its "environment" (all the other molecules in the material medium, radiation) since a molecule is nothing more than the particle interpretation of an elementary excitation in the macroscopic quantum system and hence it *necessarily* inherits properties, manifested in the molecular quantum states, from the macroscopic material medium.

## 3 Linear Response Theory of Natural Optical Activity

It is convenient to begin this section by reviewing briefly some specifics of the experimental data that any theory of natural optical activity must get to grips with. It is found empirically that when a plane polarized beam of light passes through a certain class of substances the emergent beam is elliptically polarized with the major axis of the ellipse rotated from the original plane of polarization. This is natural optical activity and the substances that behave in this way are said to be optically active or circularly dichroic. Suppose one has an optically active substance that rotates the plane of polarization to the right (dextrorotatory form). Then it is found that there are always two other modifications of the same substance, one of which rotates the plane of polarization to the left to exactly the same extent (laevorotatory form), and one of which shows no rotation (racemic form). These three substances, which are referred to as the $d$-, $\ell$- and $d\ell$-isomers respectively exhibit identical chemical properties with respect to *achiral* reagents, and they are only distinguished by their response to chiral perturbations (polarized light, other chiral substances). They are said to be optical isomers, and the active pair ($d$-, $\ell$-) are called enantiomers[19, 41)].

In some cases loss of optical activity occurs at the solid → liquid phase transition: the dichroism in these cases is thought of as a property of the crystal as a whole. In other substances melting and vaporization do not destroy optical activity which may therefore be exhibited by the gaseous, liquid and solid phases of matter. It is sometimes found however that increasing temperature and/or the presence of "catalysts" eventually eliminate the dichroism without change in the chemical constitution of the substance, that is, the only property altered is the chirality: this is the phenomenon of racemization. A typical example is optically active 3-methyl-2-pentanone which under the action of alkali loses its optical activity, and upon cooling, the optical activity does not return[42]. Subsequent chemical analysis reveals a quantitative conversion to the racemic $d\ell$-isomer. In many other cases however loss of optical activity is not observed on any human timescale, except by degradation of the originally chiral substance.

Now, what can be said theoretically about optical activity? It may be taken that the theoretical framework to be used is quantum theory, and to the extent that temperature, (T), effects are interesting (racemization), statistical mechanics also. This means that the quantum mechanical formalism is based on the specification of a density matrix, or mixed state, $\hat{\xi}$, for the optically active medium, although at $T = 0$ a wavefunction, pure state, description is appropriate, and doubtless more familiar in this context. Irreversible thermally induced racemization indicates that an optically active medium cannot be in a state of thermodynamic equilibrium, that is, the irreversible loss of dichroism is to be associated with the process of the system reaching thermodynamic equilibrium. Hence $\hat{\xi}$ $(T > 0)$ for the optically active medium must be a metastable state[23]: the theoretical significance of this remark is that although one can define the usual macroscopic variables like density, pressure, temperature, the density matrix $\hat{\xi}$ for a resolved optically active medium *cannot* be simply the Gibbs state, $\exp(-\hat{\mathcal{H}}/kT)$ for a system with Hamiltonian $\hat{\mathcal{H}}$ in thermodynamic equilibrium at temperature T. We shall assume that optical isomerism is generic, and that racemization is often not observed in practice simply because the decay rate of $\hat{\xi}$ is too small (but not strictly zero).

The general statement that connects quantum theory to experimental observations such as optical activity can be put as follows. We subject the material medium to some experimental procedure using a probe light beam, and observe a certain outcome after the probe has passed through the system, for example by analysing the scattered light. If we define the density matrix $\hat{\xi}_{TOT}$ for the combined system of material medium + probe, and characterize the measurement by some operator $\hat{\mathscr{E}}$, then the probability that the specified outcome of the measurement is observed, is given by,

$$\langle \hat{\mathscr{E}} \rangle_{\hat{\xi}_{TOT}} \equiv \mathrm{Tr}\,[\hat{\xi}_{TOT} \hat{\mathscr{E}}] \tag{3.1}$$

In the particular case of optical activity I can describe the experiment by choosing $\hat{\mathscr{E}}$ to be the electric polarization operator $\hat{\mathbf{P}}(\mathbf{x}, \ell)$ for the material medium, while $\hat{\xi}_{TOT}$ refers to the combined system of matter interacting with the polarized light beam. Note that $[\hat{\xi}_{TOT}, \hat{\mathsf{P}}] = 0$, where $\hat{\mathsf{P}}$ is the space-inversion operator (see[40] and Sect. 4). Then the mean value,

$$\langle \hat{\mathbf{P}}(\mathbf{x}, \ell) \rangle_{\hat{\xi}_{TOT}} \equiv \mathrm{Tr}\,[\hat{\xi}_{TOT} \hat{\mathbf{P}}(\mathbf{x}, \ell)], \tag{3.2}$$

where the trace is taken over the degrees of freedom of the matter system, can be related to the electromagnetic radiation that arrives at the detector[43].

The coupling between the medium and the light beam is described by the interaction Hamiltonian,

$$\hat{\mathsf{H}}(t)^{\text{int}} = - \int d^3\mathbf{x}\hat{\mathbf{P}}(\mathbf{x}, t) \cdot \hat{\mathbf{E}}(\mathbf{x}, t)^{\perp} - \int d^3\mathbf{x}\hat{\mathbf{M}}(\mathbf{x}, t) \cdot \hat{\mathbf{B}}(\mathbf{x}, t) + \dots \tag{3.3}$$

where $\hat{\mathbf{M}}(\mathbf{x}, t)$ is the magnetization operator for the medium, $\hat{\mathbf{E}}^{\perp}$ and $\hat{\mathbf{B}}$ are the operators describing the transverse electric displacement and magnetic fields respectively, and $+ \dots$ indicates that other interaction terms of no interest here have been dropped[44, 45]. Notice that making a truncated multipole expansion of the polarization fields involves a non trivial assumption about the spatial localization of the elementary excitations in the material system. Thus at this stage in the argument it is preferable to keep (3.3) completely general. The polarization operators are explicitly time-dependent in a Heisenberg representation for the field-free system

$$\hat{\mathsf{X}}(t) = e^{i\mathcal{H}t/\hbar}\hat{\mathsf{X}}(0)e^{-i\mathcal{H}t/\hbar} \tag{3.4}$$

The interaction between the matter and the light beam is weak and I compute the state $\hat{\xi}_{\text{TOT}}$ using perturbation theory based on the complete set of exact states $\{\psi_n\}$, with energies $\hbar\omega_n$, of the chiral medium in the absence of the light beam, noting that the information yielded by the experiment can then be related to the optically active medium alone. The density matrix, $\hat{\xi}$, for the medium in the absence of the light beam can be given a spectral representation in terms of this complete set of states, by virtue of the spectral theorem,

$$\hat{\xi} = \sum_n \lambda_n |\psi_n\rangle\langle\psi_n| \tag{3.5}$$

Such a calculation is essentially a linear response theory of the electric polarization of the medium by the light beam.

In linear response theory the optical activity is obtained from the part of the generalized susceptibility involving the temporal correlations of the electric and magnetic polarization fields[46, 47]. For a system such as a normal fluid, described by a statistical operator $\hat{\xi}$ that is invariant under space and time translations, the appropriate retarded Green function is,

$$\mathcal{G}^{+}_{\hat{\mathbf{P}},\hat{\mathbf{M}}}(\mathbf{x}-\mathbf{x}', t-t') = i\,\theta(t-t')\,\mathrm{Tr}\{\hat{\xi}[\hat{\mathbf{P}}(\mathbf{x}, t), \hat{\mathbf{M}}(\mathbf{x}', t')]\} \tag{3.6}$$

where the Heaviside step function vanishes for negative arguments: using the spectral representation of $\hat{\xi}$[48], Eq. (3.5), we may write the space and time Fourier transform of $\mathcal{G}^{+}$ as,

$$\mathcal{G}^{+}_{\hat{\mathbf{P}},\hat{\mathbf{M}}}(\mathbf{k}, \omega) = \lim_{\delta \to +0} \frac{1}{\mathcal{V}} \sum_{n,m} (\lambda_m - \lambda_n) \frac{\langle\psi_m|\hat{\mathbf{P}}(\mathbf{k})|\psi_n\rangle\langle\psi_n|\hat{\mathbf{M}}(-\mathbf{k})|\psi_m\rangle}{\omega - \omega_{nm} + i\delta} \tag{3.7}$$

where $\omega_{nm} = \omega_n - \omega_m$ is an excitation frequency in the system. The operator $\hat{\mathbf{P}}(\mathbf{x})$ is real, while $\hat{\mathbf{M}}(\mathbf{x})$ is pure imaginary[45], and both are hermitian so that, for real wavefunctions,

$$\langle\psi_n|\hat{\mathbf{P}}(\mathbf{k})|\psi_m\rangle = \langle\psi_m|\hat{\mathbf{P}}(-\mathbf{k})|\psi_n\rangle^* = \langle\psi_m|\hat{\mathbf{P}}(\mathbf{k})|\psi_n\rangle \tag{3.8 a}$$

$$\langle\psi_m|\hat{\mathbf{M}}(-\mathbf{k})|\psi_n\rangle = \langle\psi_n|\hat{\mathbf{M}}(\mathbf{k})|\psi_m\rangle^* = -\langle\psi_n|\hat{\mathbf{M}}(-\mathbf{k})|\psi_m\rangle \tag{3.8 b}$$

Using these relations, and the symmetry in Eq. (3.7) due to the sum over both $n$ and $m$ it is a simple matter to rewrite $\mathcal{G}^+(\mathbf{k},\omega)$ in its Lehmann representation

$$\mathcal{G}^+(\mathbf{k},\omega) = 2\omega \int_{-\infty}^{+\infty} \frac{d\omega' \, \mathcal{S}(\mathbf{k},\omega')}{\omega_+^2 - \omega'^2} \tag{3.9}$$

where $\omega_+ = \omega + i\delta$ and the limit $\delta \to +0$ is understood, and the spectral function $\mathcal{S}(\mathbf{k},\omega')$ is defined as,

$$\mathcal{S}(\mathbf{k},\omega) = \frac{1}{\mathcal{V}} \sum_{n,m} \lambda_m \langle\psi_m|\hat{\mathbf{P}}(\mathbf{k})|\psi_n\rangle\langle\psi_n|\hat{\mathbf{M}}(-\mathbf{k})|\psi_m\rangle\delta(\omega - \omega_{nm}) \tag{3.10}$$

The retarded Green function is directly related to the generalized susceptibility

$$\mathcal{G}^+(\mathbf{k},\omega) = -2\hbar\,\chi''_{\hat{\mathbf{P}},\hat{\mathbf{M}}}(\mathbf{k},\omega) \tag{3.11}$$

in terms of which the net heat flow into the medium, assuming a monochromatic light beam of frequency $\omega$, is given by

$$\left\langle \frac{dQ}{dt} \right\rangle_\omega = \frac{\omega}{16\pi^3} \int d^3k \mathbf{E}(-\mathbf{k}) \cdot \chi''_{\hat{\mathbf{P}},\hat{\mathbf{M}}}(\mathbf{k},\omega) \cdot \mathbf{B}(\mathbf{k}) \tag{3.12}$$

where $\chi''$ is the absorptive part of the susceptibility[47]. It is easily seen from (3.7) that energy is dissipated if $\lambda_m > \lambda_n$ for $\omega_{nm} > 0$. Equation (3.12) is of course only one contribution to the power absorbed by the medium. The dominant term arises from the electric polarization self-susceptibility, $\chi''_{\hat{\mathbf{P}},\hat{\mathbf{P}}}$; however $\chi''_{\hat{\mathbf{P}},\hat{\mathbf{P}}}$ does not lead to circular dichroism since its contributions to $\langle dQ/dt\rangle_\omega$ for left and right circularly polarized light are identical. The term in $\chi''_{\hat{\mathbf{P}},\hat{\mathbf{M}}}$ does discriminate between different polarization states for the ligth beam, and by taking the difference in absorption for left and right circularly polarized light one finds, after averaging over the direction of the incident beam, the dichroism of the medium,

$$\varphi(\mathbf{k},\omega) = -\left(\frac{16\pi}{3}\right)\left(\frac{\omega}{c}\right) \operatorname{Im} \chi''_{\hat{\mathbf{P}},\hat{\mathbf{M}}}(\mathbf{k},\omega) \tag{3.13}$$

where $\chi''_{\hat{\mathbf{P}},\hat{\mathbf{M}}} = \operatorname{Tr} \chi''_{\hat{\mathbf{P}},\hat{\mathbf{M}}}$ is the scalar part of the susceptibility; then by the Krönig-Kramers theorem, the rotation angle, $\theta$, is given by the Hilbert transform of $\varphi$,

$$\theta(\mathbf{k},\omega) = \frac{1}{\pi} \int_{-\infty}^{+\infty} d\omega' \frac{\varphi(\mathbf{k},\omega')}{\omega' - \omega} \tag{3.14}$$

In order to make the connection with the usual discussion of optical rotation based on the Rosenfeld equation for a molecule[49], one must express the exact states $\{\psi_n\}$ of the chiral medium in terms of the states of the elementary excitations in the system, using the machinery of quantum field theory discussed in Sect. 2. Before considering this problem however it is instructive to consider first the role of the space-inversion operator $\hat{\mathsf{P}}$ in optical activity.

## 4 The Representation of the Space-Inversion Operator

Let $\hat{\mathsf{P}}$ be the operator describing space-inversion: the identity, $\hat{\mathsf{I}}$, and $\hat{\mathsf{P}}$ together form an Abelian group, $\mathcal{G}(\imath)$, isomorphic to $\mathfrak{Z}_2$, with the composition law:

$$(\hat{\mathsf{P}})^2 = \hat{\mathsf{I}} ; \tag{4.1}$$

by Schur's lemma *unitary* irreducible representations of this group are one-dimensional (non-degenerate). The realization of this group as a transformation of the $3n$-dimensional Euclidian space $\mathfrak{R}^{3n}$ is given as

$$\hat{\mathsf{P}}\mathbf{r} \to -\mathbf{r} , \qquad \mathbf{r} \in \mathfrak{R}^{3n} .$$

More generally, as a transformation on a function space $\mathcal{F}(\mathfrak{R}^{3n})$, the space-inversion operator acts on vectors in $\mathcal{F}$ as,

$$\hat{\mathsf{P}}\phi(\mathbf{r}) = \phi(-\mathbf{r}) , \qquad \phi \in \mathcal{F} .$$

Using the properties of integrals over all space under the change of variables $\mathbf{r} \to -\mathbf{r}$ we see that,

$$\begin{aligned} \langle\phi|\hat{\mathsf{P}}\psi\rangle &= \langle\hat{\mathsf{P}}\phi|\psi\rangle \\ \langle\hat{\mathsf{P}}\phi|\hat{\mathsf{P}}\psi\rangle &= \langle\phi|\psi\rangle \end{aligned} \tag{4.2}$$

for all $\phi$, $\psi$ in the domain $\mathfrak{D}_{\hat{\mathsf{P}}}$ of $\hat{\mathsf{P}}$. These equations tell us that $\hat{\mathsf{P}}$ *is an hermitian, isometric operator.*

Density matrices $\hat{\xi}$ can be characterized as trace-class linear operators on a Hilbert space $\mathfrak{H}$, $\hat{\xi}$: $\mathfrak{H} \to \mathfrak{H}$, such that $\hat{\xi}$ is self-adjoint, non-negative and of trace equal to one, where trace is defined by

$$\mathrm{Tr}[\hat{\xi}] = \sum_n \langle\psi_n|\hat{\xi}\psi_n\rangle \equiv 1 \tag{4.3}$$

and $\{\psi_n\}$ is any orthonormal basis of the Hilbert space $\mathfrak{H}$. By the spectral theorem[48] every state $\hat{\xi}$ has a representation in the form of Eq. (3.5),

$$\hat{\xi} = \sum_n \lambda_n|\psi_n\rangle\langle\psi_n|$$

and so for the $d$- and $\ell$-isomers we may write

$$\hat{\xi}_d = \sum_n \lambda_n^d |\psi_n^d\rangle\langle\psi_n^d| \ , \ |\psi_n^d\rangle \in \mathfrak{H}^d$$
$$\hat{\xi}_\ell = \sum_n \lambda_n^\ell |\psi_n^\ell\rangle\langle\psi_n^\ell| \ , \ |\psi_n^\ell\rangle \in \mathfrak{H}^\ell \tag{4.4}$$

where

$$\lambda_n^d = \lambda_n^\ell = \lambda_n$$

since

$$|\psi_n^\ell\rangle = \hat{\mathsf{P}}|\psi_n^d\rangle \ , \qquad \forall_n \tag{4.5}$$

i.e. $\hat{\mathsf{P}}$ interconverts the states of the pair of isomers, and $\hat{\mathsf{P}}$ is isometric. We identify the states $\{\psi_n^\ell\}$, $\{\psi_n^d\}$ with the complete, orthonormal bases used in the perturbation theory derivation of the optical rotation Eq. (3.14) for isomers $\ell$-, and $d$-respectively.

That there is something unusual about this description may be seen from the following argument. The $d$- and $\ell$-isomers share the same Hamiltonian $\hat{\mathscr{H}}$ for which $[\hat{\mathscr{H}}, \hat{\mathsf{P}}] = 0$ is true, so that $|\psi_n^\ell\rangle$ and $|\psi_n^d\rangle$ are degenerate in energy. However, since they are *distinct* physical systems we have assigned them their own Hilbert spaces $\mathfrak{H}^k$: $k = \ell, d$, and in the perturbation theory calculation for a given isomer we tacitly asserted that at T = 0 *all* the excitations of say, the $d$-isomer lie in $\mathfrak{H}^d$, i.e. the basis $\{\psi_n^k\}$: $k = \ell, d$ is complete for isomer $k$. Now consider the space-inversion operator $\hat{\mathsf{P}}$ as an operator with domain $\mathfrak{D}_{\hat{\mathsf{P}}} = \mathfrak{H}^d$; since $\mathfrak{H}^d$ is dense, $\hat{\mathsf{P}}$ has an adjoint $\hat{\mathsf{P}}^+$. The condition

$$\langle\phi|\hat{\mathsf{P}}\psi\rangle \equiv \langle\hat{\mathsf{P}}^+\phi|\psi\rangle \ , \quad \forall\psi \in \mathfrak{D}_{\hat{\mathsf{P}}} = \mathfrak{H}^d \tag{4.6}$$

defines $\hat{\mathsf{P}}^+$ as an operator with domain equal to the set of vectors $\{\phi\}$ satisfying this identity[50]. From (4.2) we see that if $\phi \in \mathfrak{D}_{\hat{\mathsf{P}}}$, then $\phi \in \mathfrak{D}D_{\hat{\mathsf{P}}^+}$; however, $\mathfrak{D}_{\hat{\mathsf{P}}} \subset \mathfrak{D}_{\hat{\mathsf{P}}^+}$, since the only condition on $\mathfrak{D}_{\hat{\mathsf{P}}^+}$ is that

$$\hat{\mathsf{P}}^+\phi(\mathbf{r}) = \phi(-\mathbf{r}) \ . \tag{4.7}$$

Thus on $\mathfrak{D}_{\hat{\mathsf{P}}}$ we have,

$$\hat{\mathsf{P}}^+\psi = \hat{\mathsf{P}}\psi \ ; \quad \psi \in \mathfrak{D}_{\hat{\mathsf{P}}} \tag{4.8}$$

and further,

$$(\hat{\mathsf{P}}^+\hat{\mathsf{P}})\psi = (\hat{\mathsf{P}}\hat{\mathsf{P}}^+)\psi = \hat{\mathsf{I}}\psi \ ; \quad \psi \in \mathfrak{D}_{\hat{\mathsf{P}}} \tag{4.9}$$

However, $\hat{\mathsf{P}}$ is *not* unitary, since the domain of its inverse $\hat{\mathsf{P}}^{-1}$ is obviously $\mathfrak{H}^\ell$, the Hilbert space of the conjugate $\ell$-isomer, and $\mathfrak{H}^\ell \subset \mathfrak{D}_{\hat{\mathsf{P}}^+}$. Hence $\hat{\mathsf{P}}^+ \neq \hat{\mathsf{P}}^{-1}$. Of course $\hat{\mathsf{P}}$ with domain $\mathfrak{D}_{\hat{\mathsf{P}}} = \mathfrak{H}^d \oplus \mathfrak{H}^\ell$ is unitary. Thus for optically active isomers we have two complete, orthonormal bases that are interconverted by the space-inversion operator. Since $\hat{\mathsf{P}}$ does not have a unitary representation on the states of a given isomer we can take the wavefunctions of the $d$-isomer to be orthogonal to those of the $\ell$-isomer,

$$\langle \psi_n^d | \psi_m^\ell \rangle = \langle \psi_n^d | \hat{\mathsf{P}} | \psi_m^d \rangle \equiv 0, \qquad \forall_{n,m} \tag{4.10}$$

and $\{\psi_n^d\}$, $\{\psi_n^\ell\}$ belong to orthogonal (incoherent) Hilbert spaces.

It is characteristic of systems displaying a spontaneously broken symmetry that the corresponding symmetry operator no longer has a unitary representation on the states of the system, even though it commutes with the underlying Hamiltonian $\mathscr{H}$. Physically it is obvious that the space inversion operator should not have a unitary representation when we restrict attention to just one of the isomers, since the $d$- and $\ell$-isomers are physically distinct (inequivalent) systems. In group theoretical terms we can say that for an achiral substance, $\mathscr{G}(\dot{\imath})$ is a symmetry group that can be given an irreducible unitary representation on the corresponding Hilbert space $\mathfrak{H}$. Chiral substances however have only the group $\mathscr{C}_1$ (which is the subgroup of $\mathscr{G}(\dot{\imath})$ consisting of the identity $\hat{\mathrm{I}}$) as their symmetry group, since the distinct optically active isomers are interconverted by the action of $\hat{\mathsf{P}}$, and together form an orbit of $\mathscr{G}(\dot{\imath})$[50].

An equivalent description of this situation is the statement that linear superpositions of the states of *both* isomers, i.e. a state of the form,

$$|\phi\rangle = \sum_n \mathrm{C}_n^d |\psi_n^d\rangle + \sum_n \mathrm{C}_n^\ell |\psi_n^\ell\rangle \tag{4.11}$$

with at least one pair of non-zero coefficients $(\mathrm{C}_n^d, \mathrm{C}_n^\ell)$ cannot be physically realized. If we think about this statement in the context of the optical rotation equations (Sect. 3) we see that this is to be expected since the rotation angle is calculated using non-degenerate perturbation theory, and with the interpretation that the complete set of states $\{|\psi_n\rangle\}$ refer to one particular isomer. If one admits all the states of both isomers to the calculation, i.e. allows superpositions of the form (4.11), $\theta$ should vanish, for in this case $\hat{\mathsf{P}}$ has a unitary representation; one could then choose a unitarily equivalent set of basis functions with definite parity quantum numbers, since $\hat{\mathsf{P}}$ commutes with the Hamiltonian $\mathscr{H}$ for the optically active system in the absence of the light beam.

An observable $\hat{\mathscr{E}}$ must be either even or odd under the space-inversion operator

$$\hat{\mathsf{P}}^{-1}\hat{\mathscr{E}}\hat{\mathsf{P}} = \pm \hat{\mathscr{E}} \tag{4.12}$$

Thus we may write its expectation value in some state $\hat{\xi}$ as,

$$\langle \hat{\mathscr{E}} \rangle_\xi = \mathrm{Tr}[\hat{\xi}\hat{\mathscr{E}}] = \pm\,\mathrm{Tr}[\hat{\mathsf{P}}\hat{\xi}\hat{\mathsf{P}}^{-1}\hat{\mathscr{E}}] = \pm\,\mathrm{Tr}[\hat{\xi}'\hat{\mathscr{E}}] \tag{4.13}$$

after using (4.12), and the invariance of the trace to cyclic permutation. The characteristic feature of an achiral state of matter is the property,

$$\hat{\xi}' = \hat{\mathsf{P}}\hat{\xi}\hat{\mathsf{P}}^{-1} \equiv \hat{\xi} \tag{4.14}$$

where $\hat{\mathsf{P}}$ and $\hat{\mathsf{P}}^{-1}$ share a common domain such that $\hat{\mathsf{P}}$ is unitary. If however we assign normalized density matrices $\hat{\xi}_d$ and $\hat{\xi}_\ell$ to $d$- and $\ell$-isomers respectively of some substance, then from Eq. (4.5) we may write

$$\hat{P}\hat{\xi}_m\hat{P}^{-1} \equiv \hat{\xi}_n\,; \qquad m \neq n = d, \ell \tag{4.15}$$

where $\hat{\xi}_m \neq \hat{\xi}_n$ for $m \neq n$, and with the understanding that the domain $\mathfrak{D}_{\hat{P}}$ of $\hat{P}$ is the set of states of isomer $m$, whereas $\mathfrak{D}_{\hat{P}^{-1}}$ is the set of states of the $n$ isomer.

Nevertheless Eq. (4.13) still says that

$$\langle\hat{\mathscr{E}}\rangle_{\hat{\xi}_d} = \pm\langle\hat{\mathscr{E}}\rangle_{\hat{\xi}_\ell} \tag{4.16}$$

and accounts for the empirical fact that the various optical isomers of a given substance can only be distinguished through odd-parity observables such as the optical rotation angle. Equally, Eq. (4.16) is the theoretical expression of the empirical fact that the rotation angles of enantiomers are equal in magnitude, but have opposite signs (Sect. 3). While for odd-parity observables, Eq. (4.13) shows that for systems satisfying (4.14), the identity

$$\langle\hat{\mathscr{E}}\rangle_{\hat{\xi}} = -\langle\hat{\mathscr{E}}\rangle_{\hat{\xi}} \equiv 0 \tag{4.17}$$

is always true, Eq. (4.7) is not sufficient to guarantee a non-zero expectation value for a chiral substance characterized by (4.15). This depends on all the other symmetry properties of the system as well; for example, if an optically active fluid is rotationally invariant, then the only non-zero odd-parity observables must be pseudoscalar operators[6]. Hence an isotropic field, achiral or chiral, cannot support a non-zero electric polarization.

Once this discussion of the space-inversion operator in the context of optically active isomers is accepted, it follows that a molecular interpretation of the optical activity equation will not be a trivial matter. This is because a molecule is conventionally defined as a dynamical system composed of a particular, finite number of electrons and nuclei; it can therefore be associated with a Hamiltonian operator containing a finite number ($3\mathcal{N}$) of degrees of freedom (variables) (Sect. 2), and for such operators one has a theorem that says the Hamiltonian acts on a single, *coherent* Hilbert space $\mathfrak{H} = \mathscr{L}^2(\mathfrak{R}^{3\mathcal{N}})$[51]. In more physical terms this means that *all* the possible excitations of the molecule can be described in $\mathfrak{H}$. In principle therefore any superposition of states in the molecular Hilbert space is physically realizable: in particular it would be legitimate to write the eigenfunctions of the usual molecular Hamiltonian, Eq. (2.14)[1–3] in the form of Eq. (4.14) with suitable coefficients $\{C_n^\ell, C_n^d \neq 0\}$. Moreover any unitary transformation of the eigenfunctions of the molecular Hamiltonian will also have a similar representation, since the eigenfunctions induce a unitary representation of the space inversion operator $\hat{P}$ and this property cannot be altered by subsequent unitary transformations. The essence of the previous argument however is that the superposition (4.14) cannot be realized, and that the space-inversion operator does not have a unitary representation on the states for a given isomer. How is this to be explained?

The conventional intuition is that we should start with the Hamiltonian for an isolated molecule, $\hat{H}$, or indeed a collection of $n$ molecules ($n < \infty$), and introduce intermolecular interactions as a perturbation $\hat{V}$ where, as usual, $\hat{V}$ is such that $[\hat{P}, \hat{V}] = 0$, so that the total Hamiltonian $\hat{H} = \hat{H} + \hat{V}$. But the process of constructing the ground state of $\hat{H}$ from that of $\hat{H}$ can be formulated as a unitary transformation to a new representation[52], and hence leaves $\hat{P}$ as a unitary operator in the new representation. More generally, if we start with any density matrix $\hat{\sigma}(0)$ at time $t = 0$ which is such that $[\hat{\mathsf{P}}, \hat{\sigma}] = 0$, then this density matrix develops in time under $\hat{H}$ as

$$\hat{\sigma}(t) = \exp(-\mathrm{i}\hat{H}t/\mathrm{h})\hat{\sigma}(0)\exp(+\mathrm{i}\hat{H}t/\mathrm{h}) \tag{4.18}$$

and if $3\mathscr{N}$, the total number of degrees of freedom, is finite this is a unitary transformation, and so $[\hat{\sigma}(t), \hat{\mathsf{P}}] = 0$. The difficulty with an exclusively molecular interpretation of chirality is that it gives no insight as to how $\hat{\mathsf{P}}$ loses the unitary property that we usually take for granted in most applications of quantum mechanics. Since the broken space-inversion symmetry is of the essence in optical activity, one must understand how this comes about, especially if one is concerned with the origin of optical activity in nature[4)].

There is one straightforward loophole that resolves this dilemma, namely to make the number of degrees of freedom in $\hat{H} \rightarrow \infty$[6]. In this limit, which must be described by quantum field theory, the existence of a Hamiltonian that generates a set of incoherent Hilbert spaces is the rule. One then speaks of the action of superselection rules that forbid transitions between states that lie in different, orthogonal Hilbert spaces. This is obviously a natural framework for the description of the quantum states of the $d$- and $\ell$-isomers of an optically active substance, which can be associated with the two different eigenvalues of an hermitian operator that commutes with all other observables[51)]: this operator says simply that the system is "right-handed" or "left-handed" and this is in accordance with the orthogonality relations

$$\left.\begin{aligned} \langle \psi_n^d | \psi_m^d \rangle &= \langle \psi_n^\ell | \psi_m^\ell \rangle = \delta_{nm} \\ \langle \psi_n^d | \psi_m^\ell \rangle &= 0 \end{aligned}\right\} \forall_{n,m} \tag{4.19}$$

No such non-trivial ("chirality") operator can exist in a quantum mechanical system with a finite number ($n$) of degrees of freedom because any operator commuting with all observables must be a multiple of the identity operator if $n < \infty$[51)]. Thus this symmetry argument leads us back naturally to a quantum field theory context for understanding optical activity.

In order to derive Rosenfeld's equation from Eq. (3.14) we must make some such argument as the following. Let us assume that the self-consistent quantum field theory (Sect. 2) has been worked through and has yielded composite $n$-particle elementary excitations that we identify with molecules. We can then define a charge and current density for a molecule containing $n$ electrons and nuclei,

6 Of course another resolution of this dilemma is to give up the condition $[\hat{\mathsf{v}}, \hat{\mathsf{P}}] = 0$; this is a real possibility if we modify our definition of a chemical substance so as to allow interactions between electrons and nuclei involving the parity violating weak neutral current in addition to the parity-invariant Coulomb potential. This will be considered further in Sect. 5. At this stage in the argument it should be clear that recourse to the weak interactions cannot give a *complete* theory of optical activity, since one must necessarily deal with the quantum mechanics of the macroscopic optically active medium, which as we shall see is already sufficient to generate spontaneously broken space-inversion symmetry. Moreover by working directly with quantum field theory one avoids all metaphysical difficulties over the meaning of the limit $n \rightarrow \infty$; this is a powerful reason for giving up the classical "building-block" view of matter

$$\hat{\mathsf{n}}(\mathbf{x}) = \sum_{k}^{n} Z_k \delta^3(\mathbf{q}_k - \mathbf{x})$$

$$\hat{\mathsf{j}}(\mathbf{x}) = \frac{1}{2}\sum_{k}^{n} Z_k \, [\hat{\mathbf{q}}_k, \delta^3(\hat{\mathbf{q}}_k - \mathbf{x})]_+ \,, \tag{4.20}$$

and the electric and magnetic polarization fields are related to these densities by,

$$\hat{\mathsf{n}}(\mathbf{x}) = -\nabla \cdot \hat{\mathbf{P}}(\mathbf{x})$$

$$\hat{\mathsf{j}}(\mathbf{x}) = \frac{d\hat{\mathbf{P}}(\mathbf{x})}{dt} + \nabla \vee \hat{\mathbf{M}}(\mathbf{k}) \tag{4.21}$$

If $n$ is such that we are dealing with an ordinary molecule, rather than a macromolecule, we can take the spatial extension of the charge density, $\langle 0|\hat{\mathsf{n}}(\mathbf{x})|0\rangle$, to be small in comparison with an optical wavelength, and use a multipole expansion for the polarization fields. We then easily find[45],

$$\hat{\mathbf{P}}(\mathbf{x}) = \sum_{A}^{\mathcal{N}} \{\hat{\mathbf{d}}_A + \ldots\}\delta(\mathbf{x} - \mathbf{R}_A)$$

$$\hat{\mathbf{M}}(\mathbf{x}) = \sum_{A}^{\mathcal{N}} \{\hat{\mathbf{m}}_A + \ldots\}\delta(\mathbf{x} - \mathbf{R}_A) \tag{4.22}$$

where $\hat{\mathbf{d}}_A$ and $\hat{\mathbf{m}}_A$ are the electric and magnetic dipole operators respectively for molecule A, and $\mathbf{R}_A$ is a c-number reference coordinate located within the molecule that may be conveniently identified with the molecular centre-of-mass. In the dipole approximation we thus obtain,

$$\hat{\mathbf{P}}(\mathbf{k}) = \sum_{A}^{\mathcal{N}} \hat{\mathbf{d}}_A e^{i\mathbf{k}\cdot\mathbf{R}_A}$$

$$\hat{\mathbf{M}}(\mathbf{k}) = \sum_{A}^{\mathcal{N}} \hat{\mathbf{m}}_A e^{i\mathbf{k}\cdot\mathbf{R}_A} \tag{4.23}$$

The crudest approximation to the density matrix for the system is obtained by assuming that there are no statistical correlations between the elementary excitations (perfect fluid), so that $\zeta$ can be written as a simple product of molecular density matrices $\zeta_A$. A better approximation is obtained if one does a quantum field theory calculation of the local field effects in the system which in a certain approximation gives the Lorentz-Lorenz correction $L(\bar{n})$ in terms of the refractive index $\bar{n}$[53]. One then writes,

$$\xi = \xi_0 + \xi_1$$

where (4.24)

$$\xi_0 = L(\bar{n}) \prod_{A=1}^{\mathcal{N}} \xi_A$$

and $\xi_1$ contains all the effects of interactions between the elementary excitations that give rise to spectral linewidths etc.[47]. In the same zero-order approximation the wavefunctions for the system can be written as a simple Hartree product of molecular wavefunctions $\{\phi_n\}$. The spectral function $\mathcal{S}(\mathbf{k}, \omega)$ then becomes,

$$\mathcal{S}_0(\omega) = \left(\frac{\mathcal{N}}{\mathcal{V}}\right) \sum_{n,m} \lambda_m \langle \phi_m | \hat{\mathbf{d}} | \phi_n \rangle \langle \phi_n | \hat{\mathbf{m}} | \phi_m \rangle \delta(\omega - \omega_{nm}) \tag{4.25}$$

where the $\omega_{nm}$ are interpreted as molecular excitation energies. Notice that with the zero order approximation for the density matrix, and the dipole approximation for the polarization fields, the $\mathbf{k}$-dependence drops out of $\mathcal{S}(\mathbf{k}, \omega)$ as does the centre-of-mass momentum of the molecules: the $\langle \phi_n \rangle$ can then be interpreted as wavefunctions for the molecular internal states. Of course once one allows the molecules to interact these simplifications no longer hold. Using (4.25), and (3.9)–(3.12) and taking $\lambda_0 = 1$, $\lambda_n = 0$ $(n \neq 0)$ in the density matrix we finally obtain the Rosenfeld equation[49],

$$\theta = \left(\frac{\mathcal{N}}{\mathcal{V}}\right) \mathrm{L}(\bar{n}) \left(\frac{32\pi^2}{3}\right) \left(\frac{\hbar\omega^2}{c}\right) \mathrm{Tr}[\boldsymbol{\sigma}(\omega)] \tag{4.26}$$

where we have defined the gyration tensor for the molecular ground state $|\phi_0\rangle$,

$$\boldsymbol{\sigma}(\omega) = \mathrm{Im} \sum_n \frac{\langle \phi_0 | \hat{\mathbf{d}} | \phi_n \rangle \langle \phi_n | \hat{\mathbf{m}} | \phi_0 \rangle}{\mathrm{E}_{0n}^2 - \hbar^2\omega^2} \tag{4.27}$$

The discussion of the parity operator can now be repeated verbatim with states $\{\psi_n\}$ replaced by the molecular states $\{\phi_n\}$, *and the understanding that the set $\{\phi_n\}$ are associated with the quasi-particle "solutions" of the quantum field theory of the chiral medium.* In the next section we discuss some models of these molecular (quasi-particle) states.

## 5 Microscopic Theories of Optical Activity

This last section is a critical discussion of the various models that have recently been proposed for the description of the wavefunctions $\{\phi_n\}$ of a resolved optically active molecule[5, 9–14, 16, 54, 55]. All these models have the underlying structure of a double-well potential, and can be considered to be descendants of the usual theory of the ammonia inversion. No attempt will be made to discuss racemization and quantum statistical mechanics $(T \neq 0)$. The difference between the models lies in the assumptions made about the origin of this double-well potential $V(\alpha)$ and the nature of the "coordinate" $\alpha$ involved. In one class of model the double-well potential is *assumed* at the outset as part of the model or justified by appeal to the Born-Oppenheimer approximation, and then the effects on the dynamics of various additional interactions are considered[5, 9, 54, 55]; in the second class of model the double well structure may be regarded as *arising from the dynamics* specified in the model[10–14, 16]. It may be noted that a similar distinction can be made in analogous models of broken space-reflection symmetry that are of interest in elementary particle and condensed matter physics[56–59].

The simplest model of the first type is the standard textbook discussion of optical activity that has recently been restated by Barron[5]. In this model one considers a double-well potential for a vibrational inversion coordinate, with a large barrier to inversion: the stationary states $\{\phi_\pm\}$ of this potential have definite parities, and are necessarily non-degenerate if the barrier height is finite. However, for large enough barrier heights the energy separation, $\varepsilon$, of pairs of $(\pm)$-states becomes very small, and it follows that if one of the degenerate pair of non-stationary states

$$\phi_R = \frac{1}{\sqrt{2}}[\phi_+ + \phi_-]$$

$$\phi_L = \frac{1}{\sqrt{2}}[\phi_+ - \phi_-]$$

with the system localized in one of the wells, can be prepared, it will have a long lifetime $\tau$. One can then invoke the usual argument (Sect. 1) about the relationship of $\tau$ to the timescale of an experiment in which optical activity can be measured[5]. This model certainly describes optical activity but begs the question as to how the broken symmetry molecular state is prepared[8]; nor is it clear how to interpret the model physically when the Born-Oppenheimer approximation is relaxed.

There are two other models that take the double-well potential as their starting point. In the first, due to Harris and Stodolsky[54, 55], one considers the effects on the states $\phi_R$, $\phi_L$ of the weak interaction mediated by the weak neutral current. In the simplest discussion one asserts that this interaction gives rise to an additional potential $\hat{V}_x$ between electrons and nuclei that does not have space-inversion symmetry $[\hat{P}, \hat{V}_x] \neq 0$. It is then shown that the effect of $\hat{V}_x$ is to lift the degeneracy of the pair $(\phi_R, \phi_L)$ by causing a splitting $\delta$, which for typical chiral molecules is $\gg$ the tunneling splitting $\varepsilon$ and this has the consequence that the stability of the handed states is greatly enhanced: an equivalent statement is that the weak neutral current has the effect of hindering dramatically the effective rate for tunneling through the barrier. It is then more natural to say that in the favourable cases with $\delta \gg \varepsilon$, the *handed* states are the stationary states of the molecular Hamiltonian augmented by the weak interaction force, and these states are weakly perturbed by tunneling. At a more fundamental level it may be noted that the potential $\hat{V}_x$ can be thought of as arising from the virtual interaction of the electron and nuclear fields with an additional quantized field, the quanta of which are massive, boson particles, so-called Z particles. Just as the Coulomb interaction between charges arises from the exchange of virtual photons between charges, and is a long-range potential because the photon mass is zero, so the exchange of virtual Z-particles between charges leads to an effective potential $\hat{V}_x$ which has a very short range because the mass of the Z-particle is so great[55].

This remark leads us naturally to the first model proposed by Davies[13], in which a "particle" moving in a double well potential ("the system") is coupled to a boson field (a quantized "reservoir") and the dynamical evolution of the coupled system is studied in the weak coupling limit. If the strength of the interaction between the potential wells is measured by a parameter $\mu$ (related to the tunneling matrix element), and the coupling of the particle to the field is determined by a coupling constant $\lambda$, then a rigorous mathematical analysis of the time evolution of the system is possible for the limiting case $\mu \to 0$,

$\lambda \to 0$. One finds that the asymptotic evolution of the system (i.e. the equilibrium states) depends crucially on the relative speed at which $\lambda$ and $\mu$ converge to zero. If we put $\mu = \lambda^\beta$ where $0 < \beta < \infty$ then the time evolution changes discontinuously at $\beta = 2$. Physically the result may be summarized by the statement that if $2 < \beta < \infty$ the tunneling interaction *between* the wells has no effect for times of order $\lambda^{-2}$, while if $0 < \beta < 2$ it has a direct effect and modifies the decay caused by the system-field interaction: thus the equilibrium states in the two cases are quite distinct. In a quite precise sense this difference arises because when $\beta$ exceeds the critical value 2, a superselection rule operates in the theory forbidding the appearance of superpositions between the degenerate eigenstates of the two non interacting potential wells[3,13]. This is, of course, just what one expects (Sect. 4) in a theory of optical activity, cf. the remarks following Eq. (4.14). Note that although the original discussions of Harris and Stodolsky[54,55], and Davies[13] seem rather different, they are actually quite similar once one appreciates that (a) the potential $V_x$ arises from the coupling of the charges in the system to a boson field and (b) Davies has to integrate out the field variables when calculating the equilibrium states of the system and so there is finally an "effective potential" involving only the particle coordinates. The main difference is that the weak neutral current interaction does not have space-inversion symmetry whereas in Davies' model the coupling of the particle to the boson field is not explicitly required to break this symmetry: the outcome of the two models is the same however, namely that for suitable choices of parameters in the models tunneling between the potential wells is inhibited by the field interaction ($\delta \gg \varepsilon$, $\mu = \lambda^\beta$ with $\beta > 2$ are similar conditions). This does not mean that the weak neutral current can be neglected since it is a real effect and there is no reason to suppose that both mechanisms are not operative; the nature of the boson field in Davies' model will be discussed shortly.

The other class of models[9–12,14,16] has the feature that the double well structure is generated by the dynamics, rather than being put in by hand, and so they can be considered as model calculations of the elementary excitations in a quantum field theory. I shall first summarize the mathematical analysis of these models following the recent treatment of Davies[16], and then comment on their physical interpretation in the light of the previous sections. These dynamical models are based on the idea of coupling a molecular Hamiltonian to a quantized boson field ('a reservoir'), and studying by *variational* rather than perturbative methods the ground-state energy and time evolution of the combined system. It may be helpful to note at the outset that symmetry breaking is found in variational solutions based on the Hartree (mean-field) approximation in a way that is reminiscent of symmetry breaking in unrestricted Hartree-Fock calculations of electronic structure[60] – see also[56–59] for the use of mean-field approximations.

The starting point is the Hamiltonian,

$$\hat{K} = \hat{H} + \hat{V} + \hat{\mathscr{F}} \tag{5.1}$$

on $\mathfrak{H} \oplus \mathfrak{F}$, where $\mathfrak{H}$ is the Hilbert space associated with the molecule ($\mathscr{L}^2(\Re^{3\mathscr{N}})$) for a $\mathscr{N}$-particle system (Sect. 2) which is described by the usual self-adjoint Hamiltonian $\hat{H}$, or some subspace spanned by the eigenstates of $\hat{H}$. The boson Fock space $\mathfrak{F}$ has single particle space $\mathfrak{F}_1$ and the reservoir Hamiltonian $\hat{\mathscr{F}}$ is the *free* Hamiltonian whose single-particle term $\hat{\mathscr{F}}_1$ is assumed to be self-adjoint, and strictly positive in the sense that

$$\langle \phi, \hat{\mathscr{F}}_1 \phi \rangle > 0 \tag{5.2}$$

for all non-zero $\phi$ in its domain. $\hat{\mathcal{F}}$ is the usual bilinear expression in terms of creation ($\hat{a}^+(\mathbf{k})$) and annihilation ($\hat{a}(\mathbf{k})$) operators for bosons with momentum **k** and frequency $\omega = |\mathbf{k}|c$, where $c$ is the particle speed. The analysis applies to any boson system with a dispersion law that makes $\omega$ a non-negative function of **k**.

In physical applications the bosons have been thought of as either phonons ($c$ = speed of sound in the substance) or photons ($c$ = speed of light). The interaction term is assumed to have the form

$$\hat{V} = \sum_{\varkappa=1}^{m} \{\hat{A}_\varkappa \hat{a}^+(f_\varkappa) + \hat{A}_\varkappa^+ \hat{a}(f_\varkappa)\} \tag{5.3}$$

where $\{\hat{A}_\varkappa\}$ are bounded operators on $\mathfrak{H}$, and $\hat{a}^+(f_\varkappa)$, $\hat{a}\,(f_\varkappa)$ are smeared field operators on $\mathfrak{F}$.

Davies[16] has given a rigorous mathematical analysis of the variational estimate of the ground state of the Hamiltonian $\hat{K}$ based on trial functions consisting of the tensor product $\chi$ of a molecular wavefunction $\phi$, and a coherent state $\psi$ for the boson field, i.e. one searches for the minimum value of

$$\langle\chi|\hat{K}|\chi\rangle \tag{5.4}$$

with

$$\chi = \phi \times \psi,\ ||\phi|| = ||\psi|| = 1\,. \tag{5.5}$$

It is easily seen that for such trial functions the minimization of the Hamiltonian $\hat{K}$, Eq. (5.1), may be replaced by the minimization of a specified nonlinear functional $\mathfrak{E}(\phi)$ of the molecular states $\phi$ alone. In the following we refer to either formulation as seems convenient. This argument also enables one to connect these field theoretical models with the earlier suggestion of mine that "molecular structure states" can be associated with those solutions of the Schrödinger equation for the full molecular Hamiltonian $\hat{H}$ that satisfy certain subsidiary conditions[3, 35], if the latter are associated with the non-linearity in the functional $\mathfrak{E}(\phi)$. As we shall see, it may happen that $\mathfrak{E}(\phi)$ has two degenerate minima and it is in this sense that the dynamics gives rise to a "double-well" structure.

In order to guarantee that $\hat{K}$ actually possesses a ground-state it is necessary to state the conditions that ensure that it is bounded below. This turns out to be the requirement that the energy of the boson field be finite, so we must have

$$\Lambda = ||\mathcal{F}_1^{-1/2} f_\varkappa|| < \infty \tag{5.6}$$

As a result of the interaction between the molecule and the boson field, the "bare" molecule becomes "dressed" with a cloud of boson particles: in the language of Sect. 2 the dressed molecule is an elementary excitation in the many-body system of molecules and boson particles. The number density of dressing boson particles is given by

$$\Gamma = ||\mathcal{F}_1^{-1} f_\varkappa|| \tag{5.7}$$

In other studies, Davies[14, 15] and Pfeifer[11, 12] (see below) have described models with the property that $\Gamma = \infty$ because the dressed molecule involves a cloud of soft (long wavelength) bosons of infinite number but finite total energy ($\Lambda < \infty$). In such cases the ground state of $\hat{K}$ may be degenerate and associated with inequivalent representations of the commutation relations by means of some superselection rule. It is very satisfactory to see that whether or not degeneracy is actually realized depends crucially on the energy separation of the low-lying eigenstates of the molecular Hamiltonian $\hat{H}$ in relation to the field energy $\Lambda$. Furthermore, the degenerate components of the ground state and hence also the minima in $\mathfrak{E}(\phi)$, are related by the space-inversion operation, and so these models are candidates for a dynamical theory of optical activity at $T = 0$.

The authors of these models[11, 12, 14–16)] have concentrated mainly on the technical, mathematical aspects of proving rigorous statements about the ground states of their models: it is however just as important to look at the physics that might be associated with these models, and I shall attempt this within the framework described in Sect. 2. Davies has consistently assumed that the reservoir is described by a free boson field which is taken to represent the "environment" of the molecule. Following Sect. 2 I interpret "environment" to mean all the elementary excitations that can couple to the elementary excitations we call "molecules". In a solid the existence of a phonon field of elementary excitations is relatively clear-cut, and a quadratic (free-field) phonon Hamiltonian is a useful approximation[61)]; in a fluid however the situation is less favourable, as may be seen from the following discussion.

In order to analyse the collective modes in a fluid we must examine the density fluctuation propagator $\mathsf{F}(\mathbf{x}, \tau)$ (Green function) which is essentially the probability amplitude that if the density is $\mathsf{n}$ at $x = 0$, it will be the same at $x' = \mathbf{x}, \tau$[28)]

$$\mathsf{F}(\mathbf{x}, \tau) = -\mathrm{i}\langle 0|\mathsf{T}\{\hat{\mathsf{n}}(\mathbf{x}, \tau)\hat{\mathsf{n}}(\mathbf{0}, 0)\}|0\rangle \qquad \tau > 0 \tag{5.8}$$

where $\hat{\mathsf{n}}$ is given by Eq. (2.17). $\mathsf{F}(\mathbf{x}, \tau)$ is evidently a special case of the two-particle Green function for the atomic or molecular field $\chi$, introduced in Sect. 2. Its space-time fourier transform has the Lehmann representation,

$$\mathsf{F}(\mathbf{k}, \omega) = \int_0^\infty \mathrm{d}\omega' \mathsf{S}(\mathbf{k}, \omega') \left[\frac{1}{\omega - \omega' + \mathrm{i}\delta} - \frac{1}{\omega + \omega' - \mathrm{i}\delta}\right] \tag{5.9}$$

where $\mathsf{S}(\mathbf{k}, \omega')$ is the *dynamical structure factor* measured in inelastic scattering experiments involving light, X-rays, neutrons etc. Suppose now that there is only a single density-fluctuation excitation at frequency $\omega(\mathbf{k})$ i.e.

$$\mathsf{S}(\mathbf{k}, \omega') = \mathsf{S}(\mathbf{k})\, \delta(\omega' - \omega(\mathbf{k}))\ , \tag{5.10}$$

then for this idealized case we have

$$\mathsf{F}(\mathbf{k}, \omega) = \mathsf{S}(\mathbf{k}) \left[\frac{1}{\omega - \omega(\mathbf{k}) + \mathrm{i}\delta} - \frac{1}{\omega + \omega(\mathbf{k}) - \mathrm{i}\delta}\right] \tag{5.11}$$

This expression is the Green function for a field quantum associated with the free field $\hat{\mathsf{n}}(\mathbf{x})$ since it describes undamped propagation, cf. Eq. (2.13). Hence if $\mathsf{S}(\mathbf{k}, \omega)$ possesses a well-developed peak in the vicinity of some $\omega(\mathbf{k})$ then we can regard this excitation as a quasi-particle for a collective mode resembling a *phonon*, which is an undamped sound-wave.

In real fluids, low-frequency light-scattering reveals a central Rayleigh peak, due to heat diffusion, and two symmetrically displaced Brillouin peaks, due to sound waves, such that

$$\omega(\mathbf{k}) = \omega(\mathbf{k}_0) + \mathrm{i}\,\gamma(\mathbf{k}_0)\,, \qquad \omega(\mathbf{k}_0) = |\mathbf{k}_0|c \tag{5.12}$$

where $c$ is the speed of sound. The lifetimes, $\gamma^{-1}$, of the phonons are not long since they tend to decay into random single-particle molecular motions. This is a typical hydrodynamic regime where equilibrium is maintained by fast collisions; thus if $\tau$ is the collision time, and $\omega$ the frequency of the collective mode, the description is valid when $\omega\tau \ll 1$. The absence of long range-order in the fluid leads to the breakdown of the picture of phonons as harmonic oscillations coupled by small anharmonicities that is valid in a solid. Thus although phonons in a liquid can be described as the long-wavelength excitations of a boson field, one never gets to the free-field limit, and a realistic description is in terms of a *self-interacting boson field* which of course interacts with the single-particle elementary excitations as in Davies' model. A useful analogy is to think of the short lifetimes of plasmons in an electronic system at energies near to the purely single-particle excitation spectrum. As far as the *phonon* interpretation of Eq. (5.1) is concerned we therefore have two possibilities: Firstly, we can take $\mathcal{F}$ to be the Hamiltonian for a free boson field in which case physically the reservoir must be identified with a *solid*, with which the molecules interact. Alternatively, if we also wish to have the possibility of regarding the phonons as elementary excitations in a fluid, and it seems to me that this interpretation is strongly preferred on physical grounds, we can no longer take $\mathcal{F}$ as a free-field Hamiltonian. It is then certainly much more difficult to make a rigorous mathematical analysis; although it has been conjectured that similar results hold for a self-coupled field[62], this has not yet been proven.

Finally, we must consider the model of molecular chirality proposed by Pfeifer[11, 12]. His model consists of a neutral molecule, composed of $\mathcal{N}$ point charges, coupled to the quantized radiation field described in the minimal coupling formalism using the Coulomb gauge for the vector potential operator $\hat{\mathbf{A}}(\mathbf{x})$. He then assumes that the ground and first excited internal states of the molecule are well separated in energy from the higher states so as to reduce the theory to a two state problem. Rather than using the electric dipole approximation he elects to drop the $\hat{\mathsf{A}}^2$ term from the Hamiltonian, at least initially, so that the model is of the form of Eq. (5.1), and is closely related to the Dicke model of a laser[63]. The field is quantized in a box of length $\mathcal{L}$, but eventually the limit $\mathcal{L} \to \infty$ is discussed, i.e. the infinite number of degrees of freedom of the field are taken seriously.

Analysis of the model then shows that "a molecule with an almost degenerate ground state is structurally unstable in the following sense: if the difference between the two lowest energy levels is below a certain critical value (determined by the molecular Hamiltonian, $\hat{\mathsf{H}}$, in question), then the coupling of the molecule to the quantized radiation field yields two strictly degenerate, symmetry broken ground states of the molecule which are separated by a super-selection rule. The classical observable underlying this invalidation of the unrestricted superposition principle is chirality and originates from the infrared singularity of the electromagnetic field. On the other hand, if this energy difference exceeds the critical value then the achiral ground state of the molecule is not changed by the interaction with the field"[11)].

The off-diagonal matrix element for transitions between the ground and excited states $\psi_0$, $\psi_1$ caused by a photon with wave-vector $\mathbf{k}_n$, polarization $\alpha_n$ is,

$$\lambda_{n,\alpha} = \mathrm{i}\left(\frac{1}{2\,\varepsilon_0|\mathbf{k}_n|c\,\mathrm{L}^3}\right)^{1/2}\sum_{\nu=1}^{\mathcal{N}}\left(\frac{\mathrm{Z}_\nu}{\mathrm{m}_\nu}\right)\left\langle\psi_0|\mathrm{Cos}\,(\mathbf{k}_n\cdot\hat{\mathbf{q}}_\nu)\hat{\mathbf{p}}_\nu|\psi_1\right\rangle \tag{5.13}$$

where $\{\hat{\mathbf{q}}_\nu, \hat{\mathbf{p}}_\nu, \nu = 1 \ldots \mathcal{N}\}$ are the canonical variables of the $\mathcal{N}$ particles in the molecule (charge $\mathrm{Z}_\nu$, mass $\mathrm{m}_\nu$, $\nu = 1, \ldots \mathcal{N}$). Then if we define the number density of photons that dress the molecule (Eq. (5.7) in explicit form),

$$\Gamma_{\mathcal{F}} = \sum_{n=1}^{\mathcal{F}}(\lambda_{n,\alpha}/\omega_n)^2 \tag{5.14}$$

where $\omega_n = |\mathbf{k}_n|c$ and the sum is over field modes, the crucial result obtained by Pfeifer[11)] is that

$$\lim_{\mathcal{L}\to\infty}\Gamma_\infty = \infty\;;\quad \text{(logarithmic singularity in } \mathcal{L}) \tag{5.15}$$

The superselection rule associated with the induced optical activity has its origin in the infra-red divergence ($\omega \to 0$) of the dimensionless quantity $\Gamma$ associated with the coupling of the molecule to an infinite number of degrees of freedom of the field. If one drops the $\hat{\mathbf{A}}^2$ term from the minimal coupling Hamiltonian one cannot follow through the variational calculation based on Eq. (5.1) since a simple scaling argument[16)] shows that the resulting Hamiltonian is not bounded from below. This is not a difficulty in Pfeifer's original two-state model[11)]. However the same sort of results as follow from Eq. (5.1) appear to be obtained if the $\hat{\mathbf{A}}^2$ term is put back into the Hamiltonian and one does a variational calculation using Hartree product states, (5.5), for the molecule + field system.

Although this is a striking conclusion, it has to be said that it confronts us with a paradox: if we consult a modern text of quantum electrodynamics[64)] we read "Infrared catastrophe is caused by the emission of an infinite number of soft photons by accelerated charged particles. The infrared catastrophe also occurs for particles with a continuous distribution of charge, provided that the total charge is not zero." This result is convincingly established if one studies quantum electrodynamics in the manifestly gauge-invariant formalism[64)], and it implies that a *neutral atom or molecule should have no infrared divergence.* Pfeifer argued against the use of the electric dipole approximation because of the resulting spurious ultra-violet singularity: however the crucial result is that the divergence in $\Gamma_\infty$, Eq. (5.15), is associated with the long-wavelength (soft) photons, and in this regime the electric dipole approximation is valid, and one could expect to use this approximation to investigate $\Gamma$ for $\omega \simeq 0$. Moreover for photon wavelengths shorter than the particles' Compton wavelength, $\hbar/\mathrm{m}_\alpha c$, even the complete non-relativistic Hamiltonian is quite untrustworthy so that ultra-violet singularities cannot be taken seriously in a non-relativistic framework. In the electric dipole approximation the gauge-invariant interaction operator is simply $\hat{\mathrm{H}}^{\mathrm{int}} = -\hat{\mathbf{d}}\cdot\hat{\mathbf{E}}^{\perp}(0)$ where as in Sect. 4, $\hat{\mathbf{d}} =$

$\sum_{\nu=1}^{\mathcal{N}} Z_\nu \hat{\mathbf{q}}_\nu$ ,is the molecular electric dipole moment operator, and $\hat{\mathbf{E}}^\perp(0)$ is the transverse displacement field evaluated at the molecular centre-of-mass: then since the fourier expansion of $\hat{\mathbf{E}}^\perp$ behaves as $(|\mathbf{k}_n|)^{1/2}$ (rather than $(|\mathbf{k}_n|)^{-1/2}$ as in the vector potential $\hat{\mathbf{A}}$), it is easily seen that the number density of photons, $\bar{\Gamma}$, calculated in the gauge-invariant formalism cannot have an infrared singularity. Recall that with the gauge-invariant formalism in the dipole approximation there is no term analogous to the $\hat{A}^2$ term in the minimal coupling Hamiltonian[44, 45]. The reader may find it surprising that I should claim that the operator $-\hat{\mathbf{d}}\cdot\hat{\mathbf{E}}^\perp$ leads to a result different from that obtained with the Coulomb gauge interaction operators, $(e/m)\hat{\mathbf{p}}\cdot\hat{\mathbf{A}} + (e^2/2m)\hat{A}^2$, when their "equivalence" is well-known in say light-scattering theory. The point is that in light-scattering one can calculate scattering amplitudes in the gauge-invariant and minimal coupling formalisms using *perturbation theory* and show that the physically important *on-energy-shell* amplitudes are identical[65–67]. In the variational calculations considered here there is no on-energy-shell condition to be satisfied.

There is also a deeper reason for this inequivalence which can be put as follows: in Sect. 2 I noted that in field theories one has to make a definite decision as to the representation of the Hilbert space on which the Hamiltonian is to be considered as an operator. This remark applies just as much to quantum electrodynamics as to the quantum field theory of matter: one can either use gauge-dependent "coordinate" representations in which the vector potential operator $\hat{\mathbf{A}}(\mathbf{x})$ is diagonal at every point $\mathbf{x}$, so that its eigenvalues $\mathbf{A}(\mathbf{x})$ can be used to parameterize the states, or alternatively one can do the same thing in a gauge-invariant fashion using either the electric or magnetic fields. Each choice has associated with it a photon Fock space. By working directly with the canonical formalism for quantum electrodynamics one can show that physical, gauge-invariant quantum states, $\psi$, have the form[68)]

$$\psi = \exp(i\hat{\varphi}/h)\phi \equiv \hat{U}\phi \tag{5.16}$$

where

$$\hat{\varphi} = \int d^3x \hat{\mathbf{P}}(\mathbf{x})\cdot\hat{\mathbf{A}}(\mathbf{x}) , \tag{5.17}$$

and $\phi$ is a gauge-dependent state calculated in the (arbitrary) gauge of the vector potential operator $\hat{\mathbf{A}}(\mathbf{x})$. As in Sect. 3, $\hat{\mathbf{P}}(\mathbf{x})$ is the electric polarization field for the matter. On the other hand, if $\hat{\mathbf{A}}$ is the Coulomb gauge vector potential, then Eqs. (5.16, 5.17), define the Power-Zienau-Woolley (PZW) Transformation[44, 45] that leads to the QED Hamiltonian, $\bar{\hat{H}}$, expressed in terms of the fields $(\hat{\mathbf{E}}^\perp, \hat{\mathbf{B}})$, rather than the vector potential $\hat{\mathbf{A}}$. The PZW Transformation is carried out by expressing the minimal coupling QED Hamiltonian $\hat{H}(\Omega)$ in terms of the transformed dynamical variables $\bar{\hat{\Omega}}$,

$$\bar{\hat{\Omega}} = \hat{U}\hat{\Omega}\hat{U}^{-1} \tag{5.18}$$

using the power series expansion of the exponential operator $\hat{U}$,

$$\hat{U} = \sum_{n=0}^{\infty} (n!)^{-1} (i\hat{\varphi}/h)^n \tag{5.19}$$

to compute the $\hat{\bar{\Omega}}$. At the same time the Fock space bases for photons with momentum $\mathbf{k}_n$ and polarization $\alpha_n$ (or $\bar{\mathbf{k}}_n$, $\bar{\alpha}_n$) are given by $|n\rangle = |\mathbf{k}_n, \alpha_n\rangle$, $|\bar{n}\rangle = |\bar{\mathbf{k}}_n, \bar{\alpha}_n\rangle$ and are formally related by $|\bar{n}\rangle = \hat{U}|n\rangle$.

It is important to recognize that in the limit $\mathscr{L} \to \infty$, $\mathscr{F} \to \infty$ used in Eq. (5.15) to establish the key result $\Gamma \to \infty$, $\hat{U}$ is *not* unitary. This can be exhibited by arranging that the annihilation operators act before the creation operators, so as to isolate the effects of virtual photons: if we write

$$\hat{\varphi} = \hat{\varphi}^{+} + \hat{\varphi}^{-} \tag{5.20}$$

where $+(-)$ refers to the creation (annihilation) operator part, then we may write[45, 69]

$$\hat{U} = \exp(i\hat{\varphi}/h) = \exp(i\hat{\varphi}^{+}/h)\exp(i\hat{\varphi}^{-}/h)\exp(-[\hat{\varphi}^{-}, \hat{\varphi}^{+}]/2h^2) \tag{5.21}$$

since the commutator is a c-number. If one now uses the expansion of the Coulomb gauge vector potential $\hat{\mathbf{A}}(\mathbf{x})$ as a fourier series in a box of side $\mathscr{L}$, it is a straightforward matter to show that in the $\mathscr{L} \to \infty$ limit the commutator diverges unless an ultra-violet cut-off is imposed, i.e. for point-like sources

$$\lim_{|\mathbf{k}| \to \infty} [\varphi^{-}, \varphi^{+}] = \infty \tag{5.22}$$

so that in this limit $\hat{U}$ contains the factor $\exp(-\infty)$. Thus when $\hat{U}$ acts on states $\{|n\rangle\}$ with only a finite number of photons (e.g. states in the photon Fock space) we find $\langle n'|\hat{U}|n\rangle = 0$, and so $\hat{U}$ does not transform the one photon Fock space $\{|n\rangle\}$ into the other photon Fock space $\{|\bar{n}\rangle\}$, since all its matrix elements vanish in the Fock space basis $\{|n\rangle\}$. This implies that $\hat{U}$ is *not* unitary: the Fock space bases $|n\rangle$ and $|\bar{n}\rangle$ associated with the gauge-dependent, minimal coupling and gauge-invariant formalisms respectively are orthogonal bases belonging to *inequivalent* representations of the Hilbert space. With respect to the Fock space $\{|n\rangle\}$ the power series expansion (5.19) does not converge to the exponential operator $\hat{U}$ by virtue of the fact that, for point-like sources, the mean square of $\hat{\varphi}$ i.e. $\langle n|\hat{\varphi}\hat{\varphi}|n\rangle$ diverges if $|n\rangle$ is a state in Fock space. These results are *not* dependent on the use of a non-relativistic representation of the polarization field $\hat{\mathbf{P}}(\mathbf{x})$ in Eq. (5.17), cf.[70].

Now, returning to Pfeifer's model of chirality we see that we have to make a choice of representation when selecting states to use in a Hartree variational calculation of the ground-state of the molecule-radiation field system. In Pfeifer's calculations the trial functions are chosen as coherent states, say $\psi$, based on the photon Fock space $\{|n\rangle\}$ in the Coulomb gauge theory: an inequivalent set of trial functions is obtained by choosing coherent states, $\bar{\psi}$, based on the gauge-invariant photon Fock space $\{|\bar{n}\rangle\}$. One then has to compare the results of two minimization calculations involving, (cf. Eq. 5.4),

$$\langle \phi \oplus \psi|\hat{K}|\phi \oplus \psi\rangle \quad \text{Coulomb gauge theory} \tag{5.23}$$

and

$$\begin{aligned} \langle \phi \oplus \bar{\psi}|\hat{K}|\phi \oplus \bar{\psi}\rangle &= \langle \phi \oplus \psi|\hat{U}\hat{K}\hat{U}^{-1}|\phi \oplus \psi\rangle \\ &\equiv \langle \phi \oplus \psi|\hat{\bar{K}}|\phi \oplus \psi\rangle \quad \text{gauge-invariant theory} \end{aligned} \tag{5.24}$$

where $\hat{\bar{K}}$ is the Hamiltonian obtained as the Power-Zienau-Woolley transform of $\hat{K}$. The difference between the minimization over $\hat{K}$ and $\hat{\bar{K}}$ is then simply the difference between using the interaction operators

$$H^{int} = -\frac{e}{m}\hat{\mathbf{p}}\cdot\hat{\mathbf{A}} + \frac{e^2}{2m}\hat{A}^2 \qquad \nabla\cdot\hat{\mathbf{A}} = 0 \tag{5.25}$$

and

$$\overline{H}^{int} = -\hat{\mathbf{d}}\cdot\hat{\mathbf{E}}^{\perp} - \ldots. \tag{5.26}$$

the first of which predicts an infra-red singularity in $\Gamma$ while the second is regular in the $\omega \to 0$ limit for $\bar{\Gamma}$, and hence cannot give a chirality transition. My opinion is that it makes better *physical* sense to choose trial functions that are manifestly gauge-invariant, for if one does not, one has to face the task of proving that any given result is not simply an artefact of a chosen gauge. The gauge invariance of the theory is an expression of the fact that the gauge group is a symmetry group for the combined matter + field Hamiltonian[68)]; choosing gauge-invariant trial functions is doing no more than choosing a symmetry-adapted basis. Hence I do not believe that quantum electrodynamics plays the role in chirality ascribed to it by Pfeifer and Primas[9–12)]. We are left then with Davies' interpretation of the model (5.1) involving phonon excitations, and the caveat that these results came from calculations in the spirit of the Rayleigh-Ritz variational principle with the assumption of a free boson field Hamiltonian, rather than the exact solutions of the field theory Hamiltonian involving self-coupled phonons.

It should be evident from this account that there is still plenty of scope for the development of dynamical theories of chirality; in particular there is a need for the unification of the models involving interactions with a boson reservoir (the phonon model), and the weak neutral current. It is also the case that we have tacitly assumed that in other respects the optically active medium is close to the Ideal Gas limit in as much that we have neglected the effects of intermolecular interactions (but see[55)]). This should be combined with a better treatment of the long-wavelength phonon modes which decay into random single-molecule motions in the hydrodynamic regime. Finally the difficult question of racemization and the theory of metastable states[23)] has yet to be touched, and this must be a major goal in the quantum theory of optical activity.

## 6 References


1. Woolley, R. G.: Adv. Phys. *25,* 27 (1976)
2. Woolley, R. G.: J. Amer. Chem. Soc. *100,* 1073 (1978)
3. Woolley, R. G.: Israel J. Chem. *19,* 30 (1980)
4. Origins of Optical Activity in Nature (ed. D. C. Walker), Elsevier Scientific Publishing Co., Amsterdam 1979
5. Barron, L. D.: J. Amer. Chem. Soc. *101,* 269 (1979)
6. Barron, L. D.: Mol. Phys. *43*, 1395 (1981)
7. Barron, L. D.: Chem. Phys. Letters *79,* 392 (1981)

8. Woolley, R. G.: Chem. Phys. Letters *79,* 395 (1981)
9. Primas, H.: in Proc. VI Internat. Colloq. Group Theoret. Methods in Physics, Lecture Notes in Physics, Springer Verlag, Heidelberg 1979
10. Primas, H.: in Quantum Dynamics of Molecules: The New Experimental Challenge to Theorists, NATO ASI *B57* (ed. R. G. Woolley), Plenum Press, 1980
11. Pfeifer, P.: Chiral Molecules – A Superselection Rule Induced by the Radiation Field (Doctoral Thesis, Swiss Federal Inst. of Techn., Diss. ETH No. 6551), 1980
12. Pfeifer, P.: in Classical, semiclassical and quantum mechanical problems in mathematics, chemistry and physics (Eds. Gustafson, K., Reinhardt, W. P.), Plenum Press, 1980: see also Pfeifer, P.: J. Phys. A. *A 14*, L 129 (1981)
13. Davies, E. B.: Ann. Inst. Henri Poincaré AXXVIII, 91 (1978)
14. Davies, E. B.: Commun. Math. Phys. *67*, 191 (1979)
15. Davies, E. B.: Commun. Math. Phys. *75,* 263 (1980)
16. Davies, E. B.: Ann. Inst. Henri Poincaré, A, in the press (1982)
17. Le Bel, J. A.: Bull. Soc. Chim. *22,* 337 (1874)
18. van't Hoff, J. H.: La Chimie dans L'Espace, Bazendijk, P. M., Rotterdam 1875
19. Fieser, L. F., Fieser, M.: Introduction to Organic Chemistry, p. 202, Heath, D. C. & Co. (1957). See also numerous papers published by L. Pasteur in the Comptes Rendues, Vols. 26–42 (1848–1856), and his "Leçons de Chimie professées en 1860" published by Société Chimique de Paris (1861) which is reprinted in English in the Alembic Club Reprint booklet No. 14
20. Heisenberg, W.: Physics and Beyond, p. 124, 243, George Allen and Unwin Ltd., 1971
21. Bouchiat, C.: J. Phys. Nucl. Phys. *G3,* 183 (1977)
22. Berry, R. S.: in Quantum Dynamics of Molecules: The New Experimental Challenge to Theorists (ed. Woolley, R. G.) NATO ASI *B57,* Plenum Press, 1980
23. Sewell, G. L.: Physics Reports *57,* 307 (1980)
24. Abrikosov, A. A., Gorkov, L. P., Dzyaloshinskii, I. E.: Methods of Quantum Field Theory in Statistical Physics, Dover Publications Inc., New York 1975
25. Leplae, L., Mancini, F., Umezawa, H.: Physics Reports *10 C,* 151 (1974)
26. Haag, R., Schroer, B.: J. Math. Phys. *3,* 248 (1962)
27. Umezawa, H.: Nuovo Cimento *40,* 450 (1965)
28. Pines, D.: The Many-Body Problem, W. A. Benjamin Inc., New York 1962
29. Anderson, P. W.: Concepts in Solids, W. A. Benjamin Inc., New York 1963
30. Landau, L. D., Lifshitz, E. M.: Course of Theoretical Physics, Vol. 4, Part 2, Relativistic Quantum Mechanics, § 103, § 122 (1974)
31. Gell-Mann, M., Low, F.: Phys. Rev. *84,* 350 (1951)
32. Woolley, R. G.: Chem. Phys. Letters *44,* 73 (1976)
33. Woolley, R. G.: Int. J. Quant. Chem., XII, S 1, 307 (1977)
34. Sutcliffe, B. T., Woolley, R. G.: Chem. Phys. Letters *45,* 393 (1977)
35. Woolley, R. G.: Chem. Phys. Letters *55,* 443 (1978)
36. Sutcliffe, B. T.: in Quantum Dynamics of Molecules: The New Experimental Challenge to Theorists (ed. Woolley, R. G.), Plenum Press, 1980
37. Wadati, M.: Physics Reports *50,* 88 (1979)
38. Goldstone, J.: Nuovo Cimento *19,* 154 (1961)
39. Goldstone, J., Salam, A., Weinberg, S.: Phys. Rev. *127,* 968 (1962)
40. Barron, L. D.: Nature *238,* 17 (1972)
41. Pauling, L., Pauling, P.: Chemistry, W. H. Freemann & Co., San Francisco 1975, p. 153
42. Geissmann, T. A.: Principles of Organic Chemistry, 2nd Ed., W. H. Freemann & Co., San Francisco 1962, p. 420
43. Atkins, P. W.: Molecular Quantum Mechanics, Oxford University Press, 1970, p. 464
44. Power, E. A., Zienau, S.: Phil. Trans. Roy. Soc. *A 251,* 427 (1959)
45. Woolley, R. G.: Adv. Chem. Phys. *33,* 151 (1975)
46. Harris, R. A.: J. Chem. Phys. *43,* 959 (1965)
47. Ben-Reuven, A.: Adv. Chem. Phys. *33,* 236 (1975)
48. Davies, E. B.: Quantum Theory of Open Systems, Academic Press, 1976, p. 4
49. Rosenfeld, L.: Zeits. Physik *52,* 161 (1928)

50. Roman, P.: Some Modern Mathematics for Physicists and other outsiders, Vol. 1, p. 85; Vol. 2, Ch. 12, 1975
51. Roman, P.: Advanced Quantum Mechanics, Addison-Wesley, 1965, p. 10
52. Born, M., Heisenberg, W., Jordan, P.: Zeits. Physik *35,* 557 (1926)
53. Bullough, R. K. et al.: Chem. Phys. Letters *2,* 293, ibid, 307 (1968)
54. Harris, R. A., Stodolsky, L.: Physics Letters *78 B,* 313 (1978)
55. Harris, R. A.: in Quantum Dynamics of Molecules: The New Experimental Challenge to Theorists, NATO ASI *B 57* (ed. Woolley, R. G.), Plenum Press, 1980: see also: Harris, R. A., Stodolsky, L.: J. Chem. Phys. *74,* 2145 (1981)
56. Jackiw, R., Rebbi, C.: Phys. Rev. *D 13,* 3398 (1976)
57. Su, W. P., Schrieffer, J. R., Heeger, A. J.: Phys. Rev. Letters *42,* 1698 (1979)
58. Su, W. P., Schrieffer, J. R., Heeger, A. J.: Phys. Rev. *B 22,* 2099 (1980)
59. Jackiw, R., Schrieffer, J. R.: preprint (1981)
60. Falicov, L. M., Harris, R. A.: J. Chem. Phys. *51,* 3153 (1969)
61. Ashcroft, N. W., Mermin, N. D.: Solid State Physics, Holt Rinehart & Winston, New York 1976, p. 451
62. Davies, E. B.: private communication (1978)
63. Poston, T., Stewart, I. N.: Catastrophe Theory and its Applications, Pitman, London 1972, Ch. 12
64. Białynicki-Birula, I., Białynicki-Birula, Z.: Quantum Electrodynamics, Pergamon, Oxford 1975, p. 226, 437
65. Healy, W. P., Woolley, R. G.: J. Phys. *B 11,* 1131 (1978)
66. Aharonov, Y., Au, C. K.: Phys. Rev. *A 20,* 1553 (1979)
67. Haller, K., Sohn, R. B.: Phys. Rev. *A 20,* 1541 (1979)
68. Woolley, R. G.: J. Phys. *A 13,* 2795 (1980)
69. Capps, R. H., Holliday, W. G.: Phys. Rev. *99,* 931 (1955)
70. Chrétien, M., Peierls, R. E.: Proc. Roy. Soc. (London) *A 223,* 468 (1954)


Received May 28, 1981
C. K. Jørgensen (editor)

# Spectral-Structural Correlations in High-Spin Cobalt(II) Complexes

**Lucia Banci, Alessandro Bencini, Cristiano Benelli, Dante Gatteschi and Claudia Zanchini**

General Chemistry Institute, Faculty of Pharmacy, University of Florence; I.S.S.E.C.C., C.N.R., Florence, Italy

The spectral and magnetic properties of simple high spin cobalt(II) complexes are reviewed with the aim of showing how it is possible to relate the experimental data to the electronic structure and the coordination geometry of the complexes. It is hoped that this may be of help to all researchers who need to characterize cobalt(II) complexes or use cobalt(II) as a spectroscopic probe. The data relative to electronic, EPR, MCD, NMR spectra as well as to magnetic susceptibility measurements are interpreted within an Angular Overlap approach. Energy level diagrams are employed to calculate the electronic spectra and then, applying the relevant perturbation $\mu$, $g$, $A$ values are calculated for different chromophores in different coordination environments. The underlying theory is sketched paying in every case much attention to show the possibility of calculating the electronic properties in low symmetry environments. An extensive compilation of available experimental data is presented. The chromophores are classified according to the coordination number and according to the donor atom set.



Structure and Bonding 52

# List of Symbols and Abbreviations

|  |  |
|---|---|
| AA | acrylamide |
| ac | acetylacetone |
| acac | acetylacetonate |
| acpz | N-acetylpyrazole |
| APDA | 2-amino-5 phenyl-1,3,4 oxadiazole |
| apy | 2,3 dimethyl 1-phenyl-3-pyrazolin-5-one |
| ATP | adenosine-5′-triphosphate |
| bipy | 2,2′ bipyridine |
| BIz | benzimidazole |
| $C_3H_2S_3$ | dithiole-1,2 thione-3 |
| dabp | 2,2′ diamino biphenyl |
| dacoda | 1,5 diazacyclooctane-N,N′-diacetate |
| dan | 1,8 diaminonaphtalene |
| DEAP | diethylacetylphosphonate |
| DEU | N,N′-diethylurea |
| dienMe | bis(2-dimethylaminoethyl)methylamine |
| dietu | diethylthiourea |
| 4 DMAP | 4-N,N′-dimethylaminopyridine |
| dpdmq | 6,7-dimethyl-2,3 di-(2-pyridyl)-quinoxaline |
| dpmq | 6-methyl-2,3 di-(2-pyridyl)-quinoxaline |
| DTMSO | 1,4 dithiane monosulfoxide |
| EG | ethylene glycole |
| en | ethylendiamine |
| epydo | 1-methyl-2-pyridone |
| ESU | ethylselenourea |
| Et | ethyl |
| $Et_3P$ | triethylphosphine |
| $Et_3PO$ | triethylphosphineoxide |
| $Et_4D$ | tetraethyldithio oxamide |
| 2-Et-py | 2-ethylpyridine |
| *s*-$Et_2$en | N-N′diethyletylenediamine |
| FUAenMe$_2$ | 2-furylidene-2-iminoethyl-N,N diethylamine |
| HdMePyne | 2-hydrazino-4,6-dimethyl pyrimidine |
| H-SAL-DPT | bis(salicylidene-3-iminopropyl) amine |
| 5-ClSALen-NEt$_2$ | 5-chlorosalicylidene-2-iminoethyl-N,N diethylamine |
| Iz | imidazole |
| LE | 1,2-di(2′-pyridyl)ethane |
| L-Histidine | L-histidinato |
| LS | di-(2-pyridyl)disulfide |
| MABenz-NEt$_2$ | N-methyl-o-aminobenzilidene-2-imonoethyl-N,N diethylamine |
| Me | methyl |
| Me(Cl-Ph)TazO | 1-methyl-3(2 chlorophenyl) triazene-1-oxide |
| Me$_4$daes | bis(2-dimethylaminoethyl)sulfide |
| Me$_4$dipyMT | 3,3′4,4′-tetramethyldipyrromethane |
| Me$_4$en | N,N,N′,N′-tetramethylethylenediamine |
| 2 MeIz | 2-methylimidazole |
| MeN-NAs | 1-(2′-pyridyl)-2-(o-dimethylarsinobenzene)-2 azaethane |
| MePh$_2$AsO | methyldiphenylarsineoxide |
| Me$_4$pn | N,N,N′,N′-tetramethyl-1,2-propylenediamine |
| Me$_2$ppaz | 1,4-dimethylpiperazinium cation |
| Me$_3$PSe | trimethylphosphine selenide |
| 2 Mepy | 2-methylpyridyne |
| 1 MePyne-2-T | 1-methylpyrimidine-2-thione |
| 1,4,6 MePyne-2-T | 1,4,6 trimethylpyrimidine-2-thione |
| 6 Mequin | 6-methylquinoline |
| Me$_4$tn | N,N,N′,N′-tetramethyltrimethylenediamine |
| Me$_6$tpt | tris-(3-dimethylaminopropyl) amine |
| Me$_6$tren | tris(2-dimethylaminoethyl) amine |
| 2-MIz | 2-methylimidazole |
| MOBenz-NEt$_2$ | o-methoxybenzilidene-2-iminoetyl-N,N-diethylamine |
| MSBenz-NEt$_2$ | o-methylthiobenzilidene-2-iminoethyl-N,N-diethylamine |
| NCNPh | phenylcyanamide |
| N-EtdAZAO | N-ethyl-1,4-diazabicyclo(2.2.2.)octonium cation |
| NMBuL | N-methyl-$\gamma$-butyrolactame |
| N-N-N | $\alpha,\alpha',\alpha''$-tri-imine-2,6-(dibenzothiazol-2-yl)pyridine |
| NN$_3$ | tris(2-diethylaminoethyl)amine |
| N$_3$As | bis(2-diethylaminoethyl)-2-diphenylarsinoethylamine |
| N,N′(4 ClB) PAI | N,N′di-(4-Chlorobenzen)-propene-1-iminato-3-imine |
| N$_3$P | bis(2-diethylaminoethyl)-2-diphenylphosphine ethylamine |
| N-NO$_2$ | 2-diethylaminoethyl-bis-(2-methoxyethyl)amine |
| N-N$_2$O | bis(2-diethylaminoethyl)-2-methoxyethylamine |
| N$_2$OP | N-2-(diphenylphosphino)ethyl -N-(2-methoxyethyl)-N′,N′-diethylethylenediamine |

| | |
|---|---|
| $NOP_2$ | bis(2-diphenylphosphino-ethyl)-2-methoxyethylamine |
| $N_2P_2$ | 2-diethylaminoethyl-bis-(2-diphenylphosphinoethyl)amine |
| $N\text{-}NS_2$ | bis(2-methylthioethyl)-2-diethylaminoethylamine |
| $N\text{-}N_2S$ | bis(2-diethylaminoethyl)-2-methylthioethylamine |
| $N_2SP$ | N-2-(diphenylphosphino)ethyl-N-(2-methylthioethyl)-N′,N′-diethylethylenediamine |
| $NP_3$ | tris(2-diphenylphosphino-ethyl)amine |
| $NS_3$ | tris(2-t-butylthioethyl)amine |
| N-tBpyle-2-C | N-t-butylpyrrole-2-carbaldimino |
| N-tBpyle-2-Aim | N-t-butylpyrrole-2-aldimino |
| OAB | *o*-aminobenzoate |
| OMPA | octamethylpyrophosphoramide |
| OPD | *o*-phenylenediamine |
| ox | oxalate |
| $PC_4P$ | 1,4-bis(diphenylphosphino) butane |
| Ph | phenyl |
| $Ph_3AsO$ | triphenylarsino oxide |
| $PhIPPh_3$ | phenyliminotriphenylphosphorane |
| $Ph_3P$ | triphenylphosphine |
| $Ph_3PO$ | triphenylphosphineoxide |
| $(Ph_2PS)_2N$ | imidotetramethyldithiophosphino-S,S′ |
| pic | 2-methylpyridinecarboxylate |
| $PN'H^+$ | 2-(diphenylphosphinomethyl)-6-methylpiridinium ion |
| POP | bis(diphenylphosphinoethyl) oxide |
| py | pyridine |
| $PyAenEt_2$ | pyridine-2-carboxalidene-2-iminoethyl-N,N-diethylamine |
| $py(Bu^s)_2$ | 2,6-diacetylpyridinebis(s-butylimine) |
| $3\,py(CO_2)$ | pyridine-3-carboxilate |
| $py(cy)_2$ | 2,6-diacetylpyridinebis(cyclohexylimine) |
| pyO | pyridine-N-oxide |
| pz | pyrazole |
| quin | quinoline |
| SAL | salicylaldeyde |
| SALen | N,N′-ethylenebis-salicylaldiminato |
| SALiprp | salicylidene-2-iminoisopropylamine |
| SU | selenourea |
| TMPT | 1,4,6-trimethylpyrimidine-2-thione |
| tren | 2,2′,2″-triaminotriethylamine |
| trik | 1,5-diphenyl-1,3,5-pentanetrionato |
| tripyr | 1′,6-diethoxycarbonyl-1,2,3,4,5,6′-hexamethyltripyrrene |
| TTDA | bis(N,N-dimethylacetoamido)-thioether |
| TU | thiourea |

# 1 Introduction

High spin cobalt(II) is one of the most studied ions in coordination chemistry. It shows a variety of coordination numbers and coordination geometries and is able to bind to a large number of different ligands. It has three unpaired electrons and can thus be studied with a host of different spectral and magnetic techniques. It is also largely used as a tool for determining the coordination environment of the metal site in metallo-enzymes and metallo-proteins[1-5]. In fact cobalt(II) is often able to substitute zinc(II) which is present in naturally occurring zinc enzymes, and several cases are also known where cobalt enzymes remain active. It is therefore necessary to have a clear knowledge of the spectral properties of cobalt(II) in various coordination environments in order to be able to obtain the required structural information.

The theory for the interpretation of the spectral properties of cobalt(II) is essentially known[6-9], but no extensive compilation of experimental data is to our knowledge available. We want to fill this gap in the literature giving a comprehensive review of the ligand field interpretation of the electronic structure of high spin cobalt(II) complexes, as well as of the most common spectral and magnetic techniques which are used to characterize cobalt complexes.

The ligand field description we want to refer to is the Angular Overlap Model[10,11], which Schäffer has thoroughly developed from an original molecular orbital derivation. Although the Angular Overlap Model is equivalent to the classic Ligand Field Model[11], it has the advantage of using parameters $e_\sigma$, $e_\pi$, . . . which have been shown to be actually related to the $\sigma$, $\pi$ . . . metal-ligand interactions[12-14].

Choosing a ligand field description of the electronic structure, we will deal only with spectral and magnetic properties which refer to the mainly metal d orbitals. Therefore we will review electronic spectra (d-d bands), magnetic susceptibility measurements, ESR spectra and MCD spectra. We will also briefly comment on NMR spectra which are determined by the nature of the ground state in a paramagnetic complex[15]. We will not go much into the theory of these techniques, since they are well known and adequately treated elsewhere, but will briefly summarize their fundamental basis.

In each case much attention will be paid to low symmetry components of the ligand field, which have been thoroughly studied in the last few years[16], and which are relevant to the interpretation of so many spectral data.

In this review we will not include large biological molecules although they are often interesting spectroscopic systems, but will refer only to simple model complexes.

Low spin cobalt(II) complexes will not be mentioned, and also the fascinating field of spin equilibria will be completely overlooked here since for both these topics exhaustive articles are available in the literature[17-24].

# 2 Energy Levels and Electronic Spectra

Every attempt to understand the spectral and magnetic properties of cobalt(II) complexes must start from the ligand field energy levels. Their dependence on the ligand field parameters and on the molecular symmetry has been described before[6, 9, 25–27] but we wish to demonstrate it again in an approach using the Angular Overlap formalism.

The Angular Overlap Model is a ligand field model which uses $e'_\lambda$ parameters[20] ($\lambda = \sigma, \pi, \delta \ldots$) for expressing the orbital energies. For d orbitals $\lambda$ can be only $\sigma$, $\pi$ and $\delta$, but it is customary to use as parameters: $e_\sigma = e'_\sigma - e'_\delta$ and $e_\pi = e'_\pi - e'_\delta$. In octahedral complexes a simple correlation exists between these parameters and the usual crystal field parameters[21]

$$10\,Dq = a\,(3\,e_\sigma - 4\,e_\pi)$$

a = 1 for octahedral and −4/9 for tetrahedral complexes.

In the AO formalism it is easy to calculate the energy levels for low symmetry complexes, using the reported matrices[29], and allowing for differences in the chemical nature of the ligands, and for differences in the bond distances. It is also possible to allow for anisotropic $\pi$ interactions as can be expected for instance in metal-pyridine bonds[12, 30]. Several excellent review and reference articles are available where the subtleties of the model are shown[11, 13, 14, 28–31].

In Fig. 1 the energies of the quartet levels in various coordination geometries are shown. Octahedral complexes, five coordinate, either square pyramidal and trigonal

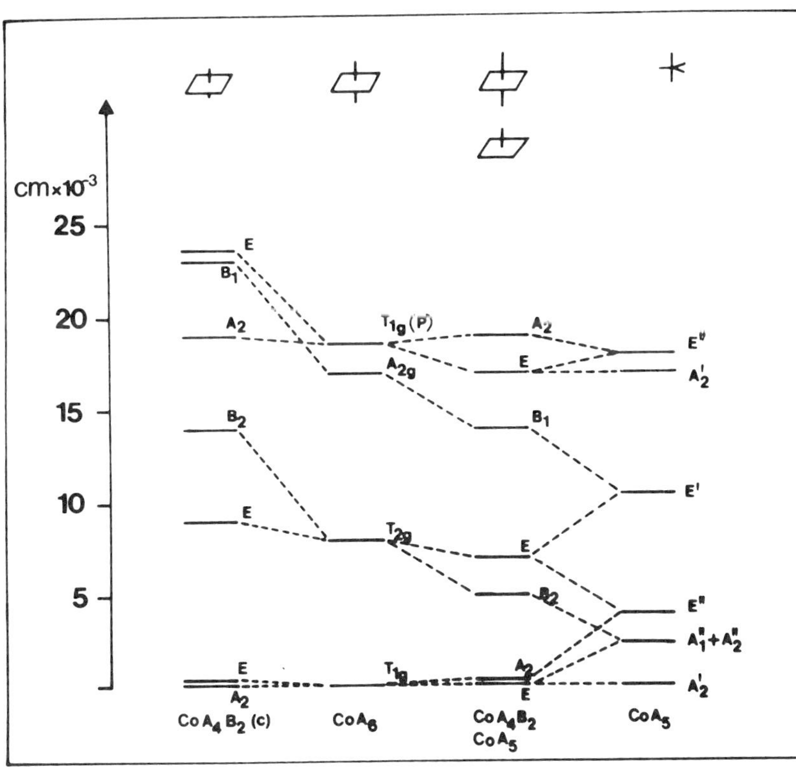


**Fig. 1.** Energies of the quartet levels in various coordination geometries

bipyramidal, are considered. Tetrahedral complexes will be taken into account separately. The purpose of Fig. 1 is not to give the exact order of the levels, which depends on the actual value of the AO parameters, but to help in providing a qualitative view of the pattern of the energy levels in various coordination environments. The ground state of tetragonal symmetry may be $^4A_2$ or $^4E$ for compressed and elongated distortions, respectively, since it depends largely on the bonding ability of the ligands as will be shown in later sections. The doublet levels have not been taken into consideration, because they are responsible only for the spin forbidden transitions and hence have been much less characterized through electronic spectroscopy.

Figure 2 shows the energy level diagrams which are relevant to $CoA_4$ tetrahedral complexes. In Fig. 2a the effect of increasing $e_\sigma$ for the A ligands in strictly $T_d$ symmetry is shown. The $e_\sigma$ values correspond to the range which in principle can be found in metal complexes. In Fig. 2b the effect of varying the $e_\pi/e_\sigma$ ratio is shown. Both positive and negative values of the ratio were taken into consideration. In the Angular Overlap Model, positive $e_\pi$ parameters correspond to antibonding effects[28]. The variation of the energy levels is rather small, as previously observed[33]. This is an important result since it shows that in tetrahedral complexes the fitting of spectral properties using only $e_\sigma$ parameters is often possible, at least in first approximation.

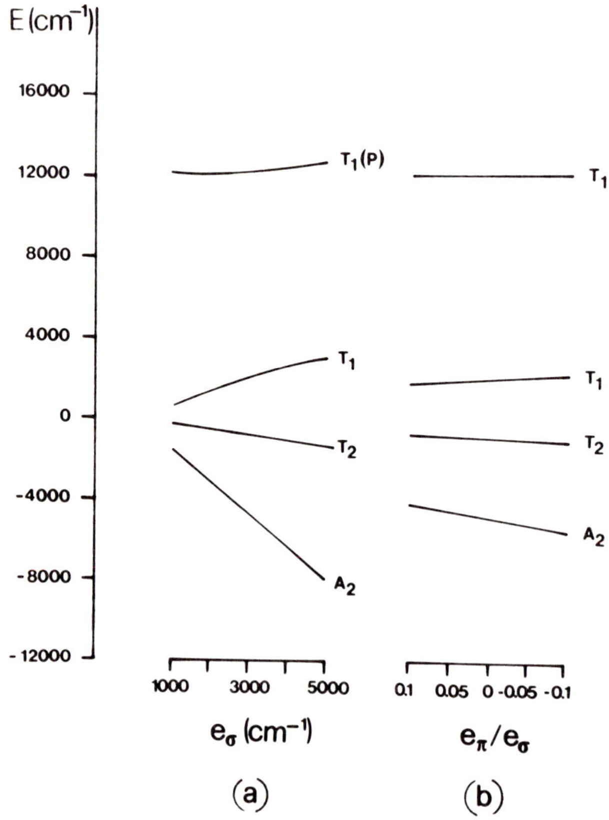


**Fig. 2a, b.** Energy level diagrams for $CoA_4$ tetrahedral complexes: (**a**) the effect of varying $e_\sigma$ ($e_\pi = 0$); (**b**) the effect of varying $e_\pi/e_\sigma$ ($e_\sigma = 3000\ cm^{-1}$)

In Fig. 3, the energy levels of pseudotetrahedral complexes of formula $CoA_2B_2$ are shown. In Fig. 3 a, the angle between the $C_2$-axis and the metal ligand bond directions is varied. For the sake of simplicity the four ligands were considered as equal and the levels are labelled according to $D_{2d}$ symmetry. The largest splitting is observed for $^4T_2$ while the two $^4T_1$ levels are much less affected. In Fig. 3 b the $e_\sigma$ values of ligands A and B are allowed to vary, and in this case relevant splittings are predicted also for $^4T_1$ (P).

Analogously to Fig. 2 b a change in the $e_\pi/e_\sigma$ ratio is of only minor importance (Fig. 3 c) and similar results are observed if the $e_\pi/e_\sigma$ ratios of individual ligands are relaxed.

In Fig. 4 similar calculations for tetrahedral $CoAB_3$ complexes are shown. In Fig. 4 a the A-Co-B angle $\beta$, is varied: it is apparent that $^4T_1$ (F) is the most split, while $^4T_1$ (P) and $^4T_2$ are less affected. Similar results are obtained also for the dependence on $e_\sigma^A/e_\sigma^B$ (Fig. 4 b), while the $e_\pi/e_\sigma$ ratios are rather unimportant for the level splittings (Fig. 4 c).

The electronic spectra of tetrahedral complexes consist in general of two groups of bands, one in the near infrared, the other in the visible region. They can be assigned,

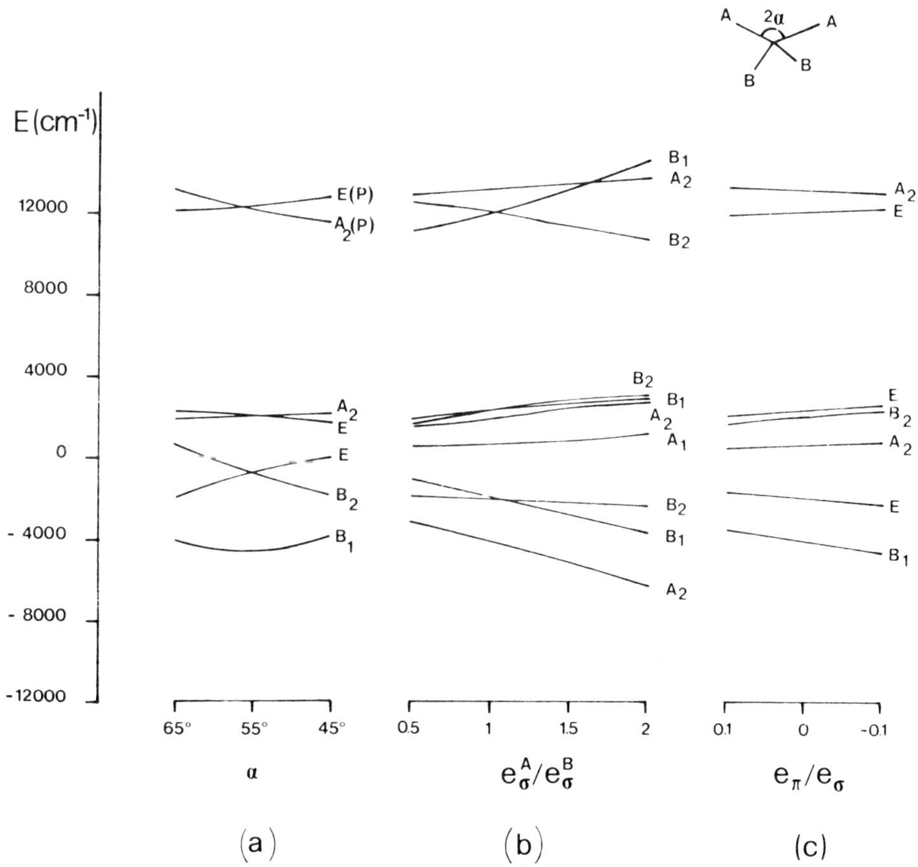


**Fig. 3a–c.** Energy level diagrams for $CoA_2B_2$ complexes: (**a**) the effect of varying $\alpha$ ($e_\sigma^A = e_\sigma^B = 3000\ cm^{-1}$, $e_\pi = 0$) [$D_{2d}$]; (**b**) the effect of varying $e_\sigma^A/e_\sigma^B$ ($e_\sigma^B = 3000\ cm^{-1}$, $e_\pi = 0$, $\alpha = 65°$) [$C_{2v}$]; (**c**) the effect of varying $e_\pi/e_\sigma$ ($e_\pi^A = e_\pi^B$, $e_\sigma^A = e_\sigma^B = 3000\ cm^{-1}$, $\alpha = 65°$) [$D_{2d}$]

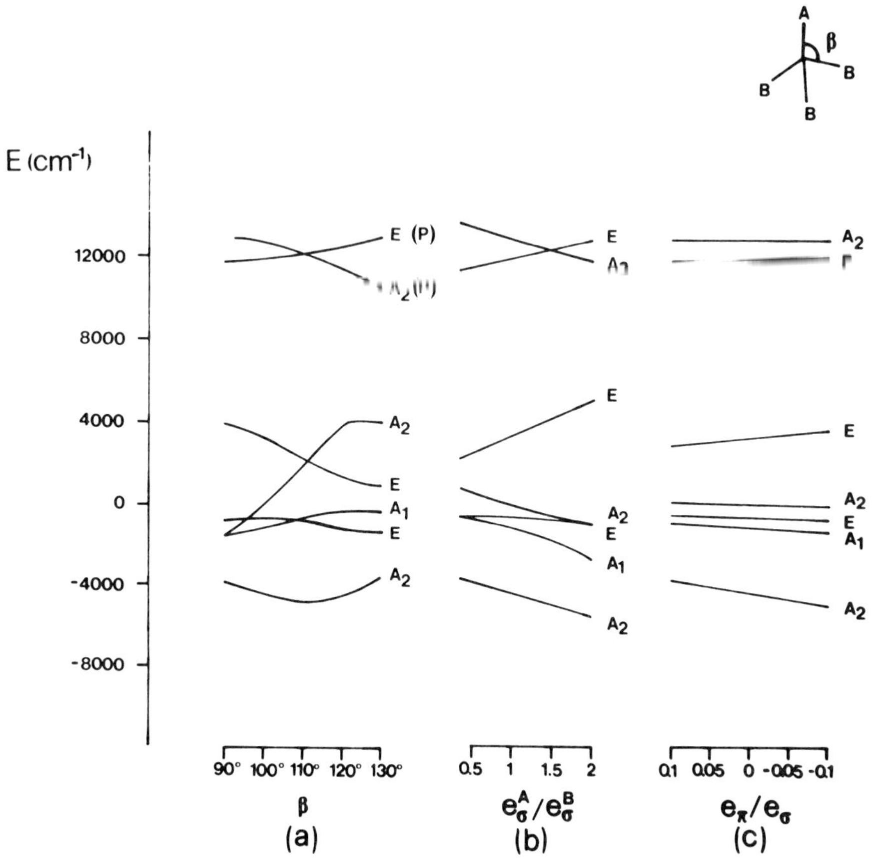


**Fig. 4a–c.** Energy level diagrams for $CoAB_3$ complexes [$C_{3v}$]: (**a**) the effect of varying $\beta$ ($e_\sigma^A = e_\sigma^B = 3000\ cm^{-1}$, $e_\pi = 0$); (**b**) the effect of varying $e_\sigma^A/e_\sigma^B$ ($e_\sigma^B = 3000\ cm^{-1}$, $e_\pi = 0$, $\beta = 100°$); (**c**) the effect of varying $e_\pi/e_\sigma$ ($e_\pi^A = e_\pi^B$, $e_\sigma^A = e_\sigma^B = 3000\ cm^{-1}$, $\beta = 100°$)

using the notation of $T_d$ symmetry, as ${}^4A_2 \rightarrow {}^4T_1$ (F) and ${}^4A_2 \rightarrow {}^4T_1$ (P) respectively. The former transition is found in the range from 5,000 to 11,000 $cm^{-1}$ at most, while the latter is observed between 13,500 and 19,500 $cm^{-1}$. A few exceptions will be mentioned in Sect. 4. The ${}^4A_2 \rightarrow {}^4T_2$ transition is usually below 5,000 $cm^{-1}$ and therefore difficult to detect.

The molar extinction coefficients of the near infrared bands are typically 50–100, while those of the visible transitions are 300–1000.

In Fig. 5d the calculated energy levels for trigonal bipyramidal $CoA_5$ chromophores ($D_{3h}$ symmetry) are shown. The ${}^4P$ levels are not largely split. Also the ${}^4A_2''$, ${}^4A_1''$ and ${}^4E''$ (F) levels remain quite close to the ground state ${}^4A_2'$. ${}^4E'$ is the most sensitive level to variations in the ligand field strength. The influence of the $e_\pi$ parameters is not large, as in the case of tetrahedral complexes.

In Fig. 6 the more general case of $CoA_2B_3$ trigonal bipyramidal complexes is considered. The most important differences as compared to the $CoA_5$ chromophores are that

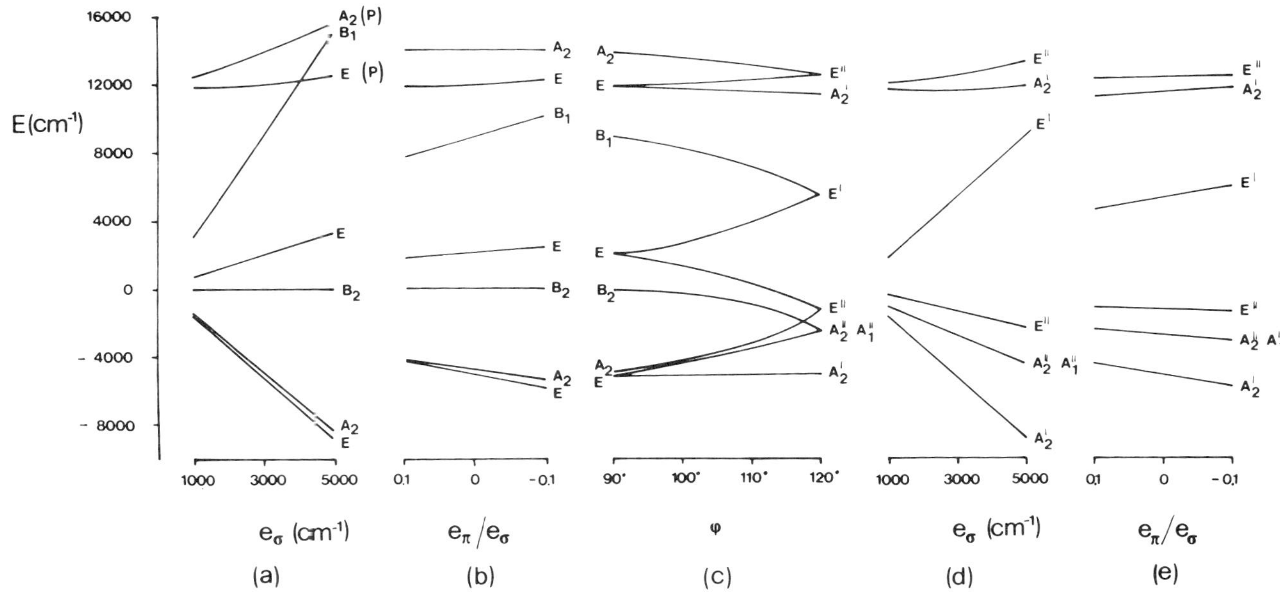


**Fig. 5 a–e.** Energy level diagrams for $CoA_5$ complexes: (**a**) and (**b**) for regular square pyramidal geometry, (**d**) and (**e**) for regular bipyramidal geometry; (**a**) the effect of varying $e_\sigma$ ($e_\pi = 0$); (**b**) the effect of varying $e_\pi/e_\sigma$ ($e_\sigma = 3000\ cm^{-1}$); (**c**) the effect of varying $\varphi$ (see text) ($e_\sigma = 3000\ cm^{-1}$, $e_\pi = 0$); (**d**) the effect of varying $e_\sigma$ ($e_\pi = 0$); (**e**) the effect of varying $e_\pi/e_\sigma$ ($e_\sigma = 3000\ cm^{-1}$)

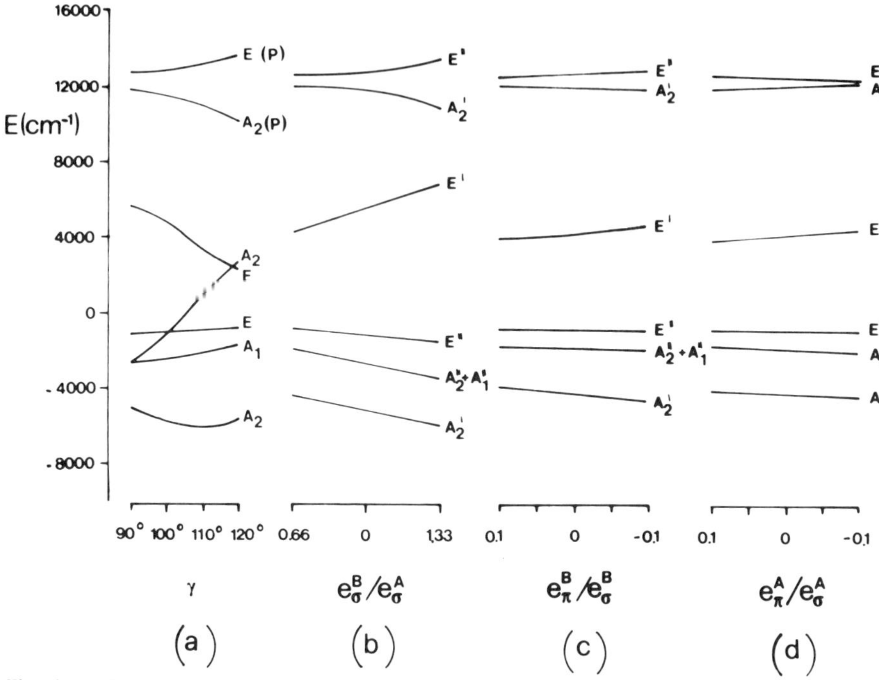


**Fig. 6a–d.** Energy level diagrams for trigonal bipyramidal $CoA_2B_3$ complexes: (**a**) the effect of varying $\gamma$ (see text) ($e_\sigma^A = e_\sigma^B = 3000\ cm^{-1}$, $e_\pi = 0$) [$C_{3v}$]; (**b**) the effect of varying $e_\sigma^B/e_\sigma^A$ ($e_\sigma^A = 3000\ cm^{-1}$, $e_\pi = 0$, $\gamma = 90°$); (**c**) the effect of varying $e_\pi^B/e_\sigma^B$ ($e_\sigma^B = 2000\ cm^{-1}$, $e_\pi^A/e_\sigma^A = 0.05$, $e_\sigma^A = 3000\ cm^{-1}$, $\gamma = 90°$); (**d**) the effect of varying $e_\pi^A/e_\sigma^A$ ($e_\sigma^A = 3000\ cm^{-1}$, $e_\pi^B/e_\sigma^B = 0.05$, $e_\sigma^B = 2000\ cm^{-1}$, $\gamma = 90°$)

the $^4P$ levels split to some extent for elongated case, and that the accidental degeneracy of $^4A_2''$ and $^4A_1''$ is removed when the A-Co-B angle $\gamma$ is allowed to vary from 90°. In particular the energy of the excited $^4A_2$ level increases markedly for $\gamma > 90°$. (The labelling in this case is that of $C_{3v}$ symmetry).

Trigonal bipyramidal spectra are characterized by a band in the near infrared region, from 5,000 to 8,000 $cm^{-1}$ ($^4A_2' \rightarrow {}^4E''$ in $D_{3h}$), a second one at 10,000–13,000 $cm^{-1}$ and two more intense bands at 16,000 $cm^{-1}$ ($^4A_2' \rightarrow {}^4A_2'(P)$) and 18,000–22,000 $cm^{-1}$ ($^4A_2' \rightarrow {}^4E''(P)$) respectively. The band at 16,000 $cm^{-1}$ has been considered by Ciampolini as diagnostic for trigonal bipyramidal spectra[34]. The molar extinction coefficients for the $^4F \rightarrow {}^4P$ transitions are: 100–300 for $^4A_2'$ and 100–600 for $^4E''$. The $^4F \rightarrow {}^4F$ transitions are usually less intense, the molar extinction coefficients ranging from 20 to 200.

In Fig. 5c the energy variation of the energy levels on passing from a trigonal bipyramid [$D_{3h}$] to a tetragonal pyramid [$C_{4v}$] is shown. This can be performed by changing one of the A-Co-A angles $\varphi$ from 90° to 120°[33]. The pattern is not much affected by the variation of the $e_\pi$ parameters.

The energy level diagrams for square pyramidal $CoA_5$ complexes are visualised in Figs. 5a, b. The ground level is quasi-degenerate $^4A_2 + {}^4E$, the splitting depending essentially on the $e_\pi$ bonding ability of the ligands (see Sect. 3). The next levels are $^4B_2$ and $^4E$, which have an octahedral $^4T_{2g}$ parentage. The $B_1$, E (P) and $A_2$ (P) levels are of comparable energy.

The variation of the $e_\pi/e_\sigma$ ratio affects mainly the relative energies of the ground $^4E$ and $^4A_2$ levels and that of $^4B_1$.

In Fig. 7 the more general case of $CoAB_4$ square pyramidal complexes is shown. The splitting of the $^4E$ and $^4B_2$ levels increases as the $e_\sigma^A/e_\sigma^B$ ratio decreases, i.e. when the square pyramid is elongated. The reverse effect is observed when the A-Co-B angle $\delta$ varies from 90° to 110°. At the same time the splitting of the P levels decreases and the energy of $^4B_1$ decreases.

Genuine examples of square pyramidal cobalt(II) complexes are relatively rare[36, 37]. In the few well documented cases bands are seen at 5,000, 7,000, 11,000, 17,000 and 20,000 $cm^{-1}$. The molar extinction coefficients increase on passing from the F-F to the F-P transitions. For the former $\varepsilon$ as low as 7 is observed while for the latter values as high as 320 were reported.

Figure 8 shows the energy levels for octahedral $CoA_6$ chromophores. Varying the $e_\pi/e_\sigma$ ratio induces larger energy changes of the ground state $^4T_{1g}$ and of the excited $^4A_{2g}$ levels, $^4T_{1g}$ (P) and $^4T_{2g}$ being less affected. Trigonal distortion components affect in

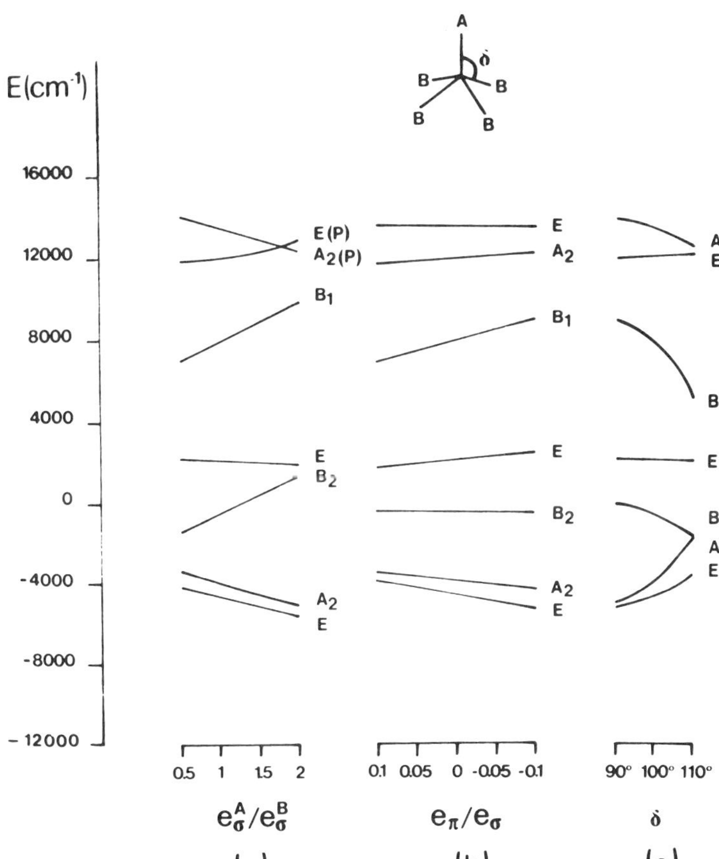


**Fig. 7a–c.** Energy level diagrams for $CoAB_4$ complexes: (**a**) the effect of varying $e_\sigma^A/e_\sigma^B$ ($e_\sigma^B = 3000\ cm^{-1}$, $e_\pi = 0$, $\delta = 100°$); (**b**) the effect of varying $e_\pi/e_\sigma$ ($e_\sigma^A = e_\sigma^B = 3000\ cm^{-1}$, $\delta = 100°$); (**c**) the effect of varying $\delta$ ($e_\sigma^A = e_\sigma^B = 3000\ cm^{-1}$, $e_\pi = 0$)

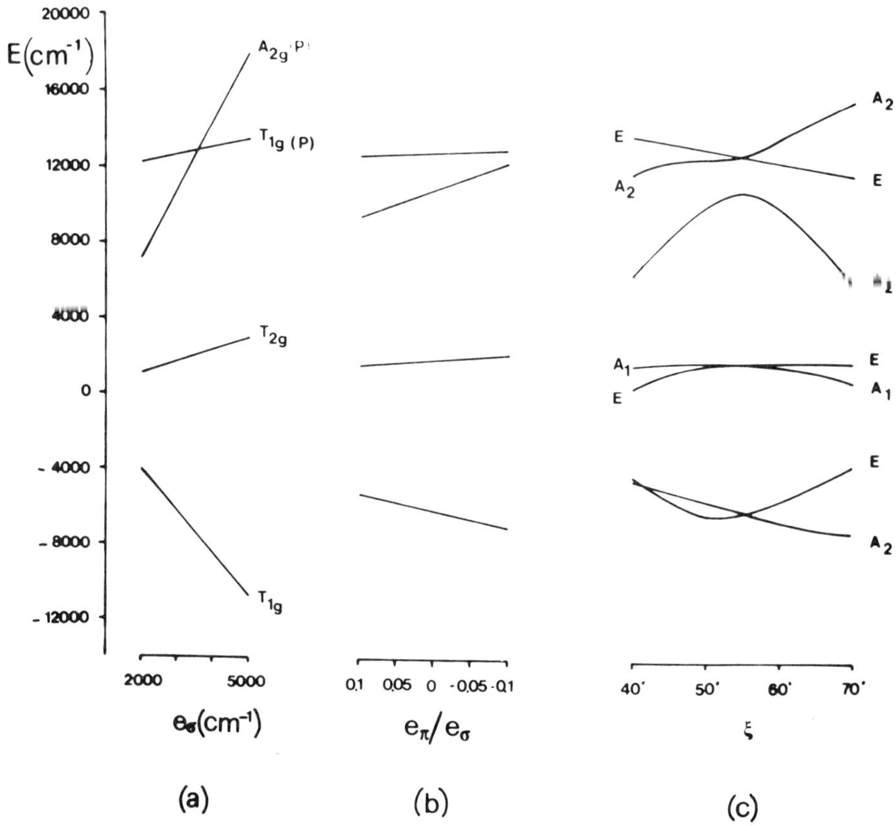


**Fig. 8a–c.** Energy level diagrams for regular octahedral $CoA_6$ complexes, (**a**) and (**b**), and for trigonally distorted chromophores, (**c**) : (**a**) the effect of varying $e_\sigma$ ($e_\pi = 0$); (**b**) the effect of varying $e_\pi/e_\sigma$ ($e_\sigma = 3000\ cm^{-1}$); (**c**) the effect of varying $\xi$ (see text) ($e_\sigma = 3000\ cm^{-1}$, $e_\pi = 0$) [$C_{3v}$]

particular the ground and the excited $^4T_{1g}$ levels (Fig. 8c). For large deviations of the angle $\xi$ between the $C_3$ axis and a metal ligand bond from the octahedral value $^4A_2$ is stabilized as the ground level.

Figure 9 considers tetragonally distorted chromophores. Whether $^4A_2$ or $^4E$ is of minimum energy depends on the $\pi$ bonding abilities of the ligands, and the overall pattern is similar to that discussed for square pyramidal complexes.

In general two main bands are seen in the spectra, namely the $^4T_{1g} \rightarrow {}^4T_{2g}$ and $^4T_{1g} \rightarrow {}^4T_{1g}$ (P) transitions. The $^4T_{1g} \rightarrow {}^4A_{2g}$ band is often hidden by the $^4T_{1g}$ (P) transition and is of lower intensity since in the strong field limit it is a two-electron jump. The lowest energy transition falls in the near infrared region, ranging from 6,500 to 10,500 $cm^{-1}$. The $^4T_{1g} \rightarrow {}^4T_{1g}$ (P) transition is typically ranging from 15,500 to 22,000 $cm^{-1}$. The molar extinction coefficients of the $^4T_{1g} \rightarrow {}^4T_{2g}$ transition are in the range 5–40 while those of $^4T_{1g} \rightarrow {}^4T_{1g}$ (P) are 15–60.

The complexes in which cobalt has a coordination number larger than six are too rare to justify a complete treatment here. They will be discussed briefly in Sect. 4.

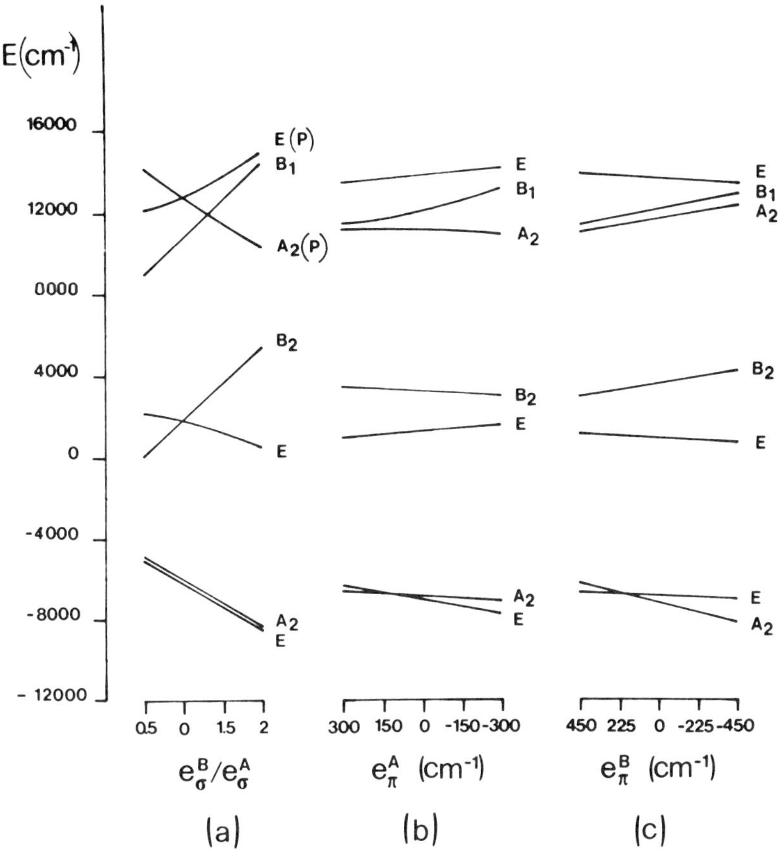


**Fig. 9a–c.** Energy level diagrams for tetragonally distorted $CoA_4B_2$ complexes: (**a**) the effect of varying $e_\sigma^A/e_\sigma^B$ ($e_\pi = 0$); (**b**) the effect of varying $e_\pi^A$ ($e_\sigma^A = 3000\ cm^{-1}$, $e_\pi^B/e_\sigma^B = 0.05$, $e_\sigma^B = 4500\ cm^{-1}$); (**c**) the effect of varying $e_\pi^B$ ($e_\sigma^B = 4500\ cm^{-1}$, $e_\pi^A/e_\sigma^A = 0.05$, $e_\sigma^A = 3000\ cm^{-1}$)

# 3 Ground States and Related Spectral and Magnetic Properties

The spectral properties depend on the ground and the excited states of the metal complexes, but for many techniques the role of the ground state is certainly more apparent. From the point of view of the ground level octahedral and square pyramidal complexes are similar, in the sense that they have orbitally (quasi) degenerate ground states, while tetrahedral and trigonal bipyramidal complexes have orbitally non-degenerate ground levels.

In the octahedral case, spin orbit coupling splits the ground $^4T_{1g}$ level yielding a Kramers doublet of lowest energy[6]. When lower symmetry components are operative it is very important to know the relative energy order of the three orbital split states. Since the parent $^4T_{1g}$ state can be considered to a good approximation to originate from the $t_{2g}^5$

$e_g^2$ configuration[6] the three orbital components can be described in the hole formalism as $(x^2 - y^2, z^2, n)$, where $n = xy, xz, yz$. Therefore when the symmetry of the complex is lower than octahedral, the relative energies of the split components of $^4T_{1g}$ will depend on the energies of the xy, xz and yz orbitals to a good approximation. Since the latter are $\pi$ antibonding orbitals in octahedral coordination, the relative energies will depend only on the $\pi$ bonding ability of the ligands. Therefore it can be anticipated that the splitting of $^4T_{1g}$ will be small in any case, typical values being 1,000 $cm^{-1}$. The same considerations apply to five coordinate square pyramidal complexes as well.

If the chromophore is orthoaxial, i.e. if the bond angles are equal to 90°, the energies of the $t_{2g}$ orbitals are given[38] by:

$$E(xy) = 4e_{\pi\perp}^{eq}$$

$$E(xz) = E(yz) = 2e_{\pi\parallel}^{eq} + 2e_{\pi}^{ax}$$

where $e_{\pi\parallel}^{eq}$ and $e_{\pi\perp}^{eq}$ refer to the $\pi$ bonding interactions with equatorial ligands, and $e_{\pi}^{ax}$ to that with the axial ligands. $\parallel$ and $\perp$ refer to the $\pi$ interactions in and perpendicular to the equatorial plane, respectively.

In the case of tetrahedral and trigonal bipyramidal coordinations the orbitally non-degenerate quartet ground state is split by spin orbit coupling into two Kramers doublets. The energy separation is generally of the order of 10 $cm^{-1}$ so that it is relevant to the interpretation of the magnetic susceptibility at low temperature and to EPR spectra.

## *3.1 Magnetic Susceptibility*

The magnetic susceptibility of an assembly of identical isooriented molecules is a second rank tensor which relates an external magnetic field, $\underline{H}$, and the resulting magnetization, $\underline{M}$, according to

$$\underline{M} = \chi \underline{H}$$

This means that different values of the susceptibility will in general be measured along different directions. In other terms $\chi$ can be represented by an ellipsoid, the principal axes of which are called the principal axes of the tensor[39, 40]. The corresponding values of $\chi$ are called the principal values. It is common practice to refer to molar magnetic susceptibility.

The molar magnetic susceptibility is usually related to the electronic structure of a transition metal complex through the well known Van Vleck equation[41] which presupposes the knowledge of the orientation of the principal axes. For the molecules lacking the symmetry necessary to determine a priori these directions a more general expression was derived[42]:

$$\chi = N \sum_i \left[ \sum_k \frac{\langle i|\hat{\mu}|k\rangle\langle k|\hat{\mu}|i\rangle}{kT} - \sum_j \frac{2\langle i|\hat{\mu}|j\rangle\langle j|\hat{\mu}|i\rangle}{E_i^0 - E_j^0} \right] \cdot \frac{e^{-E_i^0/kt}}{\sum_i e^{-E_i^0/kT}} \tag{1}$$

where $|i\rangle$ and $|k\rangle$ are thermally populated states degenerate in the absence of magnetic field, $|j\rangle$ are excited states, $E_i^0$ and $E_j^0$ are their energies in the absence of the magnetic field, and N is the Avogadro's number, $\hat{\mu}$ is the operator $\mu_B$ B $(\hat{L} + g_e \hat{S})$.

Equation (1) has the same limitations of the Van Vleck equation in that it assumes that the energies of the levels in a magnetic field can be expressed through a series expansion in the magnetic field and retaining only the terms up to second order. The $|i\rangle$, $|k\rangle$, and $|j\rangle$ functions are obtained through some ligand field calculation, usually with the Angular Overlap Model. In the calculation spin orbit coupling must be included. Since the axes which are used in relation (1) are arbitrary ones, the magnetic susceptibility $\chi$ which is calculated is in general non-diagonal. Matrix diagonalization techniques yield the principal $\chi$ values and the orientation of the principal axes in the reference frame.

A more general approach[43] to the calculation of $\chi$ which does not rely on the assumption of the Van Vleck equation uses the general definition (2)

$$\chi = -\frac{N}{H}\left\langle\frac{\delta E}{\delta H}\right\rangle_T \tag{2}$$

where $\left\langle\frac{\delta E}{\delta H}\right\rangle_T$ represents the canonical average at temperature T. $\delta E/\delta H$ is computed by direct diagonalization of the Hamiltonian matrix using the experimental field strength. In principle this calculations should be useful for low temperature work in the presence of small exchange coupling.

The magnetic susceptibility which can be calculated through (1) is the molecular susceptibility while the one which can be experimentally measured is the crystal susceptibility: the two are identical only when the site symmetry of the metal ion is equal to the point symmetry of the crystal space group. When the site symmetry is lower the molecular susceptibility can be obtained only through some indirect procedure[39].

In general however only measurements on powdered samples are performed, which yield the average magnetic susceptibility. In this case there is no way for obtaining the principal molecular values and directions.

An important feature of the magnetic susceptibility is that it is temperature dependent. The effective magnetic moment

$$\mu_{eff} = 2.828\,(\chi T)^{1/2} \tag{3}$$

is usually much less so, and it is used for a quick characterization of the magnetic properties of the ions. Typical values of the magnetic moments of Co(II) complexes at room temperature are shown in Table 1. The spin only value for a quartet is 3.78 $\mu_B$. Deviations are caused by unquenched orbital contributions.

The highest values are observed for octahedral[44–46] and square pyramidal complexes, in agreement with the (quasi) degenerate nature of the ground state. However, relatively large orbital contributions are observed also for trigonal bipyramidal and tetrahedral complexes.

For the (quasi) degenerate ground states the value of the magnetic moment depends largely on the splitting of the $^4T_{1g}$ level, a larger splitting yielding lower $\mu$ values. The sign of the splitting determines the sign of the magnetic anisotropy. For an axial case two

**Table 1.** Typical values of the room temperature effective magnetic moments of cobalt(II) complexes in various coordination environments

| Coordination | $\mu$ ($\mu_B$) |
|---|---|
| Octahedral | 4.77–5.40 |
| Square pyramidal | 4.28–5.07 |
| Trigonal bipyramidal | 4.26–5.03 |
| Tetrahedral | 3.98–4.82 |

different values of $\chi$ will be measured, one parallel to the symmetry axis ($\chi_{\|}$) and one orthogonal to it ($\chi_{\perp}$). If the orbital singlet lies lowest $\chi_{\|} < \chi_{\perp}$, while the reverse is true if the orbital doublet is of minimum energy.

Also the temperature-dependence of the magnetic moment in magnetically dilute complexes can be used as a tool for identifying the stereochemistry of the complexes since for square pyramidal and octahedral complexes, the thermal population of the excited levels decreases markedly with temperature and generally a decrease of the magnetic moment is observed also in the range 77–300 K. For observing similar effects in tetrahedral complexes, it is necessary to decrease the temperature further. It seems to be a general rule that a larger temperature-dependence is observed for those compounds which have the highest room temperature moments. Figure 10 shows the magnetic moments calculated from equation (1) at 305, 105 and 5 K for $CoA_6$ chromophore. The $\mu$'s values are little sensitive to the variation of the AOM parameters, that is distinct increase of the value of $e_\sigma$ induces only a small decrease of the $\mu$'s. The magnetic

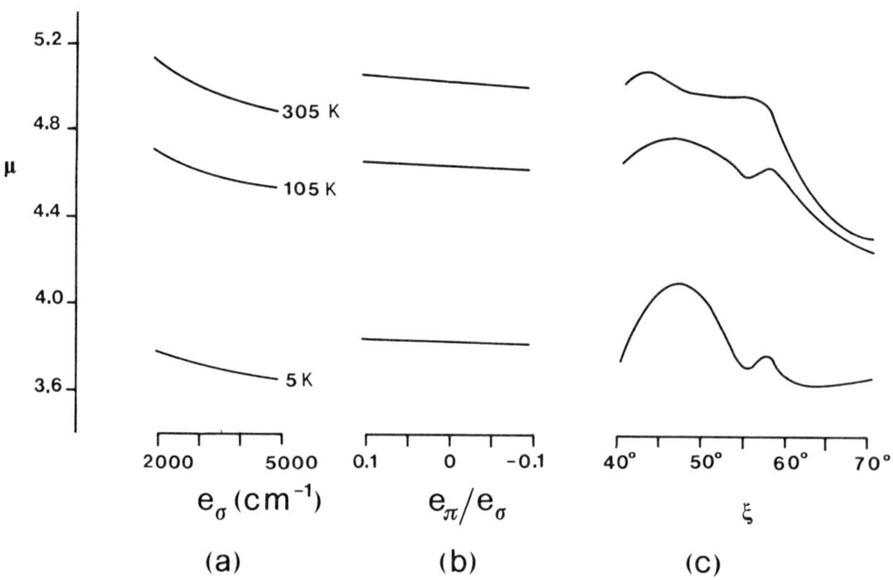


**Fig. 10 a–c.** Magnetic moments calculated (see text) at 305, 105 and 5 K for regular $CoA_6$ chromophores, (**a**) and (**b**), and for trigonally distorted geometry (**c**): (**a**) the effect of varying $e_\sigma$ ($e_\pi = 0$); (**b**) the effect of varying $e_\pi/e_\sigma$ ($e_\sigma = 3000\ cm^{-1}$); (**c**) the effect of varying $\xi$ (see text) ($e_\sigma = 3000\ cm^{-1}$, $e_\pi = 0$)

moments are even less affected by the variation of $e_\pi$. A more complicated behavior is calculated when the angle $\xi$ (where $\xi$ is the angle between the $C_3$-axes of the octahedron and the bond direction) is varied, because the lowest energy levels intersect in the respective range (see Fig. 8).

Figure 11 shows the magnetic moments calculated from equations (1,3) at 305, 105 and 5 K, for $CoA_4B_2$ chromophores. Varying the $e_\sigma^B/e_\sigma^A$ ratio moderately affects the magnetic moment, in the sense that an increase of the ratio leads to decreasing values of $\mu$. A change in the $e_\pi^A$ value does not affect $\mu$ strongly at high temperatures. A somewhat larger effect is induced by the variation of $e_\pi^B$, which causes a decrease of $\mu$ when $e_\pi$ becomes more negative (i.e. more $\pi$ bonding).

The calculated $\mu$'s for square pyramidal $CoA_5$ and $CoAB_4$ complexes do not differ substantially from those of octahedral complexes (Fig. 12). An increase in the ligand field strength causes a decrease in $\mu$. Also a decrease of $\mu$ is observed on making $e_\pi$ more negative. Similar results are calculated allowing for different ligands in $CoAB_4$ chromophores. If the B-Co-A angle $\delta$ is allowed to vary from 100° a rather complicated behavior is observed, both in the room temperature moments and in their temperature-dependence.

Figure 13 shows the calculated $\mu$'s for trigonal bipyramidal $CoA_5$ and $CoA_2B_3$ chromophores. From the comparison with Fig. 12, it is apparent that the $\mu$ values for trigonal bipyramidal complexes are smaller than those of square pyramidal ones, and that the former are more sensitive to ligand field strength variations. Also the $\mu$ value varia-

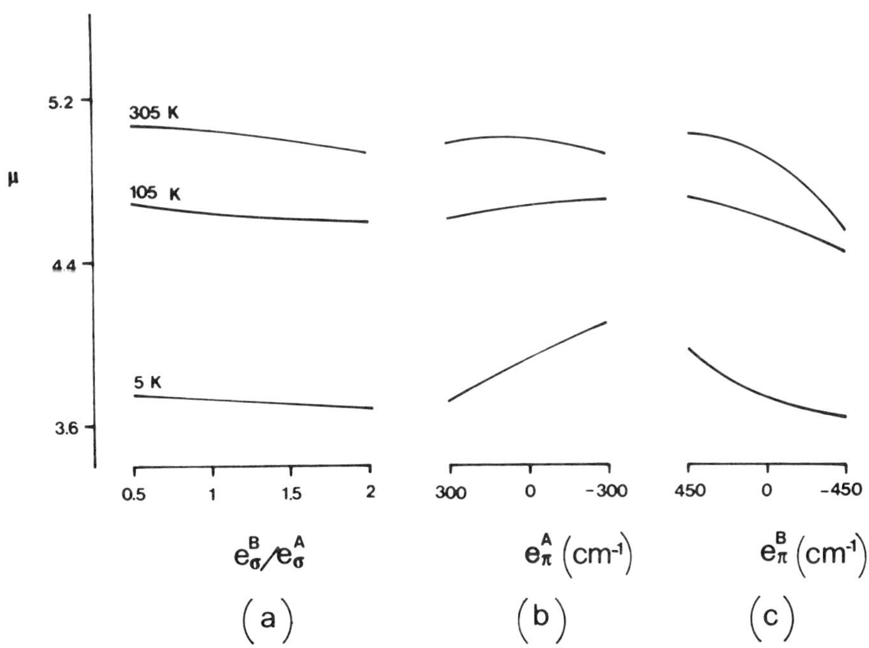


**Fig. 11 a–c.** Magnetic moments calculated (see text) at 305, 105 and 5 K for tetragonally distorted $CoA_4B_2$ chromophores: (**a**) the effect of varying $e_\sigma^B/e_\sigma^A$ ($e_\sigma^A$ = 3000 cm$^{-1}$, $e_\pi$ = 0); (**b**) the effect of varying $e_\pi^A$ ($e_\sigma^A$ = 3000 cm$^{-1}$, $e_\pi^B/e_\sigma^B$ = 0.05, $e_\sigma^B$ = 4500 cm$^{-1}$); (**c**) the effect of varying $e_\pi^B$ ($e_\sigma^B$ = 4500 cm$^{-1}$, $e_\pi^A/e_\sigma^A$ = 0.05, $e_\sigma^A$ = 3000 cm$^{-1}$)

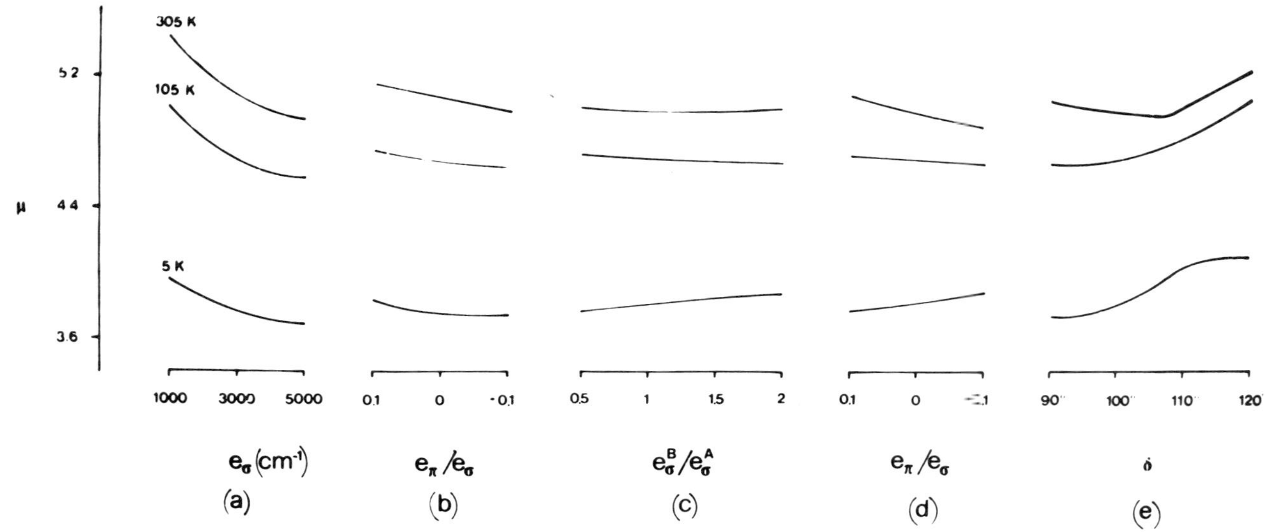


**Fig. 12 a–e.** Magnetic moments calculated (see text) at 305, 105 and 5 K for square pyramidal $CoA_5$, (**a**) and (**b**), and $CoAB_4$, (**c**), (**d**) and (**e**), complexes: (**a**) the effect of varying $e_\sigma$ ($e_\pi = 0$); (**b**) the effect of varying $e_\pi/e_\sigma$ ($e_\sigma = 3000$ cm$^{-1}$); (**c**) the effect of varying $e_\sigma^B/e_\sigma^A$ ($e_\sigma^A = 3000$ cm$^{-1}$, $e_\pi^A = e_\pi^B = 0$, $\delta = 100°$); (**d**) the effect of varying $e_\pi/e_\sigma$ ($e_\sigma^A = e_\sigma^B = 3000$ cm$^{-1}$, $\delta = 100°$); (**e**) the effect of varying $\delta$ (see text) ($e_\sigma^A = e_\sigma^B = 3000$ cm$^{-1}$, $e_\pi = 0$)

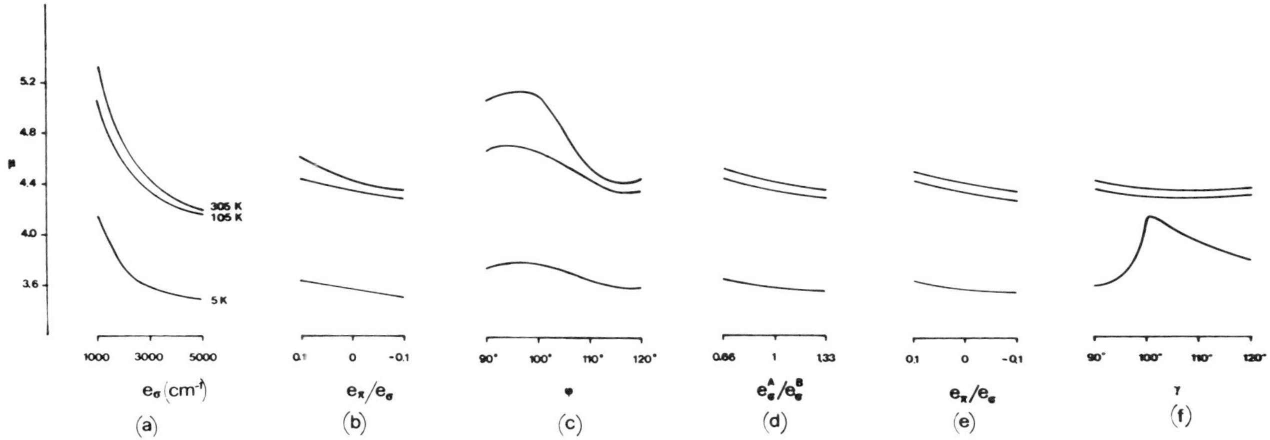


**Fig. 13 a–f.** Magnetic moments calculated (see text) at 305, 105 and 5 K for trigonal bipyramidal $CoA_5$ complexes, (**a**) and (**b**), for the transition from a trigonal bipyramid to a square pyramid $CoA_5$, (**c**), and for trigonal bipyramidal $CoA_2B_3$ complexes, (**d**), (**e**) and (**f**): (**a**) the effect of varying $e_\sigma$ ($e_\pi = 0$); (**b**) the effect of varying $e_\pi/e_\sigma$ ($e_\sigma = 3000\ cm^{-1}$); (**c**) the effect of varying $\varphi$ (see text) ($e_\sigma = 3000\ cm^{-1}$, $e_\pi = 0$); (**d**) the effect of varying $e_\sigma^A/e_\sigma^B$ ($e_\sigma^B = 3000\ cm^{-1}$, $e_\pi = 0$, $\gamma = 90°$); (**e**) the effect of varying $e_\pi/e_\sigma$ ($e_\sigma^A = e_\sigma^B = 3000\ cm^{-1}$, $\gamma = 90°$); (**f**) the effect of varying $\gamma$ (see text) with $e_\sigma^A = e_\sigma^B = 3000\ cm^{-1}$, $e_\pi = 0$

tion between 305 and 105 K si smaller for trigonal bipyramidal complexes. This is obvious from the inspection of Fig. 13c in which the variation of the angle $\varphi$ from 90° to 120° when passing from a square pyramid to a trigonal bipyramid is shown. The pattern of variation of $\mu$ is not much affected by changing the $e_\pi/e_\sigma$ ratio. In trigonal bipyramidal $CoA_2B_3$ complexes the variation of the A-Co-B angle $\gamma$ induces a change in the sign of the zero field splitting at $\gamma = 100°$. This is the reason for the cusp in the low temperature moment of Fig. 13f, the higher temperature ones being substantially unaffected.

Figure 14 shows the calculated $\mu$'s for tetrahedral $CoA_4$ complexes. The pattern is essentially similar to that of trigonal bipyramidal complexes though in this case the temperature-dependence is much smaller. A larger temperature-dependence is calculated for distorted tetrahedral $CoA_2B_2$ and $CoAB_3$ chromophores, however. The effect of the distortion on the temperature-dependence is well documented by Figs. 14e and 14h were a minimum difference $\mu_{305} - \mu_5$ is achieved at the tetrahedral angle. It is also apparent that a much higher temperature-dependence is predicted for angles $\beta < 109.47°$ as compared to $\beta > 109.47°$. On the other hand a more symmetric behavior is anticipated for $CoA_2B_2$ chromophores when the angle A-Co-A, $\alpha$, is varied.

## 3.2 EPR Spectra

High-spin cobalt(II) has three unpaired electrons. In every coordination geometry, excited levels are quite close to the ground state (Figs. 1, 2), however, so that fast relaxation of the electron spin is occurring[7,47]. As a consequence the EPR spectra are too broad to be detected at room temperature, and even at liquid nitrogen temperature only few examples of resolved high-spin cobalt(II) spectra are known. In general the electron spin lattice relaxation time becomes sufficiently long only at temperatures below 30 K. Therefore most of the reports deal with liquid helium temperature measurements.

In principle the spectra should be interpreted using an S = 3/2 spin hamiltonian

$$\hat{H} = \underline{B} \cdot \mathbf{g} \cdot \underline{S} + \underline{S} \cdot \mathbf{D} \cdot \underline{S} + \underline{I} \cdot \mathbf{A} \cdot S + \ldots\ldots \quad (4)$$

where **g**, **D**, and **A** are tensors. The first term refers to the electronic Zeeman interaction, the second is the so called spin-spin interaction which determines the zero field splitting of the electronic energy levels, and the third term is the hyperfine parameter, caused by the electron spin-nuclear spin interaction. Other terms may be added, such as nuclear quadrupole effects for instance, but their relevance to the interpretation of the electronic structure of the complexes is not large, so that they will be neglected.

Spin orbit coupling is responsible for the zero field splitting of the electronic energy levels. If the symmetry is lower than cubic, the maximum degeneracy of the electronic levels is two, corresponding to Kramers doublets. At sufficiently low temperatures only the lowest Kramers doublet is populated so that the spectra can in general be interpreted using a fictitious S = 1/2 spin hamiltonian. This means that only one transition is observed for every orientation of the molecule in the external magnetic field, in accordance with the simple relation

$$h\nu = g'\mu_B B \quad (5)$$

where $\nu$ is the spectrometer frequency.

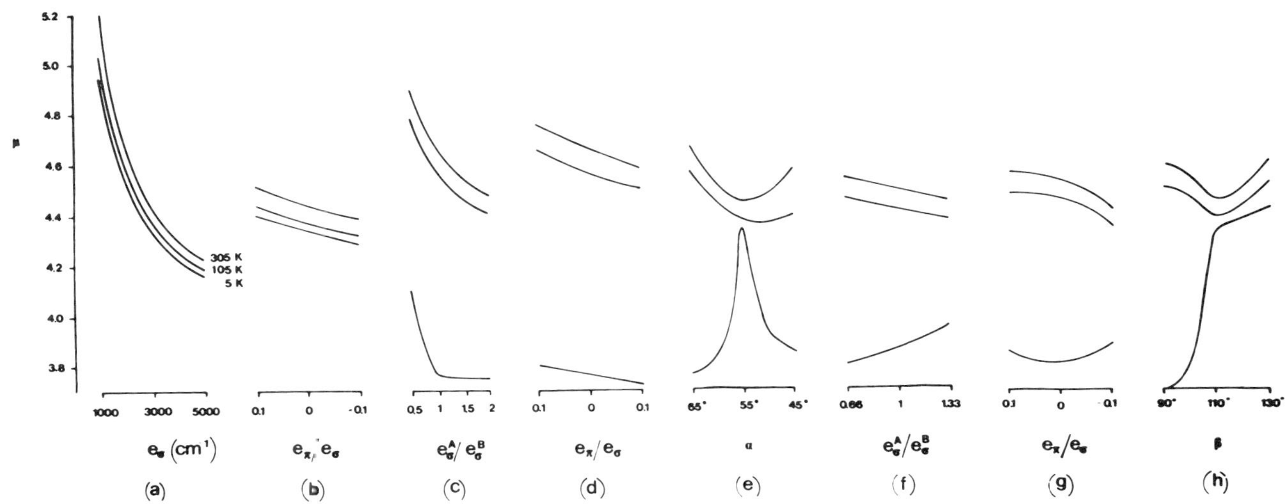


**Fig. 14a–h.** Magnetic moments calculated (see text) at 305, 105 and 5 K for regular tetrahedral $CoA_4$ complexes, (**a**) and (**b**), for pseudotetrahedral $CoA_2B_2$ complexes, (**c**), (**d**) and (**e**), and for pseudotetrahedral $CoAB_3$ complexes, (**f**), (**g**) and (**h**): (**a**) the effect of varying $e_\sigma$ ($e_\pi = 0$); (**b**) the effect of varying $e_\pi/e_\sigma$ ($e_\sigma = 3000$ cm$^{-1}$); (**c**) the effect of varying $e_\sigma^A/e_\sigma^B$ ($e_\sigma^B = 3000$ cm$^{-1}$, $e_\pi = 0$, $\alpha = 65°$); (**d**) the effect of varying $e_\pi/e_\sigma$ ($e_\pi^A = e_\pi^B$, $e_\sigma^A = e_\sigma^B = 3000$ cm$^{-1}$, $\alpha = 65°$); (**e**) the effect of varying $\alpha$ (see Fig. 3) ($e_\sigma^A = e_\sigma^B = 3000$ cm$^{-1}$, $e_\pi = 0$); (**f**) the effect of varying $e_\sigma^A/e_\sigma^B$ ($e_\sigma^B = 3000$ cm$^{-1}$, $e_\pi = 0$, $\beta = 100°$); (**g**) the effect of varying $e_\pi/e_\sigma$ ($e_\pi^B = e_\pi^A$, $e_\sigma^B = e_\sigma^A = 3000$ cm$^{-1}$, $\beta = 100°$); (**h**) the effect of varying $\beta$ (see Fig. 4) ($e_\sigma^A = e_\sigma^B = 3000$ cm$^{-1}$, $e_\pi = 0$)

Since, however, the state is not a true spin 1/2, the g′ values may be very different from 2. In the limit of axial symmetry, it is easy to show that $g'_{\|} = g_{\|}$, $g'_{\perp} = 2g_{\perp}$, where $g_{\|}$ and $g_{\perp}$ are the true Zeeman parameters of the S = 3/2 state.

The reasonce field in this case is $B_r = h\nu/2g\mu B$ so that $g' = 2g$.

The most general approach to the g values of high-spin cobalt(II) is very similar to that outlined in the previous section for the magnetic susceptibility. When the energy levels are calculated in the Angular Overlap Model, including spin orbit coupling, a Kramers doublet is found as the ground state. The g values can be easily calculated as the first order Zeeman splitting of the two levels[44, 48]. Since in the most general case the symmetry of the metal ion is $C_1$, the principal axes of g are not fixed, but must be calculated. This means that the $\mathbf{g}^2$ tensor will not be diagonal in the general axis frame within which the energy levels have been calculated. In this hypothesis

$$\mathbf{g}^2 = \sum_i \sum_k \langle i|\hat{\mu}|k\rangle\langle k|\hat{\mu}|i\rangle \tag{6}$$

the sums being over the two components of the Kramers doublet; the other symbols have the same meaning as in eq. (1). Diagonalization of the $\mathbf{g}^2$ tensor will yield the principal $g^2$ values, from which the g values and the principal directions can be obtained. With the same procedure it is possible to calculate the cobalt hyperfine splitting, using the relation:

$$A^2_{\alpha\beta} = \frac{W_{\alpha\beta}}{\sum_{m=1}^{S+1} m^2} \tag{7}$$

where $A^2_{\alpha\beta}$ is the $\alpha\beta$ element of the $\mathbf{A}^2$ tensor and

$$W_{\alpha\beta} = \sum_{i'} \sum_{k'} \langle i'|\tilde{S}_\alpha|k'\rangle\langle k'|\tilde{S}_\beta|i'\rangle \tag{8}$$

$\tilde{S}_\alpha$ and $\tilde{S}_\beta$ contain contributions of spin and orbital operators[49].

In order to analyze the relation between the observed g values and the electronic structure of the cobalt(II) ion, it is convenient to consider separately the octahedral and square pyramidal complexes which have (quasi) degenerate ground states and the trigonal bipyramidal and tetrahedral complexes. Also it is convenient to resort to approximate models which will be useful for learning which g values are expected in the various coordination environments. It is understood that only a more detailed analysis similar to the one outlined above allows one to fully understand the spin hamiltonian parameters.

In the case of tetrahedral and trigonal bipyramidal complexes, the $^4A_2$ ground level is split by spin orbit coupling into two Kramers doublets. The two can be meaningfully labelled as $M_s = \pm 1/2$ and $M_s = \pm 3/2$ only in the limit of axial symmetry, while for lower symmetries this is only loosely valid. However we will use it for the sake of simplicity. The separation between the two levels is given by $\delta = 2(D^2 + 3E^2)^{1/2}$ where D and E are the zero field splitting parameters. The ratio $E/D = \lambda$ is such that its variation between 0 and 1/3 is able to cover all the possible distortions of the ligand field[47]. For $\lambda = 0$ the symmetry is axial, while $\lambda = 1/3$ corresponds to the maximum rhombic splitting. The g′ values in the effective S = 1/2 spin hamiltonian are related to the true ones in the S = 3/2 spin hamiltonian according to[51)]:

$$
\begin{array}{ll}
M_s = \pm 1/2 & M_s = \pm 3/2 \\
g'_x = g_x \left(1 + \frac{1 - 3\lambda}{\sqrt{1 + 3\lambda^2}}\right) & g'_x = g_x \left(1 - \frac{1 - 3\lambda}{\sqrt{1 + 3\lambda^2}}\right) \\
g'_y = g_y \left(1 + \frac{1 + 3\lambda}{\sqrt{1 + 3\lambda^2}}\right) & g'_y = g_y \left(1 - \frac{1 + 3\lambda}{\sqrt{1 + 3\lambda^2}}\right) \\
g'_z = g_z \left(1 - \frac{2}{\sqrt{1 + 3\lambda^2}}\right) & g'_z = g_z \left(1 + \frac{2}{\sqrt{1 + 3\lambda^2}}\right)
\end{array}
\tag{9}
$$

The ranges of g′ values to be expected[50)] using relations (9), and assuming g values in the range 2–2.3 are shown in Fig. 15.

For octahedral and square pyramidal complexes, the theory outlined by Abragam and Pryce[56)] for axial symmetry predicted a relation between $g_{\parallel}$ and $g_{\perp}$ values as shown in Fig. 16. It was extended also to lower symmetries[50)] and the range of g values to be expected is given in Fig. 17. It is apparent that it is not possible to recognize the coordination environment of the cobalt complexes using only the g values as a criterion. As a matter of fact, only for quasi octahedral complexes are the g values significantly different from those of other symmetries, but distorted chromophores tend to have the same limiting g values.

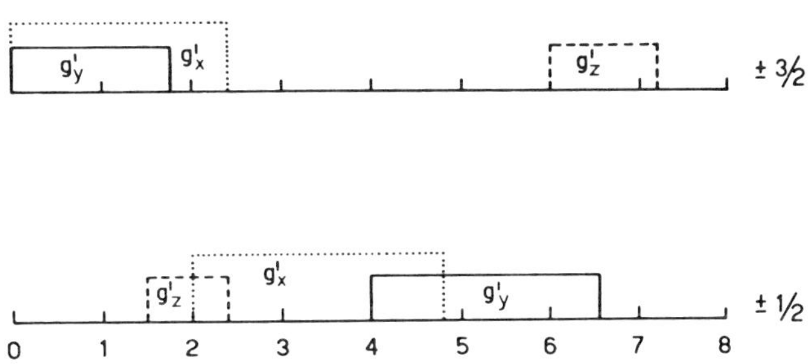


**Fig. 15.** Range of g′ values to be expected for ± 3/2 and ± 1/2 Kramers doublets in distorted tetrahedral cobalt(II) complexes

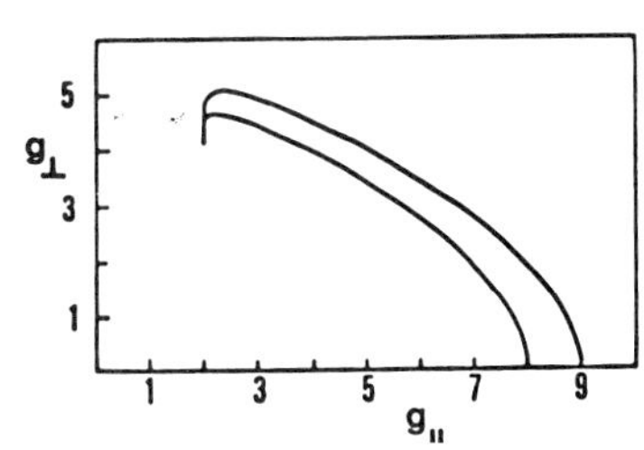


**Fig. 16.** Relation between $g_{\parallel}$ and $g_{\perp}$ for cobalt(II) complexes having axial symmetry

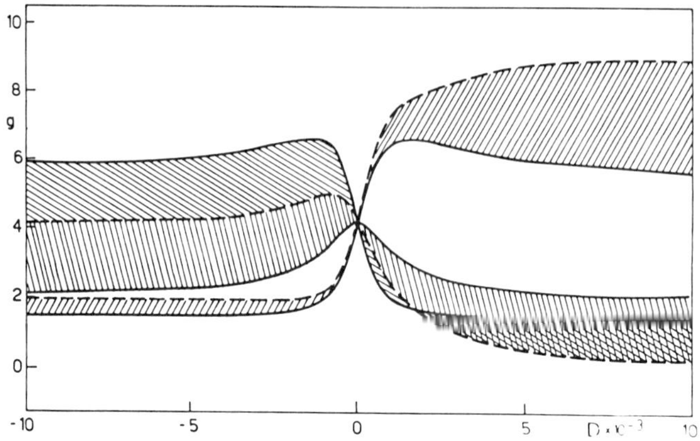

**Fig. 17.** Range of g values to be expected for distorted octahedral cobalt(II) complexes as a function of D: /// $g_z$, \\\\\\$g_x$, \\\\\\$g_y$; --- E/D = 0; — E/D = 1/3

In order to have more detailed information on the spin hamiltonian parameters, they were calculated using relation (6), on the basis of the Angular Overlap Model for various coordination environments. In Fig. 18 the true g values and the g′ values with the effective spin S = 1/2, which refer to the two lowest Kramers doublets of tetrahedral $CoA_2B_2$ chromophores, are given. The g values were calculated using a perturbation approach, together with the zero field splitting parameters, using the relation:

$$g_{ij} = g_e \delta_{ij} + \frac{\zeta}{S} \Lambda_{ij} \qquad D_{ij} = \frac{\zeta^2}{4S^2} \Lambda_{ij} \tag{10}$$

$$\Lambda_{ij} = \frac{\sum_n \langle G|L_i|n\rangle \langle n|L_j|G\rangle}{\Delta_n}$$

i and j are general components of the vectors and terms of expression (10), $\delta_{ij}$ is the Kronecker delta, $|n\rangle$ and $|G\rangle$ are an excited and the ground state respectively, $\Delta_n$ is their energy separation and $\zeta$ is the spin orbit coupling constant.

Figure 18a visualizes the effect of varying the $\alpha$ angle, when the four ligands are identical, and $e_\pi = 0$. $g_\parallel < g_\perp$ is expected for compressed tetrahedra ($\alpha > 54.74°$), while the reverse is true for elongated tetrahedra. Also the sign of the zero field splitting parameter, D, is expected to change, being positive for $\alpha > 54.74°$ and negative for $\alpha < 54.74°$.

This pattern of g′ values, however, can be changed when the A and B ligands are different. In Fig. 18b the $e_\sigma^A/e_\sigma^B$ ratio (keeping $e_\sigma^B$ fixed) is varied for $\alpha = 65°$. Within the limits of the ratio considered by us, 1/2 and 2 respectively, the direction along which the maximum zero field splitting, $|D|$, is observed moves from z, for axial symmetry, to y or x. When the change occurs, the sign of D also reverses, being negative close to the extremes, and positive close to the axial limit. The g′ values in the two Kramers doublets are seen to vary in a very complicated manner.

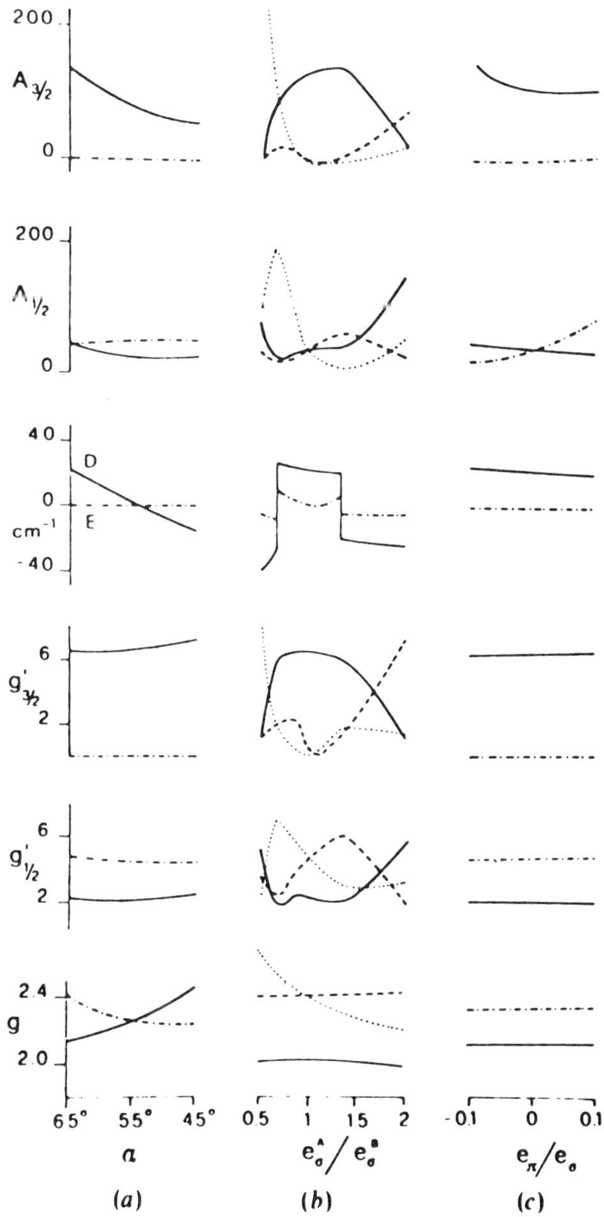


**Fig. 18 a–c.** Calculated g, g′, D, E and A values for $CoA_2B_2$ chromophores ($k = 0.85$, $\zeta = 533\ cm^{-1}$, $\varkappa = 0.350$, $P = 0.026\ cm^{-1}$). (**a**) the effect of varying $\alpha$ (with $c_\sigma^A = c_\sigma^B = 3000\ cm^{-1}$, $e_\pi^A = e_\pi^B$); (**b**) the effect of varying $e_\sigma^A/e_\sigma^B$ ($e_\sigma^B = 3000\ cm^{-1}$, $e_\pi^A = e_\pi^B = 0$, $\alpha = 65°$); (**c**) the effect of varying $e_\pi/e_\sigma$ ($e_\sigma^A = e_\sigma^B = 2000\ cm^{-1}$, $\alpha = 65°$). The A values are given in units of $10^4\ cm^{-1}$; —z, ···y, ---x

The $e_\pi/e_\sigma$ ratio does not affect the calculated g′ values, as previously observed (Fig. 18c).

The calculated A values are in general small ($< 100 \times 10^{-4}\ cm^{-1}$). Only in cases with D negative and E/D rather small is a large hyperfine splitting predicted for the largest g value. This can be easily understood considering that in this situation the ground Kramers doublet can be essentially described as $\pm 3/2$. If it were rigorously so, all the matrix elements of the hyperfine operator[47)]

$$\hat{H} = (\tilde{S}_z I_z + \tilde{S}_x I_x + \tilde{S}_y I_y)P' \tag{11}$$

would vanish, except for $\tilde{S}_z I_z$. For instance, the contribution of the Fermi contact operator is $-3\varkappa P$ along z, while it is zero along x and y. Although for $E/D = 0$ the transition probability is zero, it may become non-zero if some rhombic anisotropy is introduced.

The spin hamiltonian parameters for tetrahedral $CoAB_3$ complexes are shown in Fig. 19. For angles smaller than the tetrahedral value 109.47°, the ± 1/2 level tends to be the ground level, while the reverse is true for larger angles. This pattern may be slightly varied if the A and B ligands are allowed to be different. In particular for $e_\sigma^A > e_\sigma^B$, the ± 3/2 level tends to be stabilized, so that eventually it may become the ground state even at angles < 109.47°.

As for tetragonal tetrahedral complexes, no large effects are due to $\pi$ bonding interactions. Also for the hyperfine splitting the same considerations apply as for the above case.

For a regular trigonal bipyramidal $CoA_2B_3$ of $D_{3h}$ symmetry $g_\parallel$ is very close to the spin only value[25, 37] while $g_\perp$ is sensibly larger than this limit (Fig. 20). $A_\parallel$ is calculated to have sizeable values, while $A_\perp$ is anticipated to be small. D is positive and rather large. Relaxing the $e_\sigma^A/e_\sigma^B$ ratio determines a variation only of $g_\perp$, due to the change in the energies of the excited states. Also $A_\parallel$ is substantially unchanged, while $A_\perp$ is somewhat more sensitive. The $e_\pi/e_\sigma$ ratio slightly affects the spin hamiltonian parameters as was the case for tetrahedral complexes.

A possible geometrical distortion from the trigonal bipyramidal coordination, which preserves the trigonal symmetry and is frequently encountered, is the so-called tetrahedral distortion[36]. The A-Co-B angles, $\gamma$, become different from 90°, (Fig. 6) and one of the axial ligands moves away. Figure 20 c shows that $g_\parallel$ increases and $g_\perp$ decreases when $\gamma$ becomes larger than 90°. The reverse dependence is expected for A. D changes sign, if $\gamma$ increases.

Figure 21 shows the spin hamiltonian parameters for square pyramidal $CoAB_4$ complexes. It had already been noted by Gerloch[53] that the composition of the lowest Kramers doublet and hence the *g*-tensor is a very sensitive function of the geometrical and bonding parameters of the square pyramidal complex. For instance a variation of the $\delta$ angle (Fig. 7) from 90 to 110° should lead to an increase of $g_\parallel$ and a decrease of $g_\perp$. The hyperfine coupling constant follows the same pattern (Fig. 21 a). As compared to the trigonal bipyramidal complexes the $A_\parallel$ value tends to be larger, while $A_\perp$ is not much different.

The $e_\sigma^A/e_\sigma^B$ ratio also affects the g values; an increase in the ratio induces larger $g_\parallel$ and smaller $g_\perp$ values. A similar pattern is predicted for the A values. The $e_\pi/e_\sigma$ ratio also strongly affects the g and A values.

Figure 22 visualizes the spin hamiltonian parameters calculated for five coordinate $CoA_5$ complexes with geometries intermediate between a trigonal bipyramid ($\varphi = 120°$) and a square pyramid ($\varphi = 90°$). The variation of the calculated g values due to the two lowest Kramers doublets is shown in Fig. 22. The z direction is the unique axis in tetragonal symmetry and lies in the equatorial plane for the trigonal bipyramid, while x is the trigonal axis of the latter and lies in the equatorial plane of the former.

The g values are seen to vary in a complicated way, depending also on the magnitude of the parameters, since they strongly affect the starting values in the tetragonal limit. It

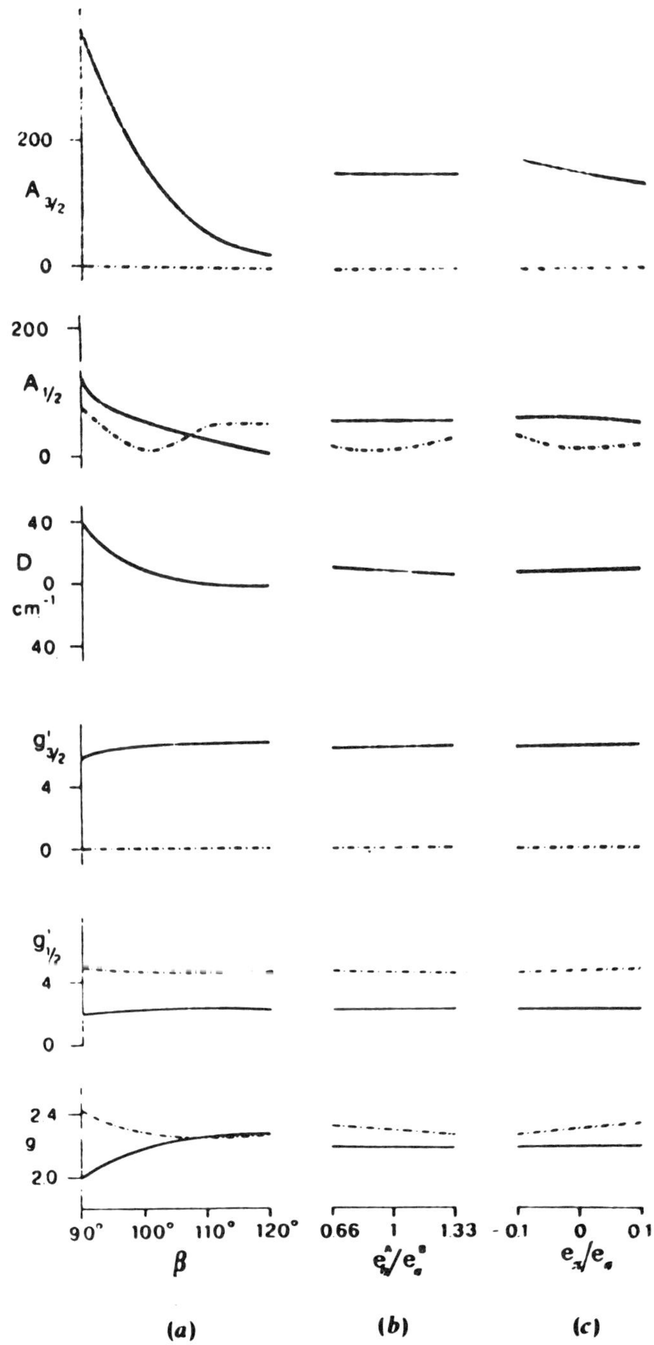


**Fig. 19 a–c.** Calculated g, g′, D and A values for $CoAB_3$ chromophores (k = 0.85, $\zeta = 533\ cm^{-1}$, $\varkappa = 0.350$, $P = 0.026\ cm^{-1}$). (**a**) the effect of varying $\beta$ ($e_\sigma^A = e_\sigma^B = 3000\ cm^{-1}$, $e_\pi^A = e_\pi^B = 0$); (**b**) the effect of varying $e_\sigma^A/e_\sigma^B$ ($e_\sigma^B = 3000\ cm^{-1}$, $e_\pi^A = e_\pi^B = 0$, $\beta = 100°$); (**c**) the effect of varying $e_\pi/e_\sigma$ ($e_\sigma^A = e_\sigma^B = 3000\ cm^{-1}$, $\beta = 100°$). The A values are given in units of $10^4\ cm^{-1}$; —z, ...y, ---x

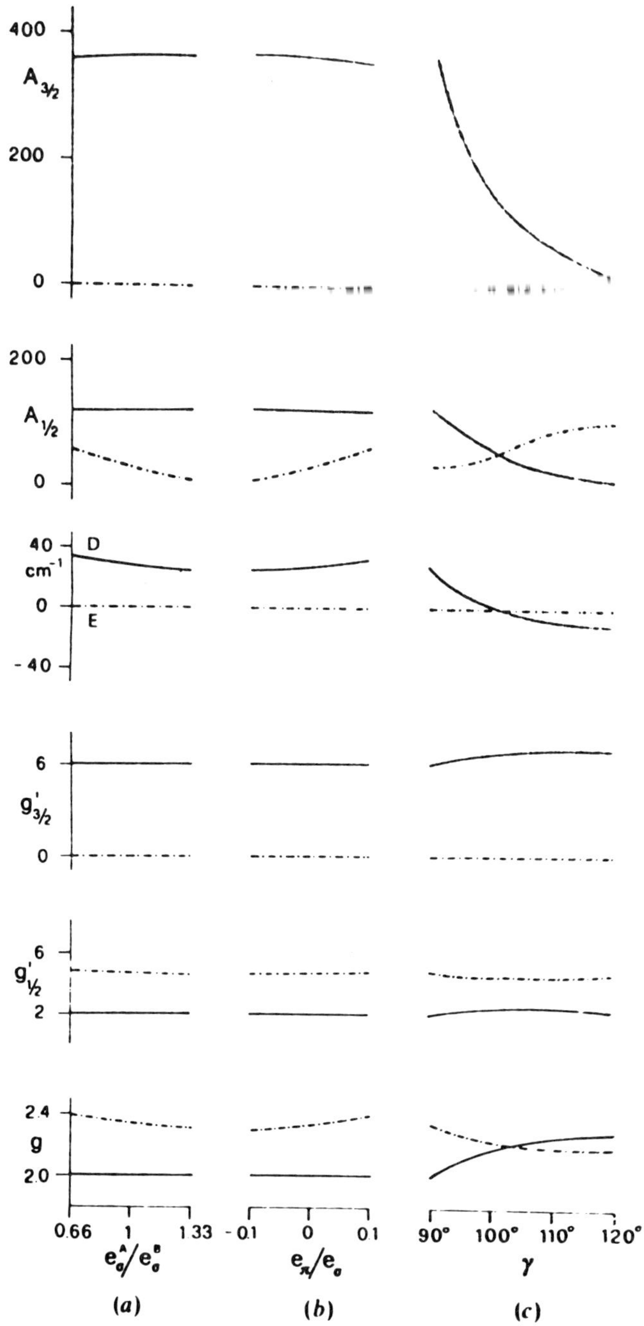


**Fig. 20a–c.** Calculated g, g′, D, E and A values for $CoA_2B_3$ chromophores (k = 0.85, $\zeta = 533\ cm^{-1}$, $\varkappa = 0.350$, $P = 0.026\ cm^{-1}$). (**a**) the effect of varying the $e^A_\sigma/e^B_\sigma$ ratio ($e^B_\sigma = 3000\ cm^{-1}$, $e^A_\pi = e^B_\pi = 0$, $\gamma = 90°$); (**b**) the effect of varying the $e_\pi/e_\sigma$ ratio ($e^A_\sigma = e^B_\sigma = 3000\ cm^{-1}$, $\gamma = 90°$); (**c**) the effect of varying $\gamma$ ($e^A_\sigma = e^B_\sigma = 3000\ cm^{-1}$, $e^A_\pi = e^B_\pi = 0$). The A values are given in units of $10^4\ cm^{-1}$; —z, ...y, ---x

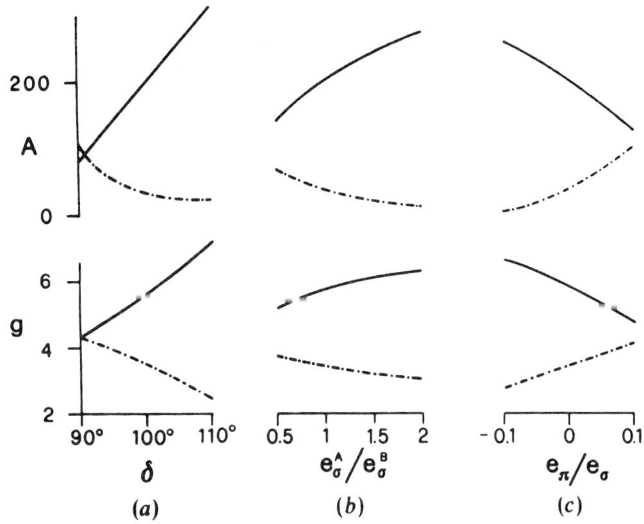


**Fig. 21 a–c.** Calculated g′ and A values for the lowest Kramers doublet for square pyramidal $CoAB_4$ chromophores (k = 0.85, $\zeta$ = 533 $cm^{-1}$, $\varkappa$ = 0.350, P = 0.026 $cm^{-1}$). The A values are given in units of $10^4$ $cm^{-1}$ (**a**) the effect of varying $\delta$ ($e_\sigma^A = e_\sigma^B$ = 3000 $cm^{-1}$, $e_\pi^A = e_\pi^B = 0$); (**b**) the effect of varying $e_\sigma^A/e_\sigma^B$ ($\delta$ = 100°, $e_\sigma^B$ = 3000 $cm^{-1}$ $e_\pi^A = e_\pi^B = 0$); (**c**) the effect of varying $e_\pi/e_\sigma$ ($\delta$ = 100°, $e_\sigma^A = e_\sigma^B$ = 3000 $cm^{-1}$). —z, ...y, ---x

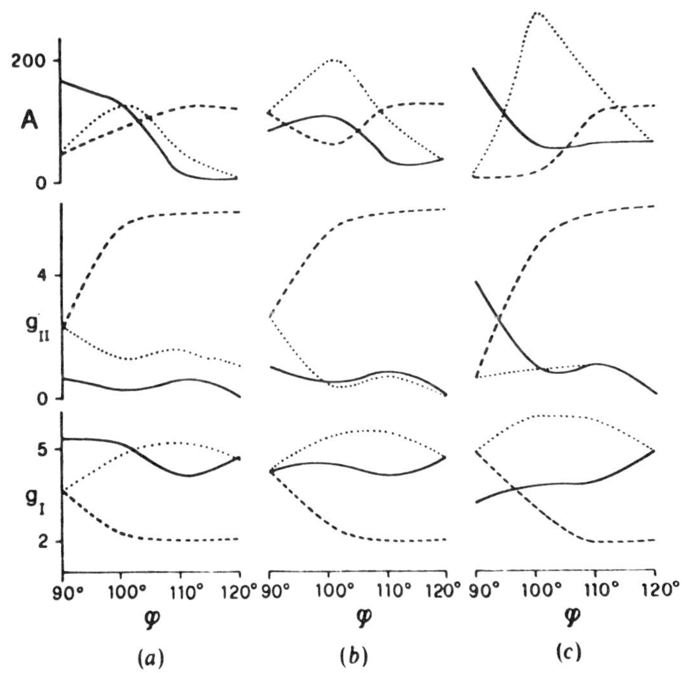


**Fig. 22 a–c.** Calculated g and A values for the two lowest Kramers doublets in $CoA_5$ chromophores with geometries intermediate between square pyramidal and trigonal bipyramidal (k = 0.85, $\zeta$ = 533 $cm^{-1}$, $\varkappa$ = 0.350, P = 0.026 $cm^{-1}$). The effect of varying the $\varphi$ angle for: (**a**) $e_\pi/e_\sigma = -0.1$; (**b**) $e_\pi/e_\sigma = 0$; (**c**) $e_\pi/e_\sigma = +0.1$. The A values are given in units of $10^4$ $cm^{-1}$, $g_I$ refers to the lowest and $g_{II}$ to the first excited Kramers doublet, —z, ...y, ---x

appears that no simple rule, of the kind suggested for five coordinate copper(II) complexes[54)], is valid in this case.

Figure 23 depicts the spin hamiltonian parameters calculated for octahedral $CoA_6$ chromophores experiencing trigonal distortions. The g values for regular octahedral complexes may range from 4.43 to 4.0 depending on the orbital reduction factor and on the strong or weak field approximation which is appropriate for the particular complex. As was observed for the magnetic moment curves (Fig. 10) the dependence of g and A on the trigonal distortion parameter, $\xi$, is rather complicated as a consequence of the intersection of the electonic energy levels.

The spin hamiltonian parameters for tetragonal $CoA_4B_2$ complexes are shown in Fig. 24. As was already stated in section 3, the g and A values are essentially determined by the $e_\pi$ values for the axial and equatorial ligands.

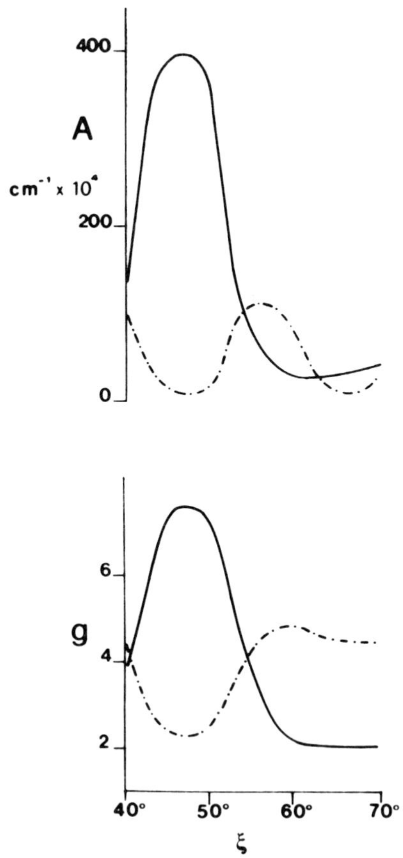


**Fig. 23.** Calculated g and A values for octahedral, trigonally distorted $CoA_6$ chromophores ($k = 0.85$, $\zeta = 533\ cm^{-1}$, $\varkappa = 0.350$, $P = 0.026\ cm^{-1}$). Dependence of g and A on $\xi$ (see text) ($e_\sigma = 3000\ cm^{-1}$, $e_\pi = 0$)

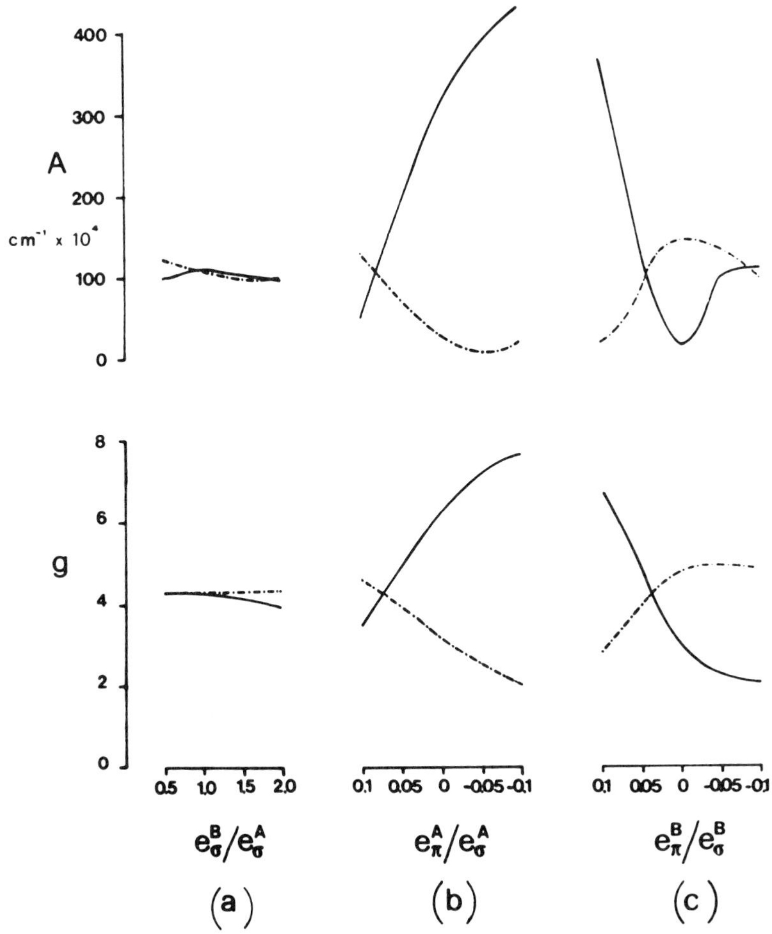


**Fig. 24a–c.** Calculated g and A values for tetragonally distorted octahedral $CoA_4B_2$ chromophores ($k = 0.85$, $\zeta = 533\ cm^{-1}$, $\varkappa = 0.350$, $P = 0.026\ cm^{-1}$): **(a)** the effect of varying $e_\sigma^B/e_\sigma^A$ ($e_\sigma^A = 3000\ cm^{-1}$, $e_\pi = 0$); **(b)** the effect of varying $e_\pi^A/e_\sigma^A$ ($e_\sigma^A = 3000\ cm^{-1}$, $e_\pi^B/e_\sigma^B = 0.05$, $e_\sigma^B = 4500\ cm^{-1}$); **(c)** the effect of varying $e_\pi^B/e_\sigma^B$ ($e_\sigma^B = 4500\ cm^{-1}$, $e_\pi^A/e_\sigma^A = 0.05$, $e_\sigma^A = 3000\ cm^{-1}$)

## *3.3 NMR Spectra*

Proton NMR spectra are only indirectly influenced by the cobalt(II) ion. A considerable signal broadening and a shift ("isotropic shift"), which is the sum of two contributions, the Fermi or contact and the pseudocontact term is observed[55]. The isotropic shift is influenced by the coordination geometry of the metal ion, but it must be admitted that the present theory is not sophisticated enough to relate an observed isotropic shift to the electronic structure of the complex[56].

Structural information can be obtained in some cases when the binding of potential donor atoms to the metal is monitored by the shift of the signals of the related protons[57].

Another method for obtaining information on the environment of the metal ion is through measurements of the nuclear spin lattice relaxation time, $T_1$. When a nucleus is under the influence of a paramagnetic center, its relaxation rate is enhanced, since the rapid fluctuations of the electronic spin provide an extremely efficient pathway to nuclear spin relaxation[58]. The theory is very complicated, but it is customary to use the Solomon-Bloembergen-Morgan equations[59] in order to analyze the observed nuclear $T_1$ values, though it is well known that it can hardly be applied to cobalt(II) complexes, where large anisotropies due to large (unquenched) orbital contributions are operative. Since it is generally accepted that the main contribution to the $T_1^{-1}$ enhancement comes from the dipolar interaction[58], the Solomon-Bloembergen-Morgan equation can be written as

$$T_{1M}^{-1} = \frac{2}{15} \frac{\gamma_I^2 g^2 S(S+1) \mu_B^2}{r^6} \left( \frac{3\tau_c}{1 + \omega_I^2 \tau_c^2} + \frac{7\tau_c}{1 + \omega_S^2 \tau_c^2} \right)$$

where all the symbols have their usual meaning.

If the metal to proton distance, r, is known it is then possible to determine the correlation time, $\tau_c$. An alternative method measures the nuclear $T_{1M}^{-1}$ enhancement at several frequencies; then the geometric factor and the correlation time can be calculated through a least squares analysis[59]. It is usually assumed that $\tau_c$ is determined by the electron spin-lattice relaxation time.

Very small electronic $T_1$ relaxation times are found for octahedral complexes ($\simeq 10^{-13}$ s), while those for tetrahedral complexes are much larger ($\simeq 10^{-11}$ s). Intermediate values have been for complexes which were considered as five coordinate[60]. It may be mentioned here that a similar trend is observed for the line widths, in the sense that the sharpest lines are observed for octahedral complexes, the broadest for tetrahedral[61].

## 3.4 MCD Spectra

Magnetic Circular Dichroism (MCD) spectra[62] use the Faraday effect, according to which a magnetic field parallel to the direction of propagation of an incident plane-polarized light wave travelling in some medium changes the plane of polarization of the electromagnetic wave. The theory of the MCD spectra has been formulated[62–65] and generally the observed ellipticity is decomposed into the sum of A, B, and C terms.

For degenerate ground states, both orbital and spin, B and C terms are possible, and this is the case for high spin cobalt(II) where the minimum degeneracy is two (Kramers doublet). In order to recognize B and C terms, it is necessary to record spectra at different temperatures, since C terms are temperature dependent, while B terms are not. This can be easily done with single crystals, but in this case the requirement is that the crystals possess an optical axis.

On the other hand, solution spectra are difficult to record at very low temperatures, due to the difficulty of having glasses of optical quality. Polymer films have been used in some cases to overcome these problems.

The sign of the ellipticity of the C terms can be calculated using the symmetry properties of the ground and the excited states, while the sign of B terms can be obtained only through detailed calculations which encompass the energy level admixtures determined by the magnetic field.

Several experimental spectra, and in some cases very detailed analyses, of high symmetry complexes, have been reported[66, 67]. Thoroughly studied examples are the tetrahalo complexes in various lattices, the hexaaquoions, and Co(II) doped into various ionic oxides.

Few reports are available for low symmetry complexes. In one of them it has been claimed that it is possible to recognize octahedral, tetrahedral, and trigonal bipyramidal complexes using MCD spectra[68]. According to this assignment the octahedral $^4T_{1g} \rightarrow {}^4T_{1g}(P)$ transition in the MCD spectrum has a peak position as in the absorption spectrum, and a negative ellipticity. The tetrahedral $^4A_2 \rightarrow {}^4T_1(P)$ transition shows both negative and positive ellipticity in the MCD spectrum, and the negative peak is centered in the absorption peak and the positive is at higher frequencies. For five coordinate complexes the split components of the transition to the P levels show negative ellipticity. In some cases a positive ellipticity is shown at higher frequencies.

# 4 Survey of Experimental Results

In the following we will explicitely refer to those species which have been studied by several techniques, and special attention will be devoted to systems for which single crystal data are available. Much more information will be reported in the Tables, without comments.

## 4.1 Tetrahedral Complexes

The energies of the electronic transitions, the effective magnetic moment and the g and A values of some relevant tetrahedral cobalt(II) complexes are shown in Table 2.

Among them the most detailed studies are available for $CoX_4^{2-}$ [68, 69] chromophores in different crystal lattices. Single crystal electronic MCD[71–75], EPR spectra[77–79] are available, as well as magnetic anisotropy measurements[78, 81–83]. The energy levels have been calculated using different approaches. In particular, Horrocks[33] used the Angular Overlap Model for reproducing all the relevant optical and magnetic data available for $CoCl_4^{2-}$. For these complexes very detailed assignments of the spin forbidden transitions are also available[84]. In $[(C_2H_5)_4N]_2CoCl_4$ the principal $g$ directions at liquid helium temperature differ from those one would expect[78] on the basis of the room temperature crystal structure. This has been attributed to a static low-temperature phase, which freezes out from a dynamic distortion operative at room temperature in analogy to the analogous copper complex[79]. The MCD spectra of the spin allowed transitions obey the rule reported in the previous section.

Other systems which have been studied in some detail and with different techniques are the trigonally distorted species $CoquinBr_3^-$ [86–89] and $CoLX_3$ (X = Cl, Br; L = N-ethyl-1,4 diazabicyclo 2.2.2. octonium)[88, 32]. For $CoquinBr_3^-$ it was found that the quinoline ligand behaves as a non-linearly ligating ligand, i.e. the interaction perpendicular to the aromatic ring is different from zero, while that parallel to the aromatic ring

**Table 2.** Spectral parameters for tetrahedral cobalt(II) complexes. The energies of the electronic transitions are labelled according to the parent tetrahedral label of the excited states. F and P refer to the free ion terms which contribute mostly to the indicated state. Don.Set is the donor set of the

| Complexes | Don.Set | Spe. | $^4T_1$(F) | ν $cm^{-1}$ ($\varepsilon_M$) | |
|---|---|---|---|---|---|
| Co(2-Et-py)$_2$(NCS)$_2$ | $N_4$ | S | 7363(129) | 8137(160) | 9204(147) |
| Co(Me$_4$dipyMT)$_2$ | $N_4$ | S | 6200sh | 8470 | 10200sh |
| Co(2MeIz)$_4$(ClO$_4$)$_2$ | $N_4$ | P | | 8800 | |
| Co(Tripyr)$_2$ | $N_4$ | S | 7700sh | 9300sh(46) | 10200(54) |
| Co(4DMAP)(NCS)$_2$ | $N_4$ | P | 7250 | 8200 | 10000 |
| Co(N-tBpyle-2-C)$_2$ | $N_4$ | C | 7500⊥ | 9300∥ | |
| Co[N,N′(4ClB)PAI]$_2$ | $N_4$ | S | | 9600(83) | |
| Co(HdMePyne)$_2$Br$_2$ | $N_4$ | S | | 7017(14) | |
| [Co(NCNPh)$_4$]$^{2-}$ | $N_4$ | S | | [illegible] | |
| [Co(1MePyne-2-T)$_4$](ClO$_4$)$_2$ [illegible] | [illegible] | P | | 10420 | |
| [illegible](NCS)$_2$ | $N_4$ | P | | 8500 | |
| Co(N-t-BPyle-2-Aim)$_2$ | $N_4$ | S | | 8330(28) | 9730(36) |
| Co(Me$_6$tpt)(BPh)$_4$ | $N_4$ | P | 8200 | 10500sh | 11300 |
| Cs$_3$CoCl$_4$ | $Cl_4$ | C | | 5500 | |
| (CoCl$_4$)$^{2-}$ (CoDEU$_6$)$^{2+}$ | $Cl_4$ | P | | 6400vb | |
| Cs$_2$(Zn, Co)Cl$_4$ | $Cl_4$ | C | 5064∥ 5348∥ | 6410⊥ 6694⊥ | 7000⊥ |
| Co(C$_9$H$_8$N)Cl$_4$ | $Cl_4$ | S | 4820(66) 5180(69) | 5470(70) | 6120(67) |
| [Co(N-EtdAZAO)Br$_3$] | $NBr_3$ | C | 4220∥ | 7270⊥ | |
| [Co(N-EtdAZAO)Cl$_3$] | $NCl_3$ | C | | 4500 | 7700 |
| [Co(Me$_2$ppaz)Br$_3$] | $NBr_3$ | P | 4700 | 6400 | 7300 |
| [Co(quin)Br$_3$]$^-$ | $NBr_3$ | C | | 5000 | 7000 |
| [Co(BIz)Br$_3$]$^-$ | $NBr_3$ | P | | 4900 | 6900 |
| Co(2-Et-py)$_2$Br$_2$ | $N_2Br_2$ | S | 5670(63) | 6692(73) | 8220sh(54) |
| Co(dpmq)I$_2$ | $N_2I_2$ | P | 5710 | 6760 | 9350 |
| Co(4-DMAP)Br$_2$ | $N_2I_2$ | P | 6200 | 7340 | 9500 |
| Co(PhIPPh$_3$)Br$_2$ | $N_2Br_2$ | P | 5000 | 5400 | 6200 |
| Co(LS)Br$_2$ | $N_2Br_2$ | P | | 6450 | 8560 |
| Co(quin)$_2$Br$_2$ | $N_2Br_2$ | P | 6150 | 7050 | 8550 |
| Co(Me$_4$en)Cl$_2$ | $N_2Cl_2$ | P | 6000sh | 7250 | 10000 |
| Co(Me$_4$tn)Cl$_2$ | $N_2Cl_2$ | P | 6550sh | 7850 | 10630 |
| Co(MSBenz-NEt$_2$)Br$_2$ | $N_2Br_2$ | P | 6000 | 7300 | 10000 |
| Co(FuAenMe$_2$)Br$_2$ | $N_2Br_2$ | P | 6000 | 7500 | 10000 |
| Co(MOBen-NEt$_2$)Br$_2$ | $N_2Br_2$ | P | 6050 | 7050 | 9250 |
| Co(NMBuL)$_4$(ClO$_4$)$_2$ | $O_4$ | S | 6300(6.5) | 8050(11) | 10500(5.8) |
| [Co(Ph$_3$AsO)$_4$](ClO$_4$)$_2$ | $O_4$ | S | 5710(66) | 6490(72) | 7410(70) |
| [Co(Ph$_3$PO)$_2$(NO$_3$)$_2$] | $O_4$ | P | | not reported | |
| Co(DTMSO)$_2$Cl$_2$ | $O_2Cl_2$ | P | | 6490 | |
| Co(Et$_3$PO)$_2$Cl$_2$ | $O_2Cl_2$ | P | | 6250 | |
| (Co, Zn)(Ph$_3$PO)$_2$Cl$_2$ | $O_2Cl_2$ | C | | | |
| Co(Ph$_3$AsO)$_2$Cl$_2$ | $O_2Cl_2$ | S | 5240(60) | 5780(65) | 6560(67) |
| Co(ESU)$_2$Cl$_2$ | $Se_2Cl_2$ | P | | 5814 | |
| Co(SU)$_2$Cl$_2$ | $Se_2Cl_2$ | NM | 5300 | 6200 | 7500 |
| Co(Me$_3$PSe)$_2$Cl$_2$ | $Se_2Cl_2$ | S | | 5050(126) | 6100sh |
| Co(acpz)Cl$_2$ | $NOCl_2$ | P | | 6060 | |
| [Co(Ph$_3$AsO)$_2$(NCS)$_2$] | $O_2N_2$ | S | 6450sh | 7140(133) | 7750sh |
| Co(CH$_3$COO)$_2$Iz$_2$ | $O_2N_2$ | NM | | not reported | |
| Co(C$_2$H$_5$COO)$_2$Iz$_2$ | $O_2N_2$ | NM | | not reported | |
| Co(C$_3$H$_7$COO)$_2$Iz$_2$ | $O_2N_2$ | NM | | not reported | |
| [Co(Me$_3$PSe)$_4$](ClO$_4$)$_2$ | $Se_4$ | P | | 5600 | 7000 |
| | | S | | 5800(170) | 7200(170) |
| Co[(Ph$_2$PS)$_2$N]$_2$ | $S_4$ | P | | 6613 | |
| Co(dietu)$_2$Cl$_2$ | $S_2Cl_2$ | C | 5600 | 7100 | 7400 |
| Co(C$_3$H$_2$S$_3$)$_2$Cl$_2$ | $S_2Cl_2$ | P | | 6940 | |
| Co(TU)$_2$Cl$_2$ | $S_2Cl_2$ | S | 5500(100) | 6200(81) | 7350(62) |
| Co(TU)$_4$(ClO$_4$)$_2$ | $S_4$ | S | | 7220(172) | |
| [Co(Ph$_3$P)$_2$(SCN)$_2$] | $P_2S_2$ | P | | not reported | |
| Co(PPh$_3$)$_2$Cl$_2$ | $P_2Cl_2$ | C | 6400 | 8000 | 10700 |
| Co(Et$_3$P)$_2$Cl$_2$ | $P_2Cl_2$ | P | | 8250 | |
| Co(PC$_4$P)Br$_2$ | $P_2Br_2$ | P | 6700sh | 8000 | 10800 |
| [Co(NOP$_2$)Cl$_2$] | $P_2Cl_2$ | P | 6300 | 7700 | 10100 |
| Co(Ph$_3$P)$_2$Br$_2$ | $P_2Br_2$ | S | 6200(110) | 7700(95) | 10200(80) |
| Co(PN′H)Cl$_3$ | $PCl_3$ | S | | 7700(48) | 8700sh |

complex; Spe. indicates the condition under which the electronic and EPR spectra were obtained: S solution, P diffuse reflectance for electronic spectra and polycrystalline powder for EPR spectra, C single crystal, NM Nujol mull. The A's are in $10^{-4}$ $cm^{-1}$.

| $^4T_1(P)$ | $\nu$ $cm^{-1}$ ($\varepsilon_M$) | | Ref. | $\mu$ | Ref. | $g_1$ ($A_1$) | $g_2$ ($A_2$) | $g_3$ ($A_3$) | Ref. |
|---|---|---|---|---|---|---|---|---|---|
| 16020(956) | 16580sh(834) | 17710sh(416) | 140 | 4.42 | 140 | | | | |
| 13300 | 14500 | 16500sh | 141 | | | | | | |
| | 17700 | 18300sh | 142 | | | | | | |
| 13400(364) | 14500(325) | 16500sh(2500) | 143 | | | | | | |
| 16470 | 16950 | 17700 | 144 | 3.98 | 144 | | | | |
| 15900∥ | 17000⊥ | | 93 | | | | | | |
| | 19300 | | 145 | | | | | | |
| 14930(264) | 16390(228) | | 146 | 4.40 | 146 | | | | |
| 15380 | 16250 | 17580 | 147 | | | | | | |
| | 18350 | | 148 | | | | | | |
| | | | 149 | | | | | | |
| 16000(58) | 17100(129) | 19610(600) | 150 | 4.74 | 150 | | | | |
| | | 20830 | | | | | | | |
| | 18200 | 19000sh | 151 | 4.41 | 151 | | | | |
| | 16000 | | 70 | 4.70 | 152 | 2.40 | | 4.60 | 153 |
| | 15150 | | 154 | | | | | | |
| | | | 69 | | | | | | |
| 14290(600) | 14930(555) 15450sh | 15870(375) | 155 | 4.60 | 155 | | | | |
| 14620 14810 | 14990⊥ 15380 | 15700 16100∥ | 89 | | | | | | |
| 15380 | 15750 | 16690 | 32 | | | | | | |
| | 15400 | | 156 | | | | | | |
| | 15000 | | 86, 88 | | | 1.60 | 2.22 | 6.31 | 87 |
| | not reported | | 149 | | | | | | |
| 15130(674) | 15700sh(543) | 16690sh(319) | 140 | 4.61 | 140 | | | | |
| 14250 | 14930 | 15870 | 157 | 4.72 | 157 | | | | |
| 15380 | 16080 | 16750 | 144 | 4.41 | 144 | | | | |
| 14925 | 15625 | 16949 | 158 | 4.70 | 158 | | | | |
| 15380 | 15870sh | 16950 | 159 | 4.50 | 159 | | | | |
| | not reported | | 149 | | | | | | |
| | 15400 | 17200 | 160 | 4.65 | 160 | | | | |
| 16260 | 17100sh | 17900sh | 161 | 4.56 | 161 | | | | |
| 15400 | 16500sh | 17500sh | 162 | 4.71 | 162 | | | | |
| | 15000/17000 | | 163 | 4.62 | 163 | | | | |
| 14700 | 16600 | 17300 | 164 | 4.61 | 164 | | | | |
| 16500(73) | 17900(84) | 18350(65) | 165 | 4.82 | 165 | | | | |
| 15480(415) | 16670(395) | 17950(260) | 166 | 4.55 | 166 | | | | |
| | 17700 | | 167 | 4.69 | 167 | | | | |
| | 16080 | | 168 | | | | | | |
| | 16105 | | 169 | 4.60 | 169 | | | | |
| | | | | 4.79 | 170 | 2.16 | 3.59 | 5.67 | 48 |
| 14810(536) | 16000sh | 16950sh | 166 | 4.60 | 166 | | | | |
| 12500 | 15267 | 16000 | 171 | 4.70 | 171 | | | | |
| 13000 | 15000 | 15800 | 172 | | | | | | |
| 12940(440) | 14750(400) | 16340(472) | 173 | 4.34 | 173 | | | | |
| | 14700 | | 174 | | | | | | |
| 15630(835) | 16390(825) | 17400sh | 166 | 4.54 | 166 | | | | |
| | not reported | | | | | 2.16 | 4.23 | 4.49 | 98 |
| 17540 | 18380 | 19050 | 98 | | | 2.06(36.5) | 3.43 | 5.37(35) | 98 |
| 17360 | 18360 | | 98 | | | | | | |
| 13300 | 13800 | 14600 | 167 | 4.40 | 167 | | | | |
| 13020(724) | 13870(762) | 14790(547) | | | | | | | |
| | 14258 | | 175 | 4.54 | 175 | | | | |
| 14000 | 15400 | 16400 | 96 | | | | | | |
| | 15290 | | 176 | 4.57 | 176 | | | | |
| 13800(435) | 15000(490) | 16200(450) | 177 | 4.44 | 177 | | | | |
| 13890(632) | 14580(690) | 15600sh | 177 | 4.52 | 177 | | | | |
| 13700 | 15500 | 16700 | 178 | 4.46 | 178 | | | | |
| 13600 | 15800 | 16700 | 96 | | | | | | |
| | 15698 | | 169 | 4.30 | 169 | | | | |
| 15000 | 16000 | 18600sh | 179 | 4.50 | 179 | | | | |
| | | 20000sh | | | | | | | |
| 13900 | 16100 | 17600sh | 180 | 4.44 | 180 | | | | |
| 13100(610) | 14900(970) | 15600(740) | 181 | 4.52 | 181 | | | | |
| 14600(420) | 15400sh | 18600sh | 182 | 4.60 | 182 | | | | |
| | 16800(410) | | | | | | | | |

plane is zero[86–88]. The sign of the $e_\pi$ parameter is negative, suggesting that the ligand is a $\pi$ acceptor. Another interesting feature of these complexes is that the EPR spectra are completely anisotropic, and also that the principal directions of $g$ are not close to any relevant direction in the molecule[86]. The fact that these data can be easily reproduced within the Angular Overlap Model is a clear proof of the power of the method.

The complex tris(3,5-dimethyl-1-pyrazolylethyl)amine cobalt(II) bis(tetraphenylborate) shows an unusual visible spectrum[90], in the sense that the energies of the electronic transitions are higher than those generally observed for tetrahedral complexes. The EPR spectra have also been recorded, which are not anomalous, however. Single crystal polarized electronic spectra yielded a reasonable assignment within $C_{3v}$ symmetry, which was substantially confirmed by the EPR data. The unusually high energy of the transition can be related[89] to the short metal-nitrogen distance[91]. It is worth noting, however, that in the bis-pyrazolyl borate complex only a slightly larger metal to nitrogen distance was observed, while the electronic spectra showed transitions at much lower energies[92].

The single crystal polarized electronic spectra of the complex bis(N-*t*-butylpyrrole-2-carbaldimino) cobalt(II) have been reported[93]. The data were interpreted using a higher symmetry than actually present but in any case the ligand field was found to be rather strong. No EPR spectra were observed, probably due to the fact that the ground Kramers doublet can be formally described as $M_s = \pm 3/2$. In fact, for an axially symmetric chromophore the transition $+3/2 \rightarrow -3/2$ is forbidden parallel to z and the two levels are not split when the static magnetic field is in the xy plane.

Horrocks reported the EPR spectra of CoSALiprp[94]. They show $g_z = 7.09$ and very small $g_x$ and $g_y$ values. These data are in agreement with a $\pm 3/2$ ground level. Other CoSAL complexes did not give any EPR spectrum at all[95].

Several studies are available on $CoA_2X_2$ complexes (X = Cl, Br). In particular the complexes with phosphine and diethylthiourea showed beautifully polarized electronic spectra with all transitions to the excited $^4T_1$ levels well resolved[96].

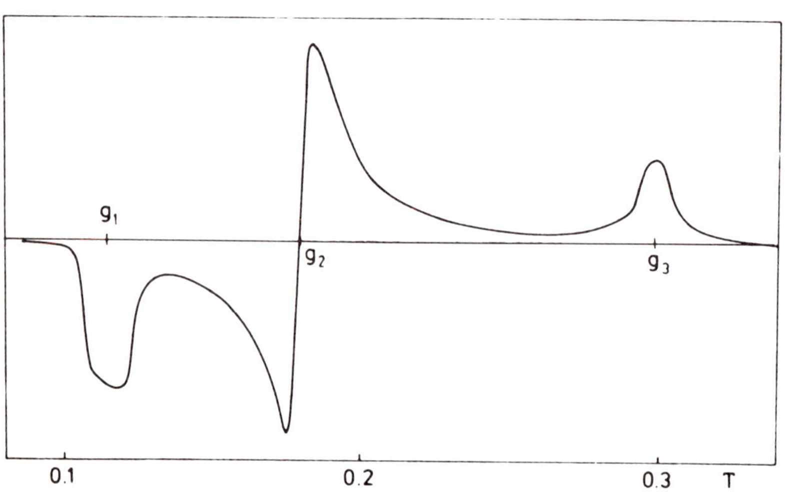


**Fig. 25.** Polycrystalline powder ESR spectra of $(Co, Zn)Cl_2(Ph_3PO)_2$ recorded at X-band frequency at 4.2 K

For the $Co(Ph_3PO)_2Cl_2$ complex also, detailed EPR[48] and MCD[97] data are available. The polycrystalline powder spectrum of the cobalt-doped zinc analogue is shown in Fig. 25. The single crystal EPR data showed that the $g_2$ value (3.59) is parallel to the $C_2$-axis of the complex, the other two being close to the projections of the metal-chlorine and metal-oxygen bonds respectively. A $^{59}Co$ hyperfine splitting is resolved parallel to the $g_3$ direction only ($53 \times 10^{-4}$ $cm^{-1}$). The electronic spectra did not show such neat polarizations as in the complexes with phosphine oxide and thiourea. The MCD spectra (Fig. 26) gave a pattern which has been assumed to be typical of tetrahedral complexes.

The complexes $CoL_2Iz_2$ (L = carboxilate; Iz = imidazole) have visible and MCD spectra similar to cobalt thermolysine and cobalt carboxipeptidase A[98]. The g values for ($Co(CH_3COO)_2Iz_2$) are almost axial, while a relatively large rhombic splitting is apparent for ($Co(C_2H_5COO)_2Iz_2$). The molar absorption coefficients of these complexes in ethanol are very small for tetrahedral complexes, but the possibility of forming octahedral species cannot be excluded.

## 4.2 Five Coordinate Complexes

The energies of the electronic transitions, the effective magnetic moment and the g and A values of some relevant five coordinate cobalt(II) complexes are given in Table 3. Genuine examples of square pyramidal chromophores are exceedingly rare. Among

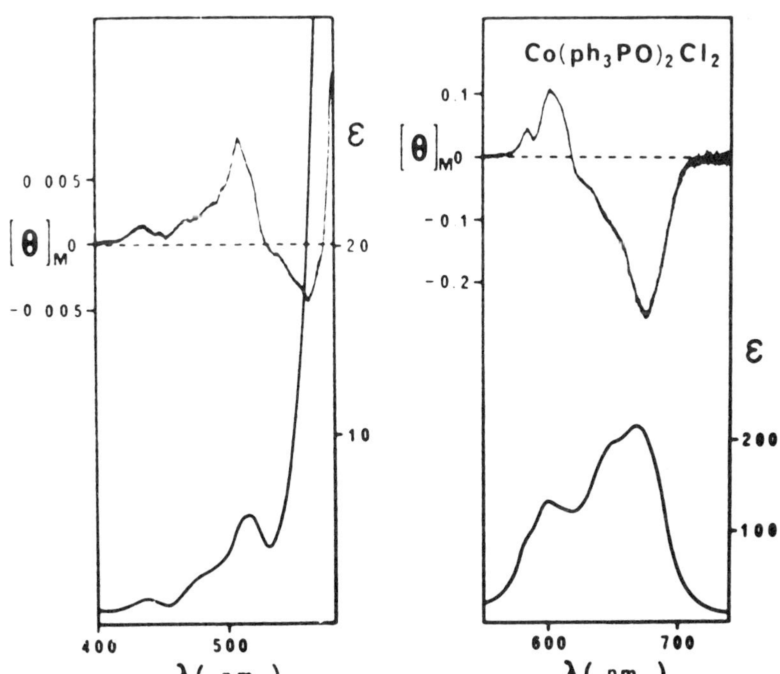


**Fig. 26.** MCD and absorption spectra of $CoCl_2$ $(Ph_3PO)_2$ in $CH_2Cl_2$ at room temperature

**Table 3.** Spectral parameters for five coordinate cobalt(II) complexes. The energies of the electronic transitions are labelled according to the symmetry labels of the excited states in $D_{3h}$ symmetry for trigonal bipyramidal, and in $C_{4v}$ symmetry for square pyramidal complexes. F and P refer to the free ion terms which contribute mostly to the indicated state. Don.Set is the donor set of the

| Complexes | Don.Set | Spe. | $^4A_1$ $^4A_2$ | $^4E''$ (F) | $^4E'$ (F) | |
|---|---|---|---|---|---|---|
| $Co(N\text{-}N\text{-}N)(NCS)_2$ | $N_5$ | P | 4400 | 5600 | 8400 | 11800 |
| $[Co(NCS)(Me_6tren)]SCN \cdot H_2O$ | $N_5$ | C | 4000 | 5800 | | 14800 |
| $Co[py(Bu^s)_2](NCS)_2$ | $N_5$ | P | | | 9600 | 12300 |
| | | S | | 5200(21) | 10200(15) | 12600(14) |
| $[Co(N_3P)(NCS)_2]$ | $N_5$ | P | | 5700 | | 12500 |
| | | S | | 5300(31) | | [illegible] |
| $[Co(N_3As)(NCS)_3]$ | $N_5$ | P | | 5600 | | 12800 |
| | | S | | 5200(32) | | 12700(36) |
| $[Co(N\text{-}N_2S)(NCS)_2] \cdot ½BuOH$ | $N_5$ | P | | 5800 | | 13000 |
| | | S | | 5100(35) | | 12600(39) |
| $[Co(tren)(NCS)]NCS$ | $N_5$ | P | | 6640 | | 13250 |
| $[Co(Me_6tren)Cl]Cl$ | $N_4Cl$ | P | | 5700 | | 12500 |
| | | S | | 5800(32) | | 12600(30) |
| $[Co(Me_6tpt)Br]BPh_4$ | $N_4Cl$ | P | | 7400 | 8500 | 11000 |
| $[Co(N_3As)Cl]BPh_4$ | $N_3AsCl$ | P | | 5800 | | 11800 |
| | | S | | 5700(30) | | 11600(57) |
| $Co(N\text{-}N\text{-}N)Cl_2$ | $N_3Cl_2$ | P | | 4800 | 9600 | |
| $Co(dienMe)Cl_2$ | $N_3Cl_2$ | P | 4000 | 5500 | 8800 | 10500 |
| $[Co(MABenz\text{-}NEt_2)Cl_2]$ | $N_3Cl_2$ | P | | 5200 | 9500 | 11000 |
| $Co[py(Bu^s)_2]Cl_2$ | $N_3Cl_2$ | P | | 5000 | 8100 | 11200 |
| $Co[py(cy)_2]Cl_2$ | $N_3Cl_2$ | P | | 5300 | 8500 | 11500 |
| $Co(PyAenEt_2)Br_2$ | $N_3Br_2$ | P | ~6000 | | ~8000 | ~12000 |
| $[Co(N\text{-}NO_2)NCS]BPh_4$ | $N_3O_2$ | P | | 6000 | 12500 | |
| $Co(MeN\text{-}N\text{-}As)Cl_2$ | $N_2AsCl_2$ | P | | 5300 | 10000 | 12500 |
| $[Co(N\text{-}NO_2)Cl]BPh_4$ | $N_2O_2Cl$ | P | | 5900 | 12100 | |
| | | S | | 6000(24) | 12300(13) | |
| $[Co(N_2P_2)Cl]BPh_4$ | $N_2P_2Cl$ | P | | 6000 | 11300 | |
| | | S | | 5900(72) | 11600(95) | |
| $[Co(N\text{-}NS_2)Cl]BPh_4$ | $N_2S_2Cl$ | P | | 5900 | 11800 | |
| | | S | | 5900(25) | 12700(43) | |
| $[Co(Me_4daes)Br_2]$ | | P | | 5300 | 9000 | 11400 |
| $[Co(N_2OP)Cl]BPh_4$ | $N_2OPCl$ | P | | 6200 | 12500 | |
| $[Co(N_2SP)Cl]BPh_4$ | $N_2SPCl$ | P | | 5800 | 12500 | |
| $[Co(Me_6tren)ClO_4]ClO_4$ | $N_4O$ | P | | 5700 | 13400 | |
| $[Co(N\text{-}N_2O)NCS]BPh_4$ | $N_4O$ | P | | 6400 | 13500 | |
| | | S | | 6100(37) | 13000(42) | |
| $[Co(Me_4daes)(NCS)_2]$ | $N_4S$ | P | | 6000 | 12500 | |
| $[Co(MSBenz\text{-}NEt_2)(NCS)_2]$ | $N_4S$ | P | | 5800 | 13000 | |
| $[Co(N\text{-}N_2S)(NCS)]BPh_4$ | $N_4S$ | P | | 5900 | 13100 | |
| | | S | | 5700(44) | 13300(85) | |
| $[Co(2\text{-}Mepy)(O_2CPh)_2]$ | $NO_4$ | P | | 7250 | 9200 | 11400 |
| $[Co(NP_3)Cl]BPh_4$ | $NP_3Cl$ | P | | 5950 | 10000 | |
| $[Co(NP_3)Br]PF_6$ | $NP_3Br$ | C | | 5700 | 10000 | |
| $[Co(NP_3)(OH)]BPh_4$ | $NP_3O$ | P | | 6500 | 10700 | |
| | | S | | 6500(115) | 10900(173) | |
| $[Co(NS_3)Br]PF_6$ | $NS_3Br$ | C | | 5200 | 11800 | |
| $[Co(NS_3)Br]^+$ | $NS_3Br$ | P | | 5400 | 11500 | |
| $[Co(C_6H_7NO)_5](ClO_4)_2$ | $O_5$ | C | | 5000 | 10500 | 12500 |
| $CoS(CH_2\text{-}CH_2\text{-}OH)_2Cl_2$ | $O_2SCl_2$ | P | | 6400 | 14500 | 14900 |
| Square Pyramidal Comp. | | | $^4B_2$ (F) | $^4E$ (F) | | $^4B_1$ (F) |
| $[Co(MePh_2AsO)_4NO_3]NO_3$ | $O_5$ | C | | 7000 | | |
| | | | 5000 | | | 11300 |
| $CoCl_2TTDA$ | $O_2SCl_2$ | P | 4790 | 6120 | | 8360 |
| $[Co(dacoda)H_2O]$ | $N_2O_3$ | S | | | | 8050(7) |
| $[Co(5\,Cl\ SALen\text{-}NEt_2)_2]$ | $N_3O_2$ | P | | 6700 | | 11400 |
| $Co(H\text{-}SAL\text{-}DPT)$ | $N_3O_2$ | P | | | 6100 | |

complex; Spe. indicates the condition under which the electronic and EPR spectra were obtained: S solution; P diffuse reflectance for electronic spectra and polycrystalline powder for EPR spectra; C single crystal. The A's are in $10^{-4}$ $cm^{-1}$.

| $^4A_2'(P)$ | $^4E'(P)$ $\nu\,cm^{-1}$ | $(\varepsilon_M)$ | Ref. | $\mu$ | Ref. | $g_1$ $(A_1)$ | $g_2$ $(A_2)$ | $g_3$ $(A_3)$ | Ref. |
|---|---|---|---|---|---|---|---|---|---|
| 17000 | 19600sh | | 183 | | | | | | |
| 16500 | 21000 | | 106 | | | | | | |
| 16400 | 18900sh | | 184 | 4.82 | 184 | | | | |
| 16400(116) | 18000(98) | | | | | | | | |
| 16100 | 10200 | 19300 20200 | 185 | 4.54 | 185 | | | | |
| 16200(174) | | 19600(147) | | | | | | | |
| 16400 | 18500 | 19400 | 185 | 4.58 | 105 | | | | |
| 16100(160) | 18500sh | 19600(140) | | | | | | | |
| 16700 | 19800 | 20800 | 186 | 4.58 | 186 | | | | |
| 16300(183) | 19600(158) | | | | | | | | |
| 16950 | 20700 | 21700 | 226 | | | | | | |
| 15600/16000 | 20000 | | 154 | 4.45 | 154 | 2.29 | | 4.25 | 187 |
| 15500/16100 | 20200(118) | | | | | | | | |
| 17000 | 18100 | | 151 | 4.26 | 151 | | | | |
| 14800 | 19200 | 21100 | 185 | 4.49 | 185 | | | | |
| 14500(90) | 18700(184) | 20600(211) | | | | | | | |
| 15600 | 17000 | 19400sh | 183 | | | | | | |
| 16200 | | 19000 | 188 | 4.60 | 188 | | | | |
| 15600 | 16700sh | 17800sh | 189 | 4.82 | 189 | | | | |
| 16500 | 17900sh | 24500sh | 184 | 4.85 | 184 | | | | |
| 16700 | 17900sh | 22900sh | 184 | 4.85 | 184 | | | | |
| ~16000 | | ~21000sh | 163 | 4.92 | 163 | | | | |
| 16700 | 20000 | 21400 | 186 | 4.55 | 186 | | | | |
| 16000 | 18500 | | 187 | | | | | | |
| 15900 | 19100 | 21400 | 106 | 4.56 | 186 | | | | |
| 16100(77) | 18500(63) | 20000(46) | | | | | | | |
| 13800 | 19400 | | 185 | 4.35 | 185 | | | | |
| 14100(168) | 19300(470) | | | | | | | | |
| 14800 | 19200 | 20800 | 186 | 4.53 | 186 | | | | |
| 15000(80) | 19600(126) | 20800(119) | | | | | | | |
| 15830 | 16700 | 18050 | 190 | 4.50 | 190 | | | | |
| 15500 | 19800 | 20800 | 191 | 4.50 | 191 | | | | |
| 14700 | 19600 | 20000 | 192 | 4.55 | 192 | | | | |
| 15600/16300 | | 20900 | 154 | | | | | | |
| 17100 | 20000 | 22000 | 186 | 4.40 | 186 | | | | |
| 16700(142) | 20000(68) | 22200(57) | | | | | | | |
| 16500 | 19400 | | 190 | 4.44 | 190 | | | | |
| 16000sh | 17500 | 18500sh | 162 | 4.58 | 162 | | | | |
| 15900 | 20400 | 21700 | 186 | 4.46 | 186 | | | | |
| 15600(167) | 20000(115) | 22700(134) | | | | | | | |
| | 18200 | | 192 | 5.03 | 187 | | | | |
| 13300 | 18200 | | 193 | | | | | | |
| 13300 | 17500 | | 107 | | | | | | |
| 13400 | 20400 | | 194 | 4.47 | 194 | | | | |
| 13800(155) | 20800(650) | | | | | | | | |
| 13800 | 18800 | | 105 | | | | | | |
| 14100 | 18900 | 19700 | 195 | | | | | | |
| 18000 | | | 103 | | | | | | |
| 16100 | 17500 | | 196 | 4.80 | 196 | 1.86(97) | 3.53(20) | 5.67(77) | 104 |
| $^4E(P)$ | | $^4A_2(P)$ | | | | | | | |
| 17000 | | 20500 | 53 | 5.07 | 53 | 0.9 | 1.3 | 8.6(462) | 99 |
| 16600 | | 20500 | | | | | | | |
| 14200 | 14700 | 17100/19400 | 197 | 4.61 | 197 | | | | |
| 18200(32) | | 22000(22.5) | 102 | 4.81 | 102 | | | | |
| 16800 | | | 198 | | | | | | |
| 15150 | | | 199 | 4.28 | 199 | | | | |

them the best characterized is $[Co(MePh_2AsO)_4NO_3]NO_3$ for which single crystal electronic[53] and EPR spectra are available[99, 100], together with magnetic anisotropy data. The complex is known to crystallize in the tetragonal system[101], and the site symmetry of the cobalt ion is $C_4$. However the anion in the axial position is disordered, so that the oxygen atom deviates from the position perpendicular to the base of the square pyramid by 15°. The actual symmetry of the complex is therefore far from axial ($g_1 = 0.9$; $g_2 = 1.3$; $g_3 = 8.6$, Fig. 27). Although $g_1$ and $g_2$ are sufficiently close to each other, they cannot be interpreted as the split components of $g_\perp$, since one, $g_1$, is very close to the direction normal to the base plane, making an angle of 12° with it. The values and the directions of g have been successfully reproduced using the Angular Overlap Model[99]. The previous quantitative assignment of the electronic spectra and of the magnetic anisotropy data, based on axial symmetry, must therefore be reconsidered.

Another interesting square pyramidal complex is $Co(dacoda)(H_2O)$[102] which is isomorphous to the nickel analogue. The latter has been reported to have an electronic spectrum similar to that of nickel carboxypeptidase. Although the same similarity does not hold for the cobalt complex, it can be used as an interesting model of cobalt enzymes which have one coordinated water molecule.

Among trigonal bipyramidal complexes the $CoO_5$ chromophore of pentakis(2-methyl pyridine-N-oxide) has been characterized through single crystal electronic[103] and EPR spectroscopy[104]. The electronic spectra have been assigned in $C_{2v}$ symmetry, and also the EPR spectra, Fig. 28, are in agreement with this view. It must be noted that sizeable $^{59}Co$ hyperfine splitting of $97 \times 10^{-4}$ and $77 \times 10^{-4}$ $cm^{-1}$ were resolved.

Complexes with the tripod ligands $NN_3$, $NP_3$ and $NS_3$ show interesting distortions in the sense that by increasing the softness of the donor atoms the axial metal-nitrogen distance grows and the complexes pass from five coordinate to capped tetrahedral[105]. The electronic spectra have been assigned on the basis of single crystal data[106–108] and

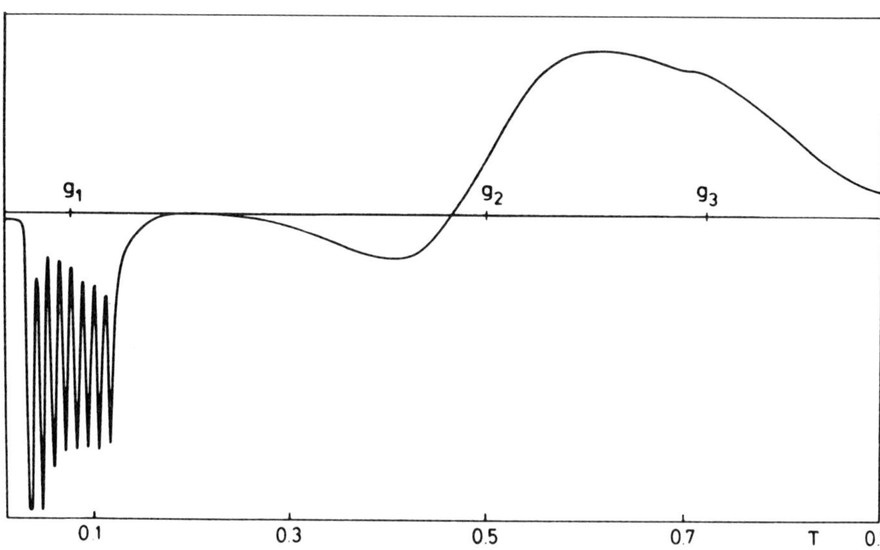


**Fig. 27.** Polycrystalline powder ESR spectra of $[(Co,Zn)(MePh_2AsO)_4NO_3]$ $NO_3$ recorded at X-band frequency at 4.2 K

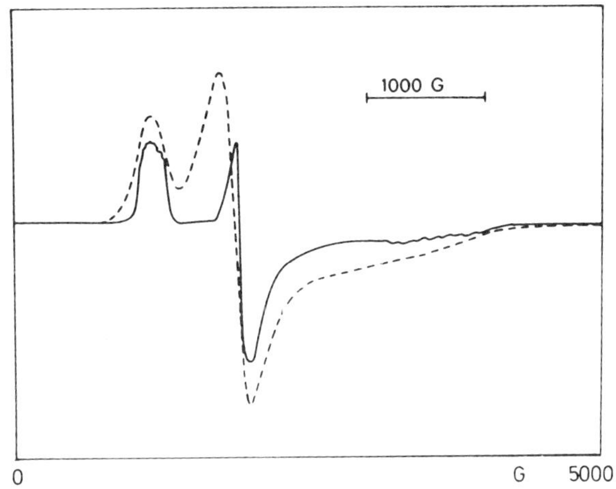


**Fig. 28.** Polycrystalline powder ESR spectra of Co(picoline N-oxide)$_5$($ClO_4$)$_2$ (---) and of (Zn, Co)(picoline N-oxide)$_5$-($ClO_4$)$_2$ (—)

the dependence of the energy levels on the geometrical distortions satisfactorily reproduced[105].

It is worth mentioning here that the Angular Overlap parameters for the $NP_3$ ligand suggest a $\pi$-back-bonding interaction with the metal ion, a result which has been later confirmed for other cobalt(II) and nickel(II) complexes[109].

EPR data are also available for the $CoNN_3Br^+$ and $CoNP_3Br^+$ derivatives[108]. They show that the ground quartet level is split into two Kramers' doublets, which can be loosely described through the spin quantum number $\pm 1/2$ and $\pm 3/2$ respectively. In the $CoNN_3Br^+$ complex the $\pm 1/2$ level is lowest in energy, while the reverse is true for the $CoNP_3Br^+$ complex. This change in the nature of the ground state has been attributed to geometrical distortion effects and the g values have been reproduced by the Angular Overlap Model.

## 4.3 *Octahedral Complexes*

The energies of the electronic transitions, the effective magnetic moment, and the g and A values of some relevant octahedral cobalt(II) complexes are collected in Table 4.

Several detailed single crystal studies are available for octahedral complexes.

The hexaaquoion has been studied through MCD spectroscopy in $Co(H_2O)_6(BrO_3)_2$[110, 111]. It was shown that the intensities of the d-d transitions are mainly determined by the cobalt-oxygen $t_{1u}$ vibrations.

Recently Ferguson[112] re-examined the polarized spectra of $CoCl_2 \cdot 6\,H_2O$ which contains octahedral $CoO_4Cl_2$ chromophores. A thorough analysis of the polarizations, and of their temperature-dependence showed that both electric and magnetic dipole mechanisms are operative. The analysis of the crystal field parameters yielded Dq = 810, Ds = −640, Dt = 30 $cm^{-1}$. Translated into Angular Overlap parameters, these values show that water is a better $\sigma$ and $\pi$ donor towards cobalt(II) than the chloride ion.

Striking examples of the sensivity of octahedral cobalt(II) towards low symmetry components are $Co(pyO)_6(ClO_4)_2$[113, 115] and $Co(apy)_6(ClO_4)_2$[116]. The crystal structures

**Table 4.** Spectral parameters for octahedral cobalt(II) complexes. The energies of the electronic transitions are labelled according to the parent symmetry of the excited states. F and P refer to the free ion terms which contribute mostly to the indicated state. Don.Set is the donor set of the complex; Spe. indicates the condition under which the

| Complexes | Don.Set | Spe. | $^4T_{2g}$(F) | $\nu$ cm$^{-1}$ ($\varepsilon_M$) | $^4A_{2g}$ cm$^{-1}$ ($\varepsilon_M$) | $^4T_{1g}$(P) | $\nu$ cm$^{-1}$ ($\varepsilon_M$) |
|---|---|---|---|---|---|---|---|
| $Co(H_2O)_6^{2+}$ | $O_6$ | S | 8100 | | 16000 | 19400 | |
| Co(OMPA) | $O_6$ | C | 6750 | 6800 | 13800 | 18400 | 19000 |
| $Co(AA)_6(ClO_4)_2$ | $O_6$ | P | 8000 | | 16000sh | 19400 | 21000sh |
| $Co(DEU)_6(ClO_4)_2$ | $O_6$ | P | 8000 | | | 19000 | 20800sh |
| $Co(Butanone)_6(InCl_4)$ | $O_6$ | P | 7570 | | 15750sh | 18890 | 21440sh |
| $Co(MeOH)_6(ClO_4)_2$ | $O_6$ | P | 8470 | | 15700sh | 19900 | |
| $Co(DTMSO)_6I_2$ | $O_6$ | P | 7850 | | 15400 | 19000 | [illegible] |
| $Co(DEAP)_3$ | $O_6$ | P | 8690 | | 17050 | 20400 | |
| $Co(acac)_2(H_2O)_2$ | $O_6$ | P | | | | | |
| $(Zn, Co)SeO_4 \cdot 6H_2O$ | $O_6$ | C | | | | | |
| $(Cd, Co)(epydo)_6(BF_4)_2$ | $O_6$ | P | 7100 | | 14700 | 18500 | 20400sh |
| $(Cd, Co)(epydo)_6(ClO_4)_2$ | $O_6$ | P | 7150 | | 14400 | 18500 | 20500sh |
| $Co(pyO)_6(ClO_4)_2$ | $O_6$ | P | 7250sh | 8300 | 17700 | 18590sh | 20410sh |
| $Co(Ph_3PO)_2(NO_2)_2$ | $O_6$ | P | 7400 | | 12700 | 17950 | |
| | | S | 7800(8.5) | | 13000(2) | 18400(80) | |
| $Co(apy)_6(ClO_4)_2$ | $O_6$ | P | 6800 | | | 18500 | 20800 |
| $KCoF_3$ | $F_6$ | C | 7150 | | 15200 | 19200 | |
| $CoCl_2$ | $Cl_6$ | C | 6600 | | 13300 | 17150 | |
| $Co(bipy)_3Br_2 \cdot 6H_2O$ | $N_6$ | C-P | 11000 | | | 22000 | |
| $Co(OPD)_3(ClO_4)_2$ | $N_6$ | P | 9800 | | 18000 | 20200 | |
| $Co(N\text{-}N\text{-}N)_2$ | $N_6$ | P | 9000 | 11800 | | 20000 | |
| $Copy_6^{2+}$ | $N_6$ | S | 9800 | | | 20400 | |
| $Co(en)_3(NO_3)_2$ | $N_6$ | C | 10000 | | | 21000 | |
| $Co(NH_3)_6^{2+}$ | $N_6$ | S | 8100 | | | 21100 | |
| $Co(Iz)_6(NO_3)_2$ | $N_6$ | P | | | | | |
| $Co(APDA)_6SO_4$ | $N_6$ | P | 8330 | | | 19050 | |
| $Co(N_2H_5)_2(SO_4)_2$ | $O_4N_2$ | P | 8500 | 9900 | 18000sh | 19900 | 21000sh |
| $Copy_2(O_2CPh)_2$ | $O_4N_2$ | P | 7600 | 9800 | 17000sh | 19500 | |
| $Co[3py(CO_2)]_2 \cdot 4H_2O$ | $O_4N_2$ | P | 8200 | | | 19400 | |
| $Co(pic)_2 \cdot 4H_2O$ | $O_4N_2$ | P | 9300 | | 19200 | 20200 | 21600 |
| $Co(acac)_2(6\,Mequin)_2$ | $O_4N_2$ | C | | | | | |
| $Co(acac)_2(py)_2$ | $O_4N_2$ | C | | | | | |
| $Co(acpz)_3(BF_4)_2$ | $O_3N_3$ | P | 9300 | | | 20400 | |
| $Co[py(Bu^s)_2](NO_3)_2$ | $O_3N_3$ | P | 8500 | | 17400 | 20400 | |
| $Co(L\text{-}Histidine)_2H_2O$ | $O_2N_4$ | C | 10000 | | 22400 | 20800 | |
| $Co(pic)_2(py)_2$ | $O_2N_4$ | P | 9200 | | 19000 | 20300 | 21500 |
| $Co(en)_2(OAB)_2$ | $O_2N_4$ | P | 9627 | | | 21980 | |
| $Co(OPD)_2(OAB)_2$ | $O_2N_4$ | P | 9434 | | 18870 | 22730 | |
| $Co(HdMePyne)_2Br_2 \cdot 2H_2O$ | $O_2N_4$ | P | 9500 | | | 22000 | |
| $Co(bipy)_2(CH_3C_6H_4SO_2)_2$ | $O_2N_4$ | P | 9800 | | | 20830 | |
| $Co(EG)_2Cl_2$ | $Cl_2O_4$ | P | 7300 | | 14400sh | 18200 | 19600 |
| $CoCl_2 \cdot 6H_2O$ | $Cl_2O_4$ | C | 8060 | 8330 | 16450 | 18600 | 22250 |
| $Co(OPD)_2Br_2$ | $Br_2N_4$ | P | 9500 | | 18600sh | 20600b | |
| $Co(s\text{-}Et_2en)_2Cl_2$ | $Cl_2N_4$ | P | 7180 | 10020 | 19370 | 20050 | 21840 |
| $Co(pz)_4Cl_2$ | $Cl_2N_4$ | P | | | | | |
| $Co(py)_4Cl_2$ | $Cl_2N_4$ | P | | | | | |
| $Co(Et_2D)_3ClO_4)_2$ | $S_6$ | P | 6850 | | | 15460 | 16810sh |
| | | S | 6940(29) | | 14710sh(60) | 15480(107) | 16670(77) |
| $Co(1\,MePyne\text{-}2\text{-}T)_3(ClO_4)_2$ | $S_3N_3$ | P | 8475 | | | 18690 | |
| $Co(1{,}4{,}6\,MePyne\text{-}2\text{-}T)_3(ClO_4)_2$ | $S_3N_3$ | P | 8475 | | | 18100 | |
| $Co(1{,}4{,}6\,MePyne\text{-}2\text{-}T)_3(CoCl_4)$ | $S_3N_3$ | P | 8300 | | | 18500 | |
| $Co(bipy)_2(CH_3C_6H_4SO_2)_2$ | $S_2N_4$ | P | 11000 | | | | |
| $[Co(N\text{-}NS_2)(NCS)_2]$ | $S_2N_4$ | P | 9300 | | | 17200 | 18300 |
| | | S | 6300(20) | 9100(21) | 12500(17) | 16100(145) | 18200sh |
| $Co(1\,MePyne\text{-}2\text{-}T)_2(H_2O)_2(ClO_4)_2$ | $S_2N_2O_2$ | P | 7810 | 10500 | 16800sh | 19000sh | |
| $Co(1\,MePyne\text{-}2\text{-}T)_2Cl_2$ | $S_2N_2Cl_2$ | P | 6500 | 10400 | | 16400 | |
| $Co[S(CH_2CH_2OH)_2]_2Cl_2$ | $S_2O_2Cl_2$ | P | | | | | |
| $Co(Me(Cl\text{-}Ph)TazO)$ | $O_2N_2Cl_2$ | P | 7300 | 11360 | 14080 | 15620 | 18860 |
| $Na_2Co(ATP) \cdot 2H_2O$ | $O_6$ | P | | | 16670 | 19610 | 21280 |

electronic and EPR spectra were obtained: S solution; P diffuse reflectance for electronic spectra and polycrystalline powder for EPR spectra; C single crystal. The A's are in $10^{-4}$ $cm^{-1}$.

| Ref. | | Ref. | $g_1$ ($A_1$) | $g_2$ ($A_2$) | $g_3$ ($A_3$) | Ref. |
|---|---|---|---|---|---|---|
| 200 | | | | | | |
| 122 | | | | | | |
| 201 | | | | | | |
| 184 | | | | | | |
| 202 | | | | | | |
| 203 | | | | | | |
| 168 | | | | | | |
| 204 | 4.83 | 204 | | | | |
| | | | 1.88(29) | 2.74(20) | 6.84(180) | 130 |
| | | | 1.76(60) | 3.6 (40) | 7.45(230) | |
| | | | 2.26(40) | 4.27(40) | 5.60(110) | 205 |
| 206 | | | 2.38(38) | 4.28(46) | 6.10(67) | 206 |
| 206 | | | 2.58 | 5.00(54) | 5.24(61) | 206 |
| 207 | | | C 2.26(186) | | 4.77(384) | 209 |
| 227 | 4.96 | 227 | | | | |
| 95 | | | C 3.06 | | 4.75(246) | 116 |
| 208 | | | | | | |
| 208 | | | | | | |
| 121 | | | 3.16 | | 4.62 | 95 |
| 209 | 4.93 | 209 | | | | |
| 183 | | | | | | |
| 210 | | | | | | |
| 120 | 5.4 | 95 | 2.6 | | 5.0 | 95 |
| 200 | | | | | | |
| | | | 3.57 | | 5.50 | 116 |
| 209 | 4.82 | 209 | | | | |
| 212 | | | | | | |
| 192 | | | | | | |
| 213 | | | | | | |
| 214 | 4.94 | 214 | | | | |
| | | | 1.90 | 4.11 | 5.67 | 130 |
| | | | 1.98 | 3.92 | 5.83 | 130 |
| 215 | | | | | | |
| 184 | 4.95 | 184 | | | | |
| 216 | | | | | | |
| 217 | 4.90 | 217 | | | | |
| 218 | | | | | | |
| 218 | | | | | | |
| 146 | 4.93 | 136 | | | | |
| 219 | 4.77 | 219 | | | | |
| 220 | | | | | | |
| 112 | | | | | | |
| 209 | 5.00 | 209 | | | | |
| 133 | 5.07 | 133 | | | | |
| | | | 2.14 | 4.55 | 5.83 | 128 |
| | | | 3.33 | | 4.46 | 128 |
| 221 | | | | | | |
| 148 | | | | | | |
| 222 | | | | | | |
| 223 | | | | | | |
| 219 | 4.13 | 219 | | | | |
| 186 | 4.87 | 186 | | | | |
| 148 | | | | | | |
| 148 | | | | | | |
| 196 | 4.84 | 196 | | | | |
| 224 | | | | | | |
| 225 | 4.99 | 225 | | | | |

of these complexes have been determined and the metal ions were found to have trigonal site symmetry. Although the bond angles are extremely close to 90°[117–119] so that the $CoO_6$ chromophores are exactly octahedral, the electronic and the EPR spectra exhibit large anisotropy effects. The g values and magnetic moments are axial with $g_\parallel < g_\perp$ and $\mu_\parallel < \mu_\perp$, respectively, indicating an orbitally non-degenerate ground state. These data can be interpreted considering that the ligands are not linearly ligating, i.e. the interaction of the oxygen donor atom with the metal ion is highly anisotropic as would be expected assuming a $sp^2$ hybridization.

Detailed studies are available also for tris-bis-chelate complexes $CoL_3$ (L: 1,2 diaminopropane[120, 95], 2,2 bipyridine[121, 95], octamethylpyrophosphoramide[122]. In all these cases the bond angles within the chelate rings are smaller than 90°, and all the complexes can be described as trigonally compressed octahedral. The $Co(en)_3^{2+}$ complexes show only a very small splitting of the electronic transitions[120], but the EPR spectra are distinctly axial. The pattern of the g values is in agreement with an orbitally non-degenerate ground state, as would be expected for $\delta$-bonding ligands in a trigonally compressed octahedron. In this case, where no $\pi$ metal ligand interaction is anticipated, the removal of octahedral symmetry is entirely due to the deviation of the N-Co-N bond angles from 90°. Similar conclusions had been reached from the analysis of the electronic and CD spectra[120]. The low symmetry splitting of the ground level is calculated to be 500 $cm^{-1}$. Also in 2,2 bipyridine[121] and in octamethylpyrophorphoramide[122] complexes no trigonal splitting of the electronic transitions is observed, although large polarization effects are discovered. These data have been interpreted as due to the presence of an orbitally non-degenerate ground state. Similar conclusions were reached from the analysis of the MCD spectra of the $Co(OMPA)_3$ complex[123].

The g values[95] of the $CoN_6$ chromophore in $Co(en)_3^{2+}$ and $Co(bipy)_3^{2+}$ are again in agreement with the previous analyses in the sense that they show $g_\parallel < g_\perp$. Matters are different for $CoIz_6^{2+}$, however, where the trigonally distorted $CoN_6$ chromophores[124] yield $g_\parallel > g_\perp$ and $\mu_\parallel > \mu_\perp$[45] in agreement with an orbitally doubly degenerate ground state.

The g-tensor of the tetragonally distorted[125, 126] cobalt(II) complexes $Copy_4Cl_2$ and $Copz_4Cl_2$ is axial in the first compound while a large rhombic component is present in the second[38, 127, 128]. The electronic spectra of $Co(pz)_4Cl_2$ show a well resolved transition to the excited quartet levels. The rhombic component of the pyrazole complex has been attributed to the deviation of the in-plane bond angle, $\varphi$, from 90°. Figure 29 demonstrates how sensitive the calculated g values are with respect to variations of $\varphi$. A $\delta\varphi$ of 5° splits $g_\perp$ by about 6 units.

Also several base adducts of ketonate ligands have been characterized[38, 40, 129, 130]. In compounds $Co(acac)_2L_2$ (L = 6 Mequin, py, water) the rhombic ligand field component, evidenced by the pattern of g values, increases on passing from 6 Mequin to py and water. The rhombic splitting has been attributed to variations in the $\pi$-bonding ability of the axial ligands[130]. A similar pattern of g and A values was found in the binuclear Co Zn $trik_2py_4$ complexes: $g_3 = 6.3$, $A_3 = 198 \times 10^{-4}\ cm^{-1}$, $g_2 = 3.5$, $A_2 = 3.8 \times 10^{-4}\ cm^{-1}$, $g_1 = 1.87$, $A_1 = 98 \times 10^{-4}\ cm^{-1}$[131].

Tetragonally distorted chromophores are present in the $Co(EtNHCH_2CH_2NHEt)_2L_2$ complexes[132] (L = Cl, Br, NCS, $NO_3$). The electronic spectra were interpreted assuming an orbitally degenerate ground state, which is what one expects for only σ-bonding in plane nitrogen ligands. The sign of the splitting depends only on the $\pi$ bonding interac-

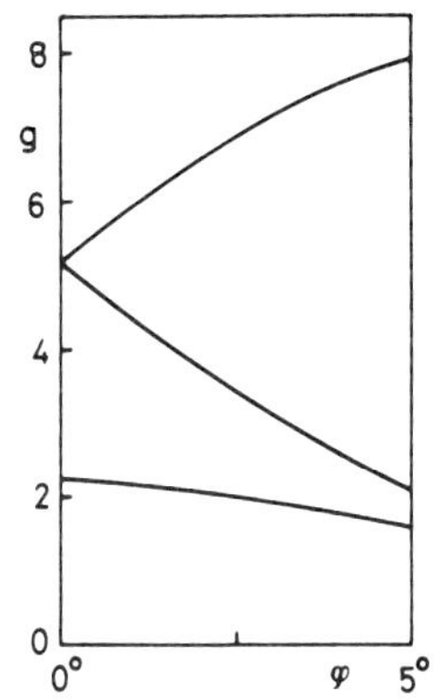


**Fig. 29.** Variations of the g values for $Co(pz)_4Cl_2$ as a function of the deviation of the in-plane bond angles from 90° (B = 800 $cm^{-1}$, $\zeta = -533$ $cm^{-1}$, $e_\sigma^{pz} = -4333$ $cm^{-1}$, $e_\pi^{pz} = -250$ $cm^{-1}$, $e_\sigma^{Cl} = -1333$ $cm^{-1}$ $e_\pi^{Cl} = 0$).

tion of the axial ligands. The EPR spectra however, although not very well defined, do not seem to confirm to this view and should be further analyzed.

The single crystal polarized electronic spectra of bis(L-histidinato) cobalt(II)[133] show that Dq of the histidine ligand is 1,100 $cm^{-1}$.

## *4.4 Coordination Numbers Larger than Six*

Detailed experimental data have been reported for the $[Co(DAPSC)Cl(H_2O)]Cl \cdot 2H_2O$ and $[Co(DAPSC)(H_2O)_2](NO_3)_2 \cdot H_2O$ complexes[134]. They are pentagonal bipyramidal[135] (Fig. 30) with the pentadentate ligand occupying the equatorial positions. The electronic spectra show a weak and broad feature in the infrared region near 6,000 $cm^{-1}$ and two bands at 15,800 and 17,700–18,500 respectively. The latter were assigned to the $^4F \rightarrow {}^4P$ transitions split by the low symmetry of the complex. The 15,800 band is assigned to a transition to an orbital singlet while the other is to an orbital doublet. The ground state was calculated to be an orbital singlet but several excited quartet spin levels are present within 4,500 $cm^{-1}$. The EPR spectra[136] were in agreement with these data.

Eight coordinate complexes have been reported with nitrate[137] and carboxylate[138] ligands. The molecular geometry is such that four shorter and four larger metal to ligand bond distances are present[139]. The electronic spectra can be interpreted essentially on the basis of tetrahedral geometry.

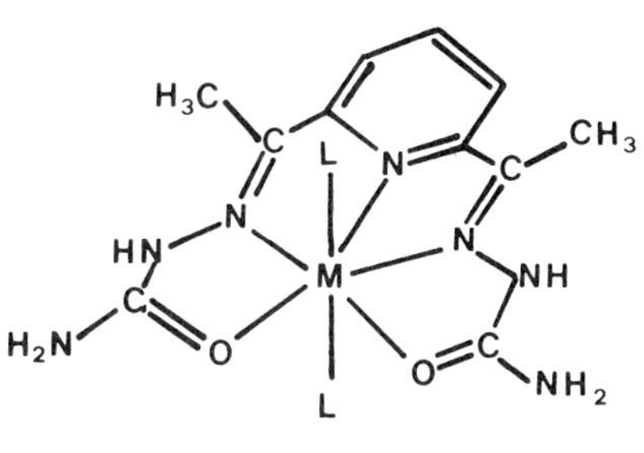


**Fig. 30.** Numbering of DAPSC donor atoms in $[Co(DAPSC)(H_2O)L]^{2+}$

# 5 References


1. Bertini, I., Luchinat, C.: In: Metal Ions in Biological System, (ed. H. Sigel) vol. 15, in press
2. Chelebowski, J. F., Coleman, J. E.: ibid. vol. 6, 2 (1976)
3. Lindskog, S.: Structure and Bonding *8*, 153 (1970)
4. Vallee, B. L.: Trends. Biochem. Sci. *1*, 88 (1976)
5. Bertini, I., Luchinat, C., Scozzafava, A.: Structure and Bonding *48*, 45 (1982)
6. Griffith, J. S.: The theory of Transition Metal Ions. University Press, Cambridge 1961
7. Abragam, A., Bleaney, B.: Electron Paramagnetic Resonance of Transition Ions. Clarendon Press, Oxford 1970
8. Ballhausen, C. J.: Introduction to Ligand Field Theory. McGraw-Hill, New York 1962
9. Carlin, R. L.: Transition Metal Chem. *1*, 1 (1965)
10. Jørgensen, C. K., Pappalardo, R., Schmidtke, H. H.: J. Chem. Phys. *39*, 1422 (1963)
11. Schäffer, C. E.: In: Wave Mechanics, (eds. W. C. Price, S. S. Chissick, T. Ravensdale) Butterworths, London 1973, p. 174
12. Glerup, J., Mønsted, O., Schäffer, C. E.: Inorg. Chem. *15*, 1399 (1976)
13. Bertini, I., Gatteschi, D., Scozzafava, A.: Isr. J. Chem. *15*, 189 (1976)
14. Gerloch, M., Slade, R. C.: Ligand Field Parameters. Cambridge University Press, London 1973
15. La Mar, G. N., Horrocks, W. De W. Jr., Holm, R. H. eds.: NMR of Paramagnetic Molecules. Academic Press, New York and London 1973
16. Donini, J. C., Hollebone, B. R., Lever, A. B. P.: Progr. Inorg. Chem. *22*, 225 (1977)
17. Daul, C., Schlapfer, C. W., Von Zelewsky, A.: Structure and Bonding *36*, 129 (1979)
18. Walker, F. A.: J. Magnetic Res. *15*, 201 (1974)
19. Malatesta, V., Mc Garvey, B. R.: Canad. J. Chem. *53*, 5791 (1975)
20. Griffith, J. S.: Disc. Faraday Soc. *26*, 81 (1958)
21. Lin, W. C.: Mol. Phys. *31*, 657 (1976)
22. Hitchman, M.: Inorg. Chem. *16*, 1985 (1977)
23. Backes, G., Reinen, D.: Z. Anorg. Allg. Chem. *418*, 217 (1975)
24. Gatteschi, D., Ghilardi, C. A., Orlandini, A., Sacconi, L.: Inorg. Chem. *17*, 3023 (1978)
25. Bencini, A., Gatteschi, D.: J. Phys. Chem. *80*, 2126 (1976)
26. Jesson, J. P.: J. Chem. Phys. *48*, 161 (1968)
27. Ciampolini, M., Bertini, I.: J. Chem. Soc. (A) 2241 (1968)
28. Schäffer, C. E.: Structure and Bonding *5*, 68 (1968)
29. Schäffer, C. E.: ibid. *14*, 69 (1973)
30. Smith, D. W.: Inorg. Chem. *17*, 3153 (1978)
31. Tennyson, J., Murell, J. N.: J. Chem. Soc. Dalton Trans. 2395 (1980)
32. Gerloch, M., Manning, M. R.: Inorg. Chem. *20*, 1051 (1981)
33. Horrocks, W. De W. Jr., Burlone, D. A.: J. Am. Chem. Soc. *98*, 6512 (1976)
34. Ciampolini, M.: Structure and Bonding *6*, 52 (1969)
35. Sacconi, L.: Pure Appl. Chem. *17*, 95 (1968)
36. Morassi, R., Bertini, I., Sacconi, L.: Coord. Chem. Rev. *11*, 343 (1973)
37. Wood, J. S.: Progr. Inorg. Chem. *16*, 227 (1972)
38. Gerloch, M., Mc Meeking, R. F., White, A. M.: J. Chem. Soc. Dalton Trans. 655 (1975)
39. Mitra, S.: Transition Metal Chem. *7*, 183 (1972)
40. Horrocks, W. De W. Jr., Hall, D. De W.: Coord. Chem. Rev. *6*, 147 (1971)
41. Van Vleck, J. H.: The Theory of Electric and Magnetic Susceptibilities. Oxford Univ. Press, London 1953
42. Gerloch, M., Mc Meeking, R. F.: J. Chem. Soc. Dalton Trans. 2443 (1975)
43. Blake, A. B.: ibid. 1041 (1981)
44. Figgis, B. N. et al.: J. Chem. Soc. (A) 2086 (1968)
45. Gerloch, M., Quested, P. N.: J. Chem. Soc. (A) 3729 (1971)
46. Gerloch, M., Quested, P. N., Slade, R. C.: ibid. 3741 (1971)
47. Bencini, A., Gatteschi, D.: Transition Metal Chem. *8*, in press
48. Bencini, A., Benelli, C., Gatteschi, D., Zanchini, C.: Inorg. Chem. *18*, 2137 (1979)
49. Bencini, A., Gatteschi, D.: J. Magn. Reson. *34*, 653 (1979)

50. Bencini, A. et al.: J. Inorg. Biochem. *14*, 81 (1981)
51. Pilbrow, J. R.: J. Magn. Reson. *31*, 479 (1978)
52. Abragam, A., Price, M. H. L.: Proc. Roy. Soc. *A 206*, 173 (1951)
53. Gerloch, M. et al.: J. Chem. Soc. (A) 3283 (1970)
54. Bencini, A., Bertini, I., Gatteschi, D., Scozzafava, A.: Inorg. Chem. *17*, 3194 (1978)
55. Jesson, J. P.: In: NMR of Paramagnetic Molecules (Eds. La Mar, J. N., Horrocks, W. De W. Jr., Holm, R. H.) Academic Press, New York and London 1973, p. 2
56. Bertini, I., Gatteschi, D.: Inorg. Chem. *12*, 2740 (1973)
57. Bertini, I., Sacconi, L.: J. Mol. Struct. *19*, 371 (1973)
58. Swift, T. J.: In: NMR of Paramagnetic Molecules (Eds. La Mar, J. N., Horrocks, W. De W. Jr., Holm, R. H.) Academic Press, New York and London 1973, p. 53
59. Bertini, I., Canti, G., Luchinat, C.: Inorg. Chim. Acta *56*, 99 (1981)
60. Bertini, I.: personal communication
61. Bertini, I., Drago, R. S. eds.: ESR and NMR of Paramagnetic Species in Biological Systems. D. Reidel Publ. Comp., Dordrecht 1980
62. Stephens, P. J.: In: Electronic States of Inorganic Compounds: New Experimental Techniques. D. Reidel Publ. Comp., Dordrecht 1975, p. 141
63. Buckingham, A. D., Stephens, P. J.: Ann. Rev. Phys. Chem. *17*, 399 (1966)
64. Schatz, P. M., Mc Caffery, A. J.: Quart. Rev. Chem. Soc. *23*, 552 (1969)
65. Denning, R. G.: In: Electronic States of Inorganic Compounds: New Experimental Techniques. D. Reidel Publ. Comp., Dordrecht 1975, p. 153
66. Briat, B.: ibid., p. 177
67. Schatz, P. M.: ibid., p. 223
68. Kaden, T. A., Holmquist, B., Valley, B. L.: Inorg. Chem. *13*, 2585 (1974)
69. Ferguson, J.: J. Chem. Phys. *39*, 116 (1963)
70. Ferguson, J.: ibid. *32*, 528 (1960)
71. Denning, R. G., Spencer, J. A.: Symposia Faraday Soc. *3*, 84 (1968)
72. Denning, R. G.: J. Chem. Phys. *45*, 1307 (1966)
73. Lo Menzo, J. A. et al.: Chem. Phys. Lett. *9*, 332 (1971)
74. Rivoal, J. C., Briat, B.: Mol. Phys. *27*, 1081 (1974)
75. Bird, B. D. et al.: Chem. Comm. 225 (1971)
76. Collingwood, J. C., Day, P., Denning, R. G.: Chem. Phys. Lett. *10*, 274 (1971)
77. Quested, P. N. et al.: Mol. Phys. *27*, 1553 (1974)
78. Van Stapele, R. P. et al.: Phys. Rev. *150*, 310 (1967)
79. Shankle, G. E. et al.: J. Chem. Phys. *56*, 3750 (1972)
80. Bencini, A., Benelli, C., Gatteschi, D.: Inorg. Chem. *19*, 1632 (1980)
81. Mitra, S.: J. Chem. Phys. *49*, 4724 (1968)
82. Cruse, D. A., Gerloch, M.: J. Chem. Soc. Dalton Trans. 1617 (1977)
83. Gerloch, M., Lewis, J., Rickards, R.: ibid. 980 (1972)
84. Bird, B. D. et al.: Phil. Trans. *A 276*, 277 (1974)
85. Stucky, G. D., Folkers, J. B., Kistenmacher, T. J.: Acta Crystallog. *23*, 1064 (1967)
86. Bertini, I., Gatteschi, D., Mani, F.: Inorg. Chim. Acta *7*, 717 (1973)
87. Bencini, A., Gatteschi, D.: Inorg. Chem. *16*, 2141 (1977)
88. Gerloch, M., Hanton, R. L.: ibid. *19*, 1692 (1980)
89. Garrett, B. B., Goedken, V. L., Quagliano, J. V.: J. Am. Chem. Soc. *92*, 499 (1970)
90. Banci, L., Benelli, C., Gatteschi, D., Mani, F.: Inorg. Chem. *21*, 1133 (1982)
91. Mani, F., Mealli, C.: Inorg. Chim. Acta *21*, 1133 (1982), *54*, L 77 (1981)
92. Guggenberger, L. J. et al.: Inorg. Chem. *12*, 508 (1973)
93. Bertini, I., Gatteschi, D., Scozzafava, A.: Inorg. Chim. Acta *13*, 145 (1975)
94. Horrocks, W. De W. Jr., Burlone, D. A.: ibid. *35*, 165 (1979)
95. Unpublished results of this laboratory
96. Tomlinson, A. A. G. et al.: J. Chem. Soc. Dalton Trans. 350 (1972)
97. Katô, M., Akimoto, K.: J. Am. Chem. Soc. *96*, 1351 (1974)
98. Horrocks, W. De W. Jr. et al.: J. Inorg. Biochem. *12*, 131 (1980)
99. Bencini, A., Benelli, C., Gatteschi, D., Zanchini, C.: Inorg. Chem. *18*, 2526 (1979)
100. Bencini, A., Benelli, C., Gatteschi, D., Zanchini, C.: J. Mol. Struct. *60*, 401 (1980)
101. Pauling, P., Robertson, G. B., Rodley, G. A.: Nature (London) *207*, 73 (1965)

102. Averill, D. F., Legg, J. I., Smith, D. L.: Inorg. Chem. *11*, 2344 (1972)
103. Bertini, I., Dapporto, P., Gatteschi, D., Scozzafava, A.: ibid. *14*, 1639 (1975)
104. Bencini, A., Benelli, C. Gatteschi, D., Zanchini, C.: ibid. *19*, 3839 (1980)
105. Bertini, I., Gatteschi, D., Scozzafava, A.: ibid. *14*, 812 (1975)
106. Bertini, I., Ciampolini, M., Gatteschi, D.: ibid. *12*, 693 (1973)
107. Bertini, I., Ciampolini, M., Sacconi, L.: J. Coord. Chem. *1*, 73 (1971)
108. Benelli, C., Gatteschi, D.: Inorg. Chem. in press
109. Gerloch, M., Hanton, R. L.: Inorg. Chem. *20*, 1046 (1981)
110. Harding, M. J., Briat, B.: Mol. Phys. *25*, 745 (1973)
111. Harding, M. J., Billardon, M., Kramer, A.: ibid. *27*, 457 (1974)
112. Ferguson, J., Wood, T. E.: Inorg. Chem. *14*, 184 (1975)
113. Carlin, R. L., O'Connor, C. J., Bathia, R. S.: J. Am. Chem. Soc. *98*, 685 (1976)
114. Mackey, D. J., Mc Meeking, R. F.: J. Chem. Soc. Dalton Trans. 2186 (1977)
115. Mackey, D. J., Evans, S. V., Mc Meeking, R. F.: ibid. 160 (1978)
116. Reedijk, J., Van der Put, P. J.: Proc. XVI I.C.C.C., Dublin 1974, 2.27 b
117. Van Ingen Schenau, A. D., Vershoor, G. C., Romers, C.: Acta Crystallog. *(B) 30*, 1386 (1974)
118. Bergendahl, T. S., Wood, J. S.: Inorg. Chem. *14*, 338 (1975)
119. Vijayan, M., Viswamitra, M. A.: Acta Crystallog. *23*, 1000 (1967)
120. Yang, C-L. M., Palmer, R. A.: J. Am. Chem. Soc. *97*, 5390 (1975)
121. Palmer, R. A., Piper, T. S.: Inorg. Chem. *5*, 864 (1966)
122. Palmer, R. A., Taylor, C. R.: ibid. *10*, 2546 (1971)
123. Evans, R. S. et al.: ibid. *15*, 3164 (1976)
124. Santoro, A. et al.: Acta Crystallogr. *(B) 25*, 842 (1969)
125. Reimann, G. W., Mighell, A. D., Maner, F. A.: ibid. *23*, 135 (1967)
126. Porai-Koshits, M. A.: Tr. Inst. Krystallogr. Akad. Nauk., SSSR *19*, (1955)
127. Bencini, A., Benelli, C., Gatteschi, D., Zanchini, C.: Inorg. Chem. *19*, 1301 (1980)
128. Bencini, A., Benelli, C., Gatteschi, D., Zanchini, C.: Inorg. Chim. Acta *45*, L 127 (1980)
129. Cotton, F. A., Eiss, R.: J. Am. Chem. Soc. *90*, 38 (1968)
130. Bencini, A., Benelli, C., Gatteschi, D., Zanchini, C.: Inorg. Chem. *19*, 3027 (1980)
131. Banci, L. et al.: ibid. *20*, 1399 (1981)
132. Lever, A. B. P., Walker, I. M., Mc Carthy, P. J.: Inorg. Chim. Acta *39*, 81 (1980)
133. Meredith, P. L., Palmer, R. A.: Inorg. Chem. *10*, 1546 (1971)
134. Gerloch, M., Morgenstern-Badarau, I., Audiere, J. P.: ibid. *18*, 3220 (1979)
135. Palenick, G. J., Wester, D. W.: ibid. *17*, 864 (1978)
136. Morgenstern-Badarau, I., Ammeter, J. H.: Proc. XXI I.C.C.C., Tolouse 1980
137. Cotton, F. A., Bergman, J. G. Jr.: J. Am. Chem. Soc. *86*, 2941 (1964)
138. Garner, D. C., Mabbs, F. E.: J. Chem. Soc. Dalton Trans. 525 (1976)
139. Bergman, J. G. Jr., Cotton, F. A.: Inorg. Chem. *5*, 1420 (1965)
140. Lever, A. B. P., Nelson, S. M.: J. Chem. Soc. (A) 859 (1966)
141. Murakami, Y., Matsuda, Y., Sakata, K.: Inorg. Chem. *10*, 1728 (1971)
142. Reedijk, J.: Rec. Trav. Chim. *91*, 507 (1972)
143. Murakami, Y., Matsuda, Y., Kobayashi, S.: J. Chem. Soc. Dalton Trans 1734 (1973)
144. Land, J. M., Stubbs, J. A., Wrobleski, J. T.: Inorg. Chem. *16*, 1955 (1977)
145. Webb, G. A., Richards, C. P.: J. Inorg. Nucl. Chem. *38*, 165 (1976)
146. Saha, N., Kar, S. K.: J. Inorg. Nucl. Chem. *41*, 1233 (1979)
147. Hollebone, B. R., Stillman, M. J.: Inorg. Chim. Acta *42*, 169 (1980)
148. Goodgame, D. M. L., Leach, A. G.: J. Chem. Soc. Dalton Trans. 1705 (1978)
149. Goodgame, D. M. L., Goodgame, M.: Inorg. Chem. *4*, 139 (1965)
150. Holm, R. H., Chakravorty, A., Theriot, L. J.: ibid. *5*, 625 (1966)
151. Dei, A., Morassi, R.: J. Chem. Soc. (A) 2024 (1971)
152. Holm, R. H., Cotton, F. A.: J. Chem. Phys. *31*, 788 (1959)
153. Henning, J. C. M.: Z. Angew. Phys. *24*, 281 (1969)
154. Ciampolini, M., Nardi, N.: Inorg. Chem. *5*, 41 (1966)
155. Cotton, F. A., Goodgame, D. M. L., Goodgame, M.: J. Am. Chem. Soc. *83*, 4690 (1961)
156. Perry, W. D., Quagliano, J. V., Vallarino, L. M.: Inorg. Chim. Acta *7*, 175 (1973)
157. Geary, W. J., Colton, D. F.: J. Chem. Soc. (A) 2457 (1971)
158. Briggs, E. M., Brown, G. W., Jiricny, J.: J. Inorg. Nucl. Chem. *41*, 667 (1979)

159. Keaton, M., Lever, A. B. P.: Inorg. Chem. *10*, 47 (1971)
160. Sacconi, L., Bertini, I., Mani, F.: ibid. *6*, 262 (1967)
161. Mani, F., Bertini, I.: Inorg. Chim. Acta *3*, 451 (1975)
162. Sacconi, L., Bertini, I., Mani, F.: Inorg. Chem. *6*, 262 (1967)
163. Zakrzewski, G., Sacconi, L.: ibid. *7*, 1034 (1968)
164. Sacconi, L., Bertini, I.: ibid. *7*, 1178 (1968)
165. Madan, S. K.: J. Inorg. Nucl. Chem. *33*, 1025 (1971)
166. Goodgame, D. M. L., Goodgame, M., Cotton, F. A.: Inorg. Chem. *1*, 239 (1962)
167. Bannister, E., Cotton, F. A.: J. Chem. Soc. 2276 (1960)
168. Fleur, A. H. M., Groeneveld, W. L.: Rec. Trav. Chim. *91*, 317 (1972)
169. Sznajder, J., Jablonski, A., Wojciechowski, W.: J. Inorg. Nucl. Chem. *41*, 305 (1979)
170. Holm, H. R., Cotton, F. A.: J. Chem. Phys. *32*, 1168 (1960)
171. Devillanova, F. A., Verani, G.: J. Inorg. Nucl. Chem. *41*, 1111 (1979)
172. Aitchen, G. B., Duncan, J. L., Mc Quillan, G. P.: J. Chem. Soc. Dalton Trans. 2103 (1972)
173. Brodie, A. M., Rodley, G. A., Wilkins, C. J.: J. Chem. Soc. (A) 2927 (1969)
174. Hill, W. E. et al.: Inorg. Chim. Acta *39*, 249 (1980)
175. Casey, A. T., Mackey, D. J., Martin, R. L.: Austral. J. Chem. *24*, 1587 (1971)
176. Petillon, F., Guerchais, J. F.: Bull. Soc. Chim. France 2455 (1971)
177. Cotton, F. A., Faut, O. D., Mague, J. T.: Inorg. Chem. *3*, 17 (1964)
178. Cotton, F. A. et al.: J. Am. Chem. Soc. *83*, 4157 (1961)
179. Sacconi, L., Gelsomini, J.: Inorg. Chem. *7*, 291 (1968)
180. Morassi, R., Sacconi, L.: J. Chem. Soc. (A) 492 (1971)
181. Cotton, F. A. et al.: J. Am. Chem. Soc. *83*, 1780 (1961)
182. Dahlhoff, W. V., Dick, T. R., Nelson, S. M.: J. Chem. Soc. (A) 2919 (1969)
183. Livingstone, S. E., Nolan, J. D.: J. Chem. Soc. Dalton Trans. 218 (1972)
184. Sacconi, L., Morassi, R., Midollini, S.: J. Chem. Soc. (A) 1510 (1968)
185. Sacconi, L., Morassi, R.: ibid. 2904 (1969)
186. Sacconi, L., Morassi, R.: ibid. 575 (1970)
187. Chiswell, B., Lee, K. W.: Inorg. Chim. Acta *7*, 509 (1973)
188. Ciampolini, M., Speroni, G. P.: Inorg. Chem. *5*, 45 (1966)
189. Sacconi, L., Bertini, I., Morassi, R.: ibid. *6*, 1548 (1967)
190. Ciampolini, M., Gelsomini, J.: ibid. *6*, 1821 (1967)
191. Morassi, R., Sacconi, L.: J. Chem. Soc. (A) 1487 (1971)
192. Catterick, J., Thornton, P.: J. Chem. Soc. Dalton Trans. 1634 (1976)
193. Sacconi, L., Bertini, I.: J. Am. Chem. Soc. *90*, 5443 (1968)
194. Orlandini, A., Sacconi, L.: Inorg. Chem. *15*, 78 (1976)
195. Ciampolini, M., Gelsomini, J., Nardi, N.: Inorg. Chim. Acta *2*, 342 (1968)
196. Sen, B., Johnson, D. A.: J. Inorg. Nucl. Chem. *34*, 609 (1972)
197. Ackerman, J., Du Preez, J. G. H., Gibson, M. L.: J. Coord. Chem *3*, 57 (1973)
198. Sacconi, L., Ciampolini, M., Speroni, G. P.: Inorg. Chem. *4*, 1116 (1965)
199. Sacconi, L., Bertini, I.: J. Am. Chem. Soc. *88*, 5180 (1966)
200. Lever, A. B. P.: Inorg. Electronic Spectroscopy, Elsevier, 1968
201. Reedijk, J.: Inorg. Chim. Acta *5*, 687 (1971)
202. Driessen, W. L., Groeneveld, W. L.: Rec. Trav. Chem. *90*, 258 (1971)
203. Van Ingen Schenau, A. D., Groeneveld, W. L., Reedijk, J.: ibid. *91*, 88 (1972)
204. Mikulski, C. M. et al.: J. Inorg. Nucl. Chem. *40*, 769 (1978)
205. Subramanian, S., Rahman, Z., Whittery, J.: J. Chem. Phys. *49*, 473 (1968)
206. Hulsbergen, F. B., Welleman, J. A., Reedijk, J.: Delft Progr. Rept. *A 1*, 137 (1976)
207. Wasson, J. R., Stoklosa, H. J.: J. Inorg. Nucl. Chem. *36*, 227 (1974)
208. Ferguson, J., Wood, D. L., Knox, K.: J. Chem. Phys. *39*, 881 (1963)
209. Kakazal, B. J. A., Melson, J. A.: Inorg. Chim. Acta *4*, 360 (1970)
210. Jones, W. C., Ball, W. E.: J. Chem. Soc. (A) 1849 (1968)
211. Sing, N. B., Sing, B.: J. Inorg. Nucl. Chem. *40*, 919 (1979)
212. Niewpoort, A., Reedijk, J.: Inorg. Chim. Acta *7*, 323 (1973)
213. Anagnostopoulos, A., Matthews, R. W., Walton, R. A.: Canad. J. Chem. *50*, 1307 (1972)
214. Larsen, E. et al.: Inorg. Chem. *11*, 2652 (1971)
215. Driessen, W. L., Everstijn, P. L. A.: Inorg. Chim. Acta *41*, 179 (1980)

216. Meredith, P. L., Palmer, R. A.: Inorg. Chem. *10*, 1546 (1971)
217. Wheeler, S. H., Zingheim, S. C., Nathan, L. C.: J. Inorg. Nucl. Chem *40*, 779 (1978)
218. Aggarwal, R. C., Chandrasekhar, V.: ibid. *41*, 1057 (1979)
219. König, E. et al.: Inorg. Chim. Acta *6*, 123 (1972)
220. Knersh, D., Groeneveld, W. L.: ibid. *7*, 81 (1973)
221. Peyronel, G. et al.: ibid. *5*, 263 (1971)
222. Goodgame, D. M. L., Leach, G. A.: ibid. *25*, L127 (1977)
223. Goodgame, D. M. L., Leach, G. A.: ibid. *32*, 69 (1979)
224. Chakravorty, A., Zacharias, P. S.: Inorg. Chem. *10*, 1961 (1971)
225. Bhattacharyya, R. G., Bhadure, I.: J. Inorg. Nucl. Chem. *40*, [illegible] (1978)
226. Ciampolini, M., Paoletti, P.: Inorg. Chem. *6*, 1261 (1967)
227. Goodgame, D. M. L., Hitchman, M. A.: ibid. *4*, 721 (1965)

# Relationships Between Structure and Low-Dimensional Magnetism in Fluorides

**Alain Tressaud and Jean-Michel Dance**

Laboratoire de Chimie du Solide du CNRS, Université de Bordeaux I, 351 Cours de la Libération, F-33405 Talence Cedex, France

The available models and experimental results used in the description of magnetic order phenomena have been recently summarized in several papers. Among the materials showing low-dimensional magnetism, fluorides are of great interest as they provide examples for most of the different theoretical models. This article emphasizes the relationships between crystallographic features and dimensionality of the magnetic interactions in fluorides. The involved compounds are characterized by rings, chains, layers and three-dimensional arrangements of fluorinated $MF_6$ octahedra. Spin-flop and zero-point spin-reduction phenomena are also considered, as many fluorides exhibit these behaviors.

# Introduction

Fluorides have long been known to provide useful materials for the checking and improvement of magnetic models[1-4]. Since they are generally strong insulators, their magnetic behavior can be thus interpretated more easily than that of oxides and sulfides, due to the absence of electronic delocalization.

Another simplification consists in the fact that among the various structural types of fluorides, the involved d-transition elements generally possess the coordination number 6. The crystallographic features can be deduced from the arrangement of ($MF_6$) octahedra[5,6]. Besides in three-dimensional (3-D) networks such as found in perovskite, rutile or pyrochlore types for instance, fluorides crystallyze in two-dimensional (2-D) layer structures, one-dimensional (1-D) chain structures and isolated unit arrangements.

3-D fluorinated materials have been used in the establishment of Néel theories of antiferro- and ferrimagnetism[1,3] and also in early studies of the relationships between the exchange constants and structural features[2]. The sign and the intensity of magnetic couplings have been derived from the Goodenough-Kanamori rules[4]. More recently, several articles dealing extensively with magnetism of low-dimensional materials have been published[7-9]. In particular, in an important contribution, De Jongh and Miedema have described the different theories and available examples[7].

This article is mainly concerned with the relationships between crystallographic features and the presence of a low-dimensional magnetism in fluorinated materials. Fluorides provide examples for most of the different theoretical models of low-dimensional magnetism and, due to the presence of simple bondings, they can contribute to a better understanding of this type of magnetism.

In the first two chapters, the different theories explaining low-dimensional magnetic behavior will be briefly considered and also the available techniques. Then, the most important part (Chap. 3) will deal with fluorinated materials showing low-dimensional magnetism in structures characterized by a 3-D crystallographic arrangement of $MF_6$ octahedra, layers, chains and rings. Spin-flop and zero-point spin-reduction will be also considered, since fluorides provide in these domains most of the conclusive examples.

# 1 Theoretical Approach

The physics of systems of less than three dimensions has focussed the attention for many years. In fact, these models were of theoretical interest only during a long period until some real materials whose behavior approximates to the theory were discovered.

## *1.1 Magnetic Models*

Starting from the classical Heisenberg-Dirac-Van Vleck (HDVV) model based on quantum mechanical interactions, all the other models can also be founded on statistical calculations.

When the number of spins is very large (i.e. in a magnetic material) the statistical thermodynamical formulae involving the partition function Z are the following:

$$\text{if} \quad Z = \sum_i e^{\frac{-E_i}{kT}},$$

$$C = \frac{\partial}{\partial T}\left(kT^2 \frac{\partial \operatorname{Log} Z}{\partial T}\right) \quad \text{(specific heat)},$$

$$M = \frac{kT}{V} \frac{\partial \operatorname{Log} Z}{\partial H} \quad \text{(magnetization)},$$

$$\chi = \frac{kT}{V} \frac{\partial^2 \operatorname{Log} Z}{\partial H^2} \quad \text{(susceptibility)},$$

where V is the volume of the system.

One can show that the Curie law is an approximation to the first term of the expanded partition function of N weakly interacting spins coupled by the Zeeman interaction:

$$\mathcal{H} = -\mu_B \vec{H} \cdot \Sigma \vec{S}.$$

Hence:

$$\chi = \frac{N \sum_{-S}^{+S} \frac{b_{S_z}^2}{kT}}{2S+1} = \frac{N\mu_B^2 g^2}{3kT} S(S+1) = \frac{C}{T}$$

The second term is the classical "temperature independent paramagnetism" $N\alpha$. Assuming a Weiss internal field $H_W = \lambda M$ which tends to align the spins, we have:

$$H_E = H_0 + \lambda M$$

where $H_E$ is the effective field and $H_0$ the applied field.

$$\frac{M}{H_E} = \frac{C}{T} = \frac{M}{H_0 + \lambda M}.$$

Thus

$$\chi = \frac{M}{H_0} = \frac{C}{T - \lambda C} = \frac{C}{T - T_C}.$$

This is the Curie-Weiss law. If the Weiss "internal field" is written in terms of HDVV interactions, then

$$\mathcal{H}_{ij} = -2JS_iS_j$$

where J is the exchange constant which can be either positive (ferromagnetism) or negative (antiferromagnetism).

### 1.1.1 The Ising and the Heisenberg Model

The first attempt to explain semi-classical interactions between magnetic spins in a lattice was made by Ising[10] who considered an atom of spin S to have (2S + 1) possible orientations in a magnetic field, the interactions being proportional to $S_i^\alpha S_j^\alpha$ which is the product of the $\alpha$ components of the spins. In this case, the Hamiltonian describing the system in an external magnetic field is

$$\mathcal{H} = -\frac{J'}{S^2}\sum_{ij} S_i^\alpha S_j^\alpha - \frac{mH}{S}\sum_i S_i^\alpha .$$

It is assumed that if S varies, the maximum of interactions remains constant and equal to J′ and the maximum magnetic moment equal to m. A simple explanation can be given for the case S = ½, i.e. when the spins have only two possibilities in the magnetic field (parallel or antiparallel).

The exact calculation of the Ising model was carried out in 1944 by Onsager[11] and involved a great number of parameters. Nevertheless, it was shown that a substantial fraction of the entropy change of the system takes place in a temperature range above the transition point (or Curie point).

Heisenberg[12] suggested a Hamiltonian of the form

$$\mathcal{H} = -2J\sum_{ij}\vec{S}_i\vec{S}_j - g\beta H\sum S_i^\alpha$$

which can be compared with the Ising Hamiltonian if it is written as

$$\mathcal{H} = -\frac{J'}{S^2}\sum_{ij}\vec{S}_i\vec{S}_j - \frac{mH}{S}\sum_i S_i^\alpha$$

with $J' = 2JS^2$ and $m = g\beta S$.

In this case, there is the possibility of interactions between the ith and jth neighbour in all three directions.

In fact, these two models can be combined by a single formula of the form:

$$\mathcal{H} = -2J\sum_{ij}[aS_i^zS_j^z + b(S_i^xS_j^x + S_i^yS_j^y)] .$$

If b = 0, a = 1 the Ising model, if b = 1, a = 1 the isotropic Heisenberg model, and if b = 1, a = 0, the planar Heisenberg model or X-Y model is valid. In the following, the values of the exchange constants were calculated, or recalculated assuming a spin Hamiltonian of this form, instead of $\mathcal{H} = -J\sum_{ij}\vec{S}_i\vec{S}_j$, occasionally used by certain authors[117, 130, 140]. Thus, an order parameter, which measures the degree of order present below the critical point, is introduced. In magnetic systems, for instance, magnetization (or sublattice magnetization in the case of an antiferromagnet) can be taken as a measure of the order parameter. In the Ising model, the order is a scalar quantity whereas in the Heisenberg or X-Y model it is represented by a two- or three-dimensional vector. The nature of this vector indicates the dimensionality of the interaction.

Of course, the dimensionality of the structure is also of great importance and will be in most cases the major factor determining the 1-, 2- or 3-D character of the magnetic properties (in the following, 1-D, 2-D and 3-D will always stand for 1-, 2- or 3-dimensional).

Finally, the last parameters are the exchange constant, which can be either positive or negative, and the value of the spin S.

### *1.1.3 Ordering Temperature*

The specific heat behavior based on the Ising and Heisenberg models (for $S = \frac{1}{2}$) for D = 1, 2, 3 is illustrated in Figs. 1 and 2. For comparison, the molecular field prediction is also shown[7, 10–16]:

– The ordering temperature is always lower than $T_c = \frac{2}{3}\frac{J}{k} z \cdot S(S + 1)$ given by the molecular field.

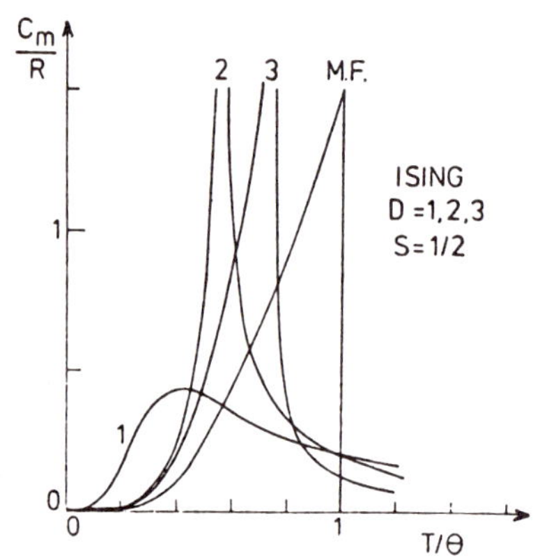


**Fig. 1.** Theoretical magnetic specific heat of the $S = \frac{1}{2}$ Ising model for 1-D, 2-D and 3-D lattices

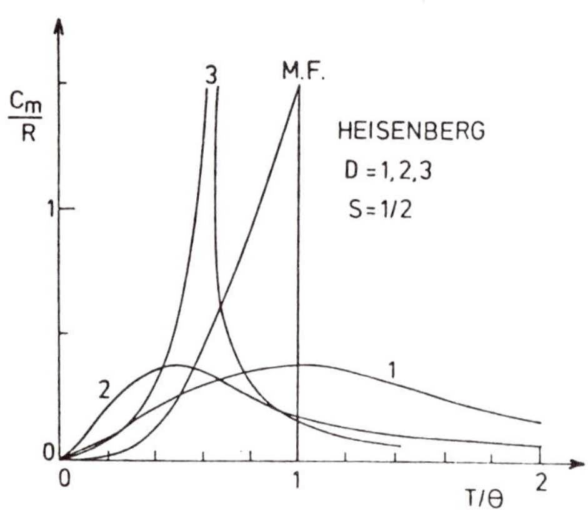


**Fig. 2.** Theoretical magnetic specific heat of the $S = \frac{1}{2}$ Heisenberg model for 1-D, 2-D and 3-D lattices

– There is a more or less pronounced tail in the specific heat curves above $T_c$, the largest being observed for the 1-D Heisenberg model.
– In the case of the Ising model, only the 1-D curve does not show any ordering temperature, whereas in the case of the Heisenberg model neither the 1-D nor the 2-D curves have an ordering temperature.

## 1.2 Dimensionality of Interactions and Structures

The Ising model assumes the magnetic interactions to be anisotropic. In fact, this phenomenon does not occur practically and the choice of an Ising Hamiltonian will be made on the basis of other factors such as the presence of a crystal field or a magnetic dipolar field both of which can polarize the spin in a certain direction of the crystal. The first case very often results from the existence of an orbital momentum[7)].

Two ions are well-known for their highly anisotropic properties. Firstly, in the rare-earth family, $Dy^{3+}$ which has a $^6H_{15/2}$ ground state. The spin-orbit interaction is stronger than the crystal field effects. The ratio $J_{\|}/J$ can be of the order of 100 ($J_{\perp} \simeq 0$), $g_{\|} = 20$ and $g_{\perp} = 0$; this is practically an ideal case. Secondly, in the transition element series, the ion $Co^{2+}$ is also characterized by anisotropic interactions (either in the tetrahedral or octahedral coordination), the anisotropy being however lower than in the case of $Dy^{3+}$. $J'/J$ is about 0.5 for this ion. Some $Fe^{2+}$ compounds also display a behavior approximating to the Ising model.

In contrast, a Heisenberg model will be chosen in the case of highly isotropic ions such as $Fe^{3+}$, $Mn^{2+}$, $Cr^{3+}$, $Gd^{3+}$ and $Ni^{2+}$.

The planar Heisenberg model is particularly appropriate when there is a strong preference for a certain direction of the moments.

Structural features have also a great influence on the dimensionality of the magnetic properties. In most cases, when a structure is characterized by arrays of magnetic ions in one or two directions (chains or layers respectively), 1-D or 2-D magnetism will result. This is not always the case, as is shown for $KCuF_3$, $BaCuF_4$ in Sect. 1.3.

## 1.3 Theoretical Calculations of Magnetic Functions

### 1.3.1 One-Dimensional (1-D) Systems

1-D Ising calculations have been tabulated by Suzuki et al.[17)] for $S = ½, 1, \frac{3}{2}$ for a zero-external field and by Van Dongen and Capel in the presence of a magnetic field[18)].

In the case of antiferromagnetic Ising chains, there is a great difference between the parallel and perpendicular components of the susceptibility. The susceptibility curves of a ferromagnetic and an antiferromagnetic interaction are illustrated in Figs. 3 and 4.

In the case of X-Y chains, otherwise known as transverse coupled chains, the perpendicular susceptibility has been calculated by Katsura[19)].

1-D Heisenberg chains have been extensively studied following the pionneering study of Bonner and Fisher[14, 20)] on the calculation of magnetic functions of the Heisenberg model of infinite chains with spins ½ and of the Heisenberg rings. The latter calculation has been extrapolated to an infinite ring, i.e. a chain. The susceptibility has the form

$$\chi = \frac{Ng^2\mu_B^2}{12\,kT}\,\frac{1+u}{1-u}\,,$$

where $u = \coth K - \dfrac{1}{K}$, with $K = \dfrac{1}{2}\dfrac{J}{kT}$

The extension to $S > \frac{1}{2}$ has been carried out by Weng[21] and the resulting susceptibility curves are shown in Fig. 5.

Extensions to more sophisticated 1-D magnetisms have been performed by Takeuchi[22] for 1-D helicoidal systems, and by Duffy and Barr[23] for alternating Heisenberg chains (two different exchange integrals).

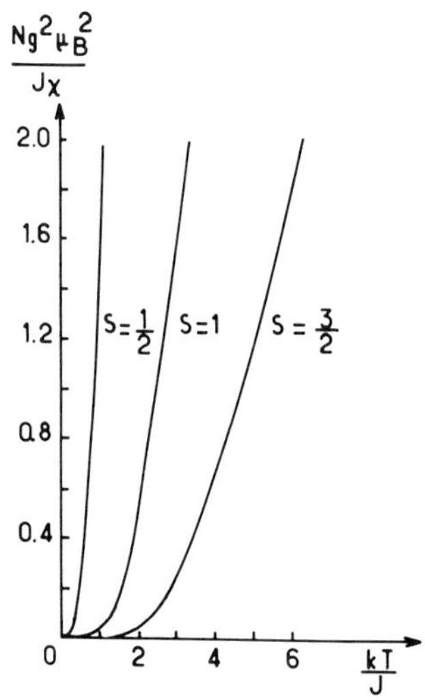


**Fig. 3.** Theoretical inverse susceptibility of ferromagnetic interactions for different spin values of an Ising model

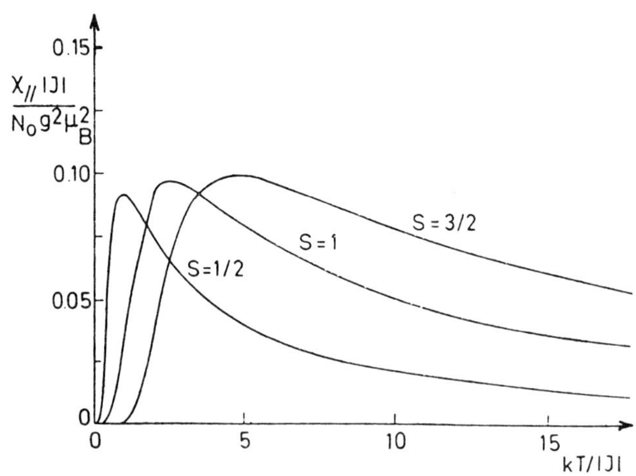


**Fig. 4.** Theoretical parallel susceptibility of an Ising model in the case of antiferromagnetic interactions

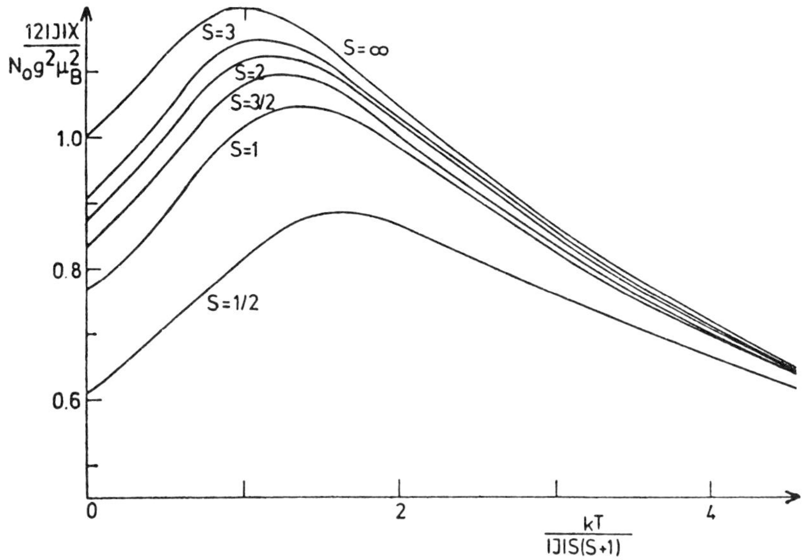


**Fig. 5.** Theoretical susceptibility of a Heisenberg chain for different spin values

A large number of publications have dealt with 1-D Heisenberg systems, among which the studies of Smith and Friedberg should be mentioned[24)]. They proposed a slight modification of the Fisher curve concerning the correction for the finite spin of the studied ion (multiplication of both $\chi$ and K by $4S(S+1)$).

Both numerical values for the different models and spins are tabulated in the review paper of De Jongh and Miedema[7)].

## *1.3.2 Two-Dimensional (2-D) Systems*

As shown in Figs. 1 and 2, when the dimensionality changes from 1 to 2, the 2-D Ising system may have an ordering temperature, in contrast to the 2-D Heisenberg model. In the Ising case, theoretical calculations on the 2-D system have been carried out by Fisher[25)] and the presence of an ordering temperature was evidenced by Griffiths[26)].

Concerning the Heisenberg model, the absence of an ordering temperature was proven by Merwin and Wagner[27)]. Nevertheless, a transition from short-range to long-range order is always possible if the system deviates from the ideal model. The transition temperature decreases to zero if the deviation from the ideal model tends towards zero. Many authors have studied the possibility for a 2-D Heisenberg model to show an ordering temperature (which is not predicted by theory). Stanley and Kaplan proposed an explanation based upon the fact that in ordinary ferromagnets a finite moment is present below $T_C$ whereas all the magnetization curves have an infinite slope at $H = 0$. The problem is the same for antiferromagnets[28)].

The value of the exchange integral in these systems can be obtained by using Green functions, the spin wave theory, the high-temperature expansion series, etc. ... The application of the expansion series is perhaps the simplest way of obtaining this value.

The partition function can be written:

$$Z = \sum_i e^{-E_i/kT} = \sum_i \left[1 - \frac{E_i}{kT} + \frac{E_i^2}{(kT)^2}\frac{1}{2!} + \frac{E_i^3}{(kT)^3}\frac{1}{3!} \cdots \right].$$

Applying this function to the expression of the susceptibility, one obtains:

$$\frac{\chi J}{Ng^2\mu_B^2} = \frac{S(S+1)\,J}{3\,kT} \sum_{n=0}^{\infty} a_n \left(\frac{J}{kT}\right)^n .$$

Rushbrooke and Wood have calculated the coefficients $a_n$ of this series for a Heisenberg ferromagnet and have extended this treatment to an antiferromagnet[29].

This series is calculated assuming that interactions occur only between nearest neighbors. In the case of a ferromagnet, it converges toward $T_C$. For an antiferromagnet, the sign of the terms of the series is alternatively positive and negative. As in the 1-D case, there is a large maximum of the susceptibility far above the 3-D ordering point (Fig. 6).

### *1.3.3 Critical Exponents*

On decreasing the temperature, the system approaches a critical state for T very close to $T_C$, say for $\frac{T - T_C}{T} < 10^{-2}$, $10^{-1}$ [25, 30].

All the thermodynamic magnetic functions are described by asymptotic expressions which are governed by exponents:

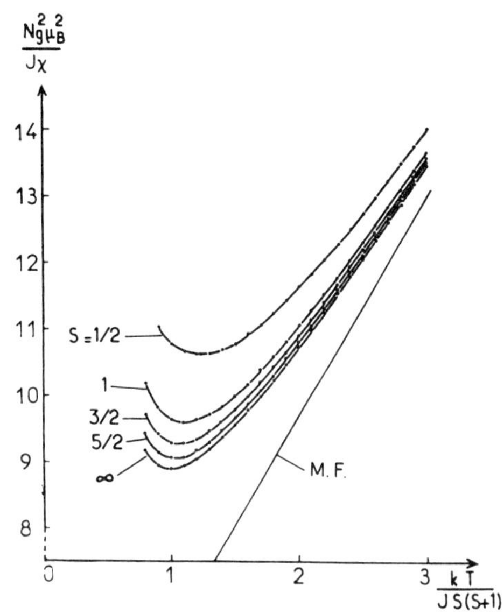


**Fig. 6.** Theoretical inverse susceptibilities in 2-D Heisenberg antiferromagnets

Specific heat:

$$C_m/R \sim A(1 - T_c/T)^{-\alpha} \quad (T \rightarrow T_c^+;\ H = 0)\ ,$$

$$C_m/R \sim A'(1 - T/T_c)^{-\alpha'} \quad (T \rightarrow T_c^-;\ H = 0)\ .$$

Spontaneous magnetization:

$$M_s(T)/M_s(0) \sim B(1 - T/T_c)^{\beta} \quad (T \rightarrow T_c^-;\ H = 0).$$

Initial susceptibility:

$$\chi T/C \sim C_0(1 - T_c/T)^{-\gamma} \quad (T \rightarrow T_c^+;\ H = 0),$$

$$\chi T/C \sim C_{0'}(1 - T/T_c)^{-\gamma'} \quad (T \rightarrow T_c^-;\ H = 0).$$

Critical isotherm:

$$H \sim D\ |M(H)|^{\delta} \quad (H \rightarrow 0;\ T = T_c)\ .$$

Inverse correlation length:

$$\varkappa \sim N(1 - T_c/T)^{\nu} \quad (T \rightarrow T_c^+;\ H = 0)\ .$$
$$\varkappa \sim N'(1 - T/T_c)^{\nu'} \quad (T \rightarrow T_c^-;\ H = 0)$$

Wave vector dependent susceptibility:

$$\chi(k) \sim k^{\eta-2} \quad (k \rightarrow 0;\ T = T_c)$$

$T^+$ or $T^-$ mean that the temperature of the system is above or below the critical temperature.

The different values of the exponents can be obtained either by exact calculations (molecular field and 2-D Ising model) or by interpolations of series expansions. Table 1 gives the critical exponents for different models.

# 2 Experimental

## 2.1 Magnetic Measurements

Most of the different techniques used for magnetic measurements are now well-known[31]. For instance, susceptibility can be measured with a Faraday balance. Magnetization measurements (vibrating sample magnetometer, extraction method, etc.) provide an indirect method of obtaining the susceptibility; in addition, the variation of magnetization with the applied field and temperature can also be determined. In some cases, this magnetization does not depend linearly on the applied field (classical Curie law), and extrapolation methods have to be used to characterize the susceptibility (weak ferromagnetism, spin orbit coupling, etc. ...). The parallel and perpendicular components of susceptibility have been determined by magnetic measurements on single crystals.

**Table 1.** Theoretical values of critical exponents [from [7]]

| Model system | | $\alpha$ | $\alpha'$ | $\beta$ | $\gamma$ | $\gamma'$ | $\delta$ | $\eta$ | $\nu$ | $\nu'$ |
|---|---|---|---|---|---|---|---|---|---|---|
| MF | | 0 (discont.) | 0 (discont.) | ½ | 1 | 1 | 3 | 0 | ½ | ½ |
| Ising | D = 3) | $\simeq$ ⅛ | ⅛–¹⁄₁₆ | 0.312 | 1.25 | 1.25–1.31 | $\simeq$ 5 | 0.03–0.05 | $\simeq$ 0.63 | – |
| XY | D = 3) | $\simeq$ 0 (logar.) | – | – | $\simeq$ 1.33 | – | – | – | – | – |
| Heisenberg | D = 3) | $\simeq$ −0.1 | – | $\simeq$ 0.36 | $\simeq$ 1.40 | – | $\simeq$ 5 | 0.03–0.04 | $\simeq$ 0.71 | – |
| Ising | D = 2) | 0 (logar.) | 0 (logar.) | ⅛ | 1.75 | 1.75 | 15 | 0.25 | 1 | 1 |

## 2.2 High Applied Fields

When the field intensity is of the same order of magnitude as the exchange energy, possible rearrangements of the spins can occur and these phenomena can be easily seen in magnetization curves.

For instance, an Ising model which is particularly anisotropic can be confirmed by a plot of $\frac{T_c(H)}{T_c(0)}$ vs. $\frac{H}{H_c}$ where $H_c(T)$ is the critical field at T. When increasing temperature, this critical field $H_c(T)$ decreases continuously until it vanishes at the critical point.

$$\frac{T_c(H)}{T_c(0)} = \left[1 - \left(\frac{H}{H_c}\right)^2\right]^{\xi}$$

where $\xi$ depends on the magnetic lattice[32]. If we consider the Heisenberg antiferromagnetic systems, the spin-flop phenomenon can be observed. This phenomenon is described in Sect. 3.5.

## 2.3 Neutron Diffraction

Neutron elastic scattering is one of the most powerful tools in the determination of the magnetic structure of an ordered material. At the onset of magnetic order additional reflections on the nuclear pattern are created, due to interactions between the electrons of the nuclei. The variation of the intensity of these magnetic reflections with temperature yields the ordering temperature as well as the spontaneous magnetization of the sublattices and thus the critical exponent $\beta$.

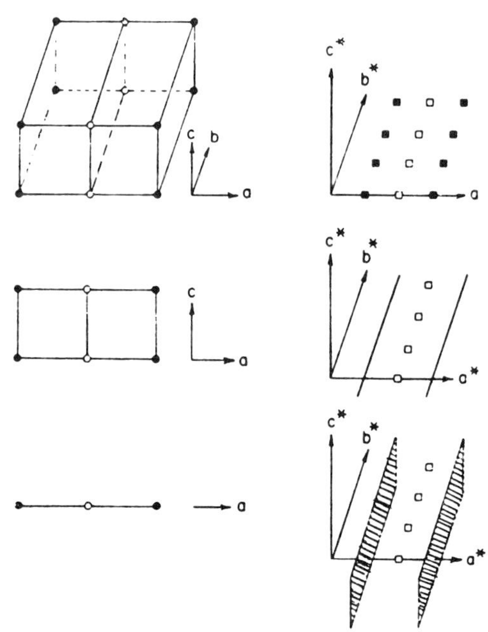


**Fig. 7.** Illustration of magnetic Bragg scatterings in 1-D, 2-D or 3-D magnetic orderings

Close to $T_c$, quasi-elastic scattering is in some cases useful and gives information on the fluctuations of regions of short-range order.

Figure 7 illustrates magnetic Bragg scatterings of low-dimensional systems[33]. Since in a 3-D material, Bragg peaks are observed, in a 2-D system lines, and in a 1-D material planes will be detected.

## 2.4 Mössbauer Resonance

Mössbauer experiments are usually performed on compounds containing $^{57}Fe$ nuclei. The variation of the hyperfine field with temperature below the ordering temperature gives the critical exponent:

$$\text{Log}\,\frac{H_{(T)}}{H_{(0)}} = D\ \text{Log}\left[1 - \frac{T}{T_N}\right]^{\beta}$$

where D is a constant.

The hyperfine field is directly proportional to the sublattice magnetization. Moreover, the appearance of the sextuplet(s) of $^{57}Fe$ in the ordered regions provides the 3-D ordering temperature. In the case of low-dimensional systems, zero-point spin reduction is often observed.

## 2.5 Nuclear Magnetic Resonance

For magnetic systems, NMR is a useful technique in the investigation of the behavior of the magnetic moments and the critical phenomena (order-disorder, critical fluctuations)[34, 35].

The magnetic dipolar and hyperfine interactions of the nucleus with the electronic moments can be expressed by:

$$\mathcal{H} = \gamma \hbar \vec{I} \cdot \vec{h}(t)\ ,$$

where $\gamma$ is the nuclear gyromagnetic ratio, $\vec{I}$ the nuclear angular momentum and $\vec{h}$ a fluctuating local field created at the nucleus by the electrons. $\vec{h}$ is the sum of a static (average) value $\langle\vec{h}\rangle$, given by the combination of local average magnetic moments ($g\beta \langle\vec{S}_i\rangle$), and a time-dependent value h(t). The first term produces a shift in the NMR resonance frequency proportional to the spontaneous magnetization in the ordered phase and to the static susceptibility in the paramagnetic phase. The fluctuating part of $\vec{h}(t)$ induces spin-spin and spin-lattice relaxation times which are useful parameters in the study of electronic spin dynamics. Hence, NMR acts as a time reference and by means of the various relaxation measurements, the spectral region of the fluctuations that can be investigated is extended to $\omega_e \simeq 10^{12}$ Hz. In low-dimensional magnetic systems, NMR measurements give valuable information on the modifications in the collective behavior due to the dimensionality restrictions for spin diffusion. For instance, cooperative effects begin in a linear chain when the temperature is of the order of the exchange interaction within the chain ($T \simeq J/k$). In real crystals, the weak interchain interactions cause a transition at $T_c$ from a 1-D to a 3-D ordered phase. The spin dynamics can be investigated between T and $T_c$ by spin lattice relaxations.

## 2.6 EPR of Low-Dimensional Systems

The EPR spectrum of magnetically condensed systems generally consists of a single "exchange-narrowed resonance" signal. The only measurable parameter in this case is the line width. Nevertheless, the variation of the line width with temperature and with angle $\theta$ ($\theta$ is the angle between the direction of the magnetic field and that of the predominant interactions) provides a great deal of information concerning the strength of the magnetic interactions and their dimensionality.

Recently, the study of EPR line widths for the determination of the spin correlations in the paramagnetic state of low-dimensional magnets has been widely extended on the basis of the pionneering publications of Richards and coworkers[36, 37].

The simplest interaction to be considered is the dipole-dipole interaction which can be divided in two parts: a secular term which has a $|3\cos^2\theta - 1|^L$ variation and a non-secular term which, when associated with the previous one, results in a variation of the form:

$$|3\cos^2\theta - 1|^2 + 10\sin^2\theta\cos^2\theta + \sin^4\theta = 1 + \cos^2\theta$$

One can see that when the exchange field is larger than the observation field ($\omega\tau \ll 1$, $\tau$ being the correlation time) the non-secular terms are not averaged out and the variation is of the $(1 + \cos^2\theta)$ type. When the exchange field is smaller than the observation field, the non-secular terms are averaged out and we have the typical $(3\cos^2\theta - 1)^2$ variation with the minimum line width at the "magic angle" $\theta = 54° 43'$. It can be shown that $L = 4/3$ for one-dimensional systems and $L = 2$ for 2-D systems[38]. For intermediate exchange fields, secular and non-secular terms must be taken into account. In most cases, the susceptibility of the compound is directly proportional to the surface of the absorption EPR signal.

# 3 Low-Dimensional Magnetism in Fluorides

## 3.1 Low-Dimensional Magnetism in 3-D Crystallographic Systems

### 3.1.1 Jahn-Teller Orderings

The cooperative Jahn-Teller effect results from the particular arrangement in a lattice of the orbital directions of certain anisotropic ions. Important distortions appear for ions in the $E_g$-ground state ($d^4$, $d^9$ in a high-spin state and $d^7$ in a low-spin state for an octahedral symmetry). In fluorides, d-transition elements generally exhibit an octahedral coordination. In the case of a $d^9$ ion, filled $d_{z^2}$ and half-filled $d_{x^2-y^2}$ orbitals lead to the formation of an elongated octahedron whose direction of elongation corresponds to the $d_{z^2}$ orbital; an opposite occupancy would give rise to compression. With a $d^4$ (high-spin state) or a $d^7$ ion (low-spin state), an elongation is also generally observed, due to the occupancy of the $d_{z^2}$

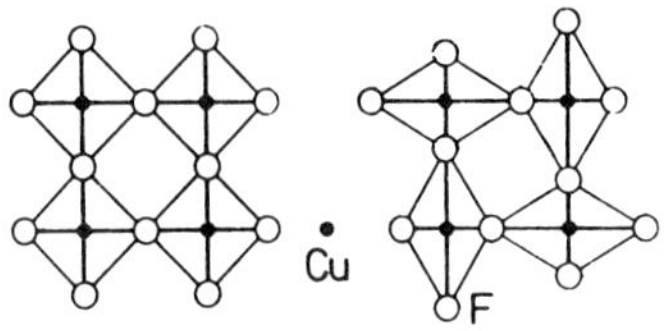


**Fig. 8.** Ferrodistortive and antiferrodistortive orderings of corner-connected ($CuF_6$) octahedra

orbital. Besides the possibility of a complicated distribution of the directions of the elongation, two simple arrangements are often found, as shown by Reinen and coworkers[39]. In the ferrodistortive ordering, all the long metal ligand distances are parallel to a certain direction while in the antiferrodistortive ordering the long bonds are alternatively directed perpendicular to each other (Fig. 8). These arrangements are favored for structures consisting of isolated or corner-connected octahedra containing the Jahn-Teller ion.

It is obvious that the magnetic couplings in the ferro- and the antiferrodistortive orderings will differ considerably. In the case of corner-shared octahedra containing $d^4$ ions, compressed octahedra of ferrodistortive order exhibit antiferromagnetic exchange whereas elongated octahedra of antiferrodistortive order favour ferromagnetism; similar arguments can be used for $d^9$ ions having elongated octahedral surroundings (the compressed case has not been reported so far).

### *3.1.2 2-D Magnetic Interactions in 3-D Structures*

#### 3.1.2.1 $KCrF_3$

The structure of $KCrF_3$ is derived from the perovskite structure by a tetragonal distortion[40]. An antiferrodistortive coupling occurs between half-filled $d_{z^2}$ orbitals and empty $d_{x^2-y^2}$ orbitals. The superexchange interaction $d^0_{x^2-y^2} - p - d^1_{z^2}$ leads to ferromagnetic layers, antiferromagnetically bound via the empty $d_{x^2-y^2}$ orbitals.

Neutron diffraction experiments have confirmed that $KCrF_3$ exhibits antiferromagnetism of type A with the spins lying in the pseudo-tetragonal (001) plane[41] ($T_N = 40$ K; $\theta_p = +5$ K[42]). The exchange constants calculated on the basis of the molecular field theory have the same order of magnitude within and between two layers: $J_1/k = 1.4$ K and $J_2/k = -2.2$ K[42].

#### 3.1.2.2. $MnF_3$

The d-transition metal trifluorides are generally derived from the $ReO_3$ structure by a rhombohedral distortion ($VF_3$ type). Due to the electronic anisotropy of high-spin $Mn^{III}$ ($t^3_{2g}\ e^1_g$), $MnF_3$ is pseudorhombohedral and shows an additional monoclinic distortion (C2/c space group)[43]. The environment of $Mn^{III}$ is characterized by three different Mn-F distances: 2.09, 1.91 and 1.79 Å. A recent Mössbauer study of $MnF_3$ doped with 2% $^{57}FeF_3$ confirmed the presence of two nonequivalent sites for Mn atoms[44]. The ordering of empty and half-filled $e_g$ orbitals here also leads to ferromagnetic layers via an antiferrodistortive mechanism. The elongation axis of ($MnF_6$) octahedra is alternatively directed along Ox and Oy (Fig. 8)[45].

Magnetic measurements have shown that $MnF_3$ is antiferromagnetic below $T_N$ = 43 K[45]. As in $KCrF_3$, the importance of ferromagnetic couplings can be deduced from the positive value of the Weiss constant ($\theta_p$ = 2 K[46]; 8 K[47]). The magnetic structure, identical with those of $KCrF_3$ and $LaMnO_3$[48], is of the A-type with ferromagnetic layers antiferromagnetically coupled. The exchange constant of nearest neighbors has been found to be positive $J_1/k = +1.7$ K while the exchange constant of second neighbors is negative ($J_2/k = -2.6$ K)[2].

#### 3.1.2.3 Ferromagnetic Layers in Difluorides

Difluoride structures are mainly derived from rutile- and fluorite-type structures. In particular, d-metal difluorides generally exhibit a rutile structure when prepared under ambient conditions. Although most of these phases show a 3-D magnetic ordering of the moments of the divalent elements, several difluorides, derived either from rutile ($CuF_2$) or exhibiting on original structure (high-pressure (HP) form of $PdF_2$; $AgF_2$) are characterized by ferromagnetic layers.

$CuF_2$ and $CrF_2$ have similar crystallographic features (distorted monoclinic cell derived from rutile). Nevertheless, while $CrF_2$ has an ordering identical with that of the other rutile phases (magnetic moments of the metal ions located in the corners of the pseudotetragonal cell antiparallel to that of the central metal[49]), $CuF_2$ shows ferromagnetic layers of $CuF_4$ squares linked by corners. These layers are shifted to form distorted $CuF_6$ octahedra and are coupled antiferromagnetically. The ferromagnetic planes are parallel to the (102) direction in the $P2_1/c$ space group[50, 51].

$AgF_2$[52, 53] and the HP form of $PdF_2$[54] are also characterized by buckled layers of $MF_4$ squares, but the bonding of the planes is different from that in the previous case. Instead of resulting from $MF_6$ octahedra linked by corners and edges, the distorted octahedra are only shared by corners. In both compounds, the ferromagnetic layers are coupled antiferromagnetically.

### *3.1.3 1-D Magnetic Interactions in 3-D Structures*

#### 3.1.3.1 $VF_2$

$VF_2$ exhibits a rutile structure and crystallizes in the tetragonal $P4_2/mnm$ space group. The structure consists of infinite chains of edge-sharing $(VF_6)^{4-}$ octahedra. These chains are directed along the c-axis and 3-D connected to each other by common corners[46, 55].

A molecular field treatment based on the hypothesis of a body-centered cubic lattice has been carried out by Smart[2]. Three main exchange constants must be taken in account:

$J_1$ (relating to nearest-neighbors within a chain),
$J_2$ (between two nearest neighbors located in two adjacent chains) and
$J_3$ (between metal ions in the ab plane), the latter constant having a smaller numerical value than the first two ones.

When the involved transition metal possesses $e_g$ electrons, the interchain constant $J_2$ is greater than the intrachain constant $J_1$. This result is due to stronger 135° M-F-M superexchange interactions (occuring via $e_g$-$p\sigma$-$e_g$ antiparallel couplings) which are responsible for the high value of $J_2$. Simultaneously, the competition within a chain between ferro- and antiferromagnetic couplings leads to low values of $J_1$, which can be either positive or negative.

Yoshimori showed that certain combinations of $J_1$, $J_2$ and $J_3$ lead to spiral magnetism[56)] and determined the ratio $J_1/J_2$ from the pitch of the spiral.

For $V^{II}$ ions ($t_{2g}^3$, $e_g^0$) the absence of $e_g$ electrons lowers the 135°-interactions and $J_2$ thus decreases to zero. The peculiar magnetic properties of $VF_2$ can be explained on the basis of almost pure 1-D helical magnetism. A broad maximum is observed for the thermal variations of the magnetic susceptibility and of the specific heat at about 40 K ($\simeq 6\ T_N$) and 27 K ($\simeq 4\ T_N$), respectively[22, 57)]. The very low anisotropy of the susceptibility (0.1%)[58)] justifies the choice of the Heisenberg model. Below 7 K, neutron diffraction experiments have revealed that in the ordered state, the spin arrangement has a wave vector Q of magnitude 1.06 $\pi/c$[58)]; The $V^{II}$ moments lie in the plane perpendicular to the c-axis and spiral with an angle of 96°. The larger exchange constant is the antiferromagnetic constant $J_1 \simeq -11$ K, and the ratio $J_2/J_1$, given by the relationship $J_2/J_1 = \cos(Qc/2)$, is about 0.1 (Table 2)[59)].

### 3.1.3.2 $KCuF_3$

The 1-D antiferromagnetic properties of $KCuF_3$, although unexpected at first sight from a crystallographic point of view, can nonetheless be easily explained on the basis of Jahn-Teller ordering (Sect. 3.1.1).

As shown by Okazaki, $KCuF_3$ exists in two stable polytypes with a tetragonal symmetry ($c/a < 1$) which may coexist over a large temperature range[60, 61)]. These polymorphs exist according to the alignment of the wave functions of $Cu^{II}$ ions (Fig. 9). The ratio of interchain (J′) to intrachain (J) exchange constants has been calculated from $kT_N/|J|$, using the Oguchi relationship[62)]. It can be seen in Fig. 9 that in the P 4 mbm space group the 1-D character is more pronounced than in the I 4/mcm group, the J′ value of the former being three times lower than that of the latter. Neutron diffraction experiments have shown that both forms display antiferromagnetism of type A with the spins lying in the ab plane[63)].

**Table 2.** Magnetic data of some $AX_2$ compounds

| Compound | Involved techniques | $T_N$(K) | Exchange constants (K) | $J_2/J_1$ | Ref. |
|---|---|---|---|---|---|
| $VF_2$ | neutron diffraction, | 7 | $J_1/k \simeq -11$; $J_2/k \simeq -1$ | 0.1 | 58 |
| | magn. measurements | | $J_1/k = -10.1$ | | 59 |
| $MnF_2$ | NMR, AFMR | 67.4 | $J_1/k = +0.2$; $J_2/k = -1.75 \pm 0.04$ | 9 | 2 |
| $MnO_2$ | $T_N/\theta$; pitch of the spiral | 84 | $J_1/k = -15$; $J_2/k = -9.5$ | 0.6 | 56 |
| $FeF_2$ | AFMR | 78.5 | $J_1/k = -0.46$; $J_2/k = -2.6$ | 6 | 2 |
| $CuF_2$ | magn. measurements | 69 | $J_1/k = -22$; $J_2/k = -34$ | 1.5 | 51 |

1-D correlations have been detected above $T_N$ by neutron diffraction, specific heat and NMR measurements mostly by Hirakawa and coworkers[64–70]. Magnetic susceptibility results have also been interpreted in terms of short-range couplings of a chain model on the basis of the calculations of Bonner and Fisher and of Oguchi. The main interaction occurs along the c-axis with $J/k = -190$ K.

The magnetic behavior is a consequence of antiferrodistortive ordering. The interaction between two half-filled $d_{x^2-y^2}$ orbitals occurs along the c-axis whilst intrachain coupling results from filled $d_{z^2}$-half-filled $d_{x^2-y^2}$ interactions in the ab plane.

The anisotropy of the susceptibility can be explained by the anisotropy of the g factor. Since only part of this anisotropy is due to the spin contribution to the magnetic moment, the Heisenberg model can serve as a basis. In addition, although the anisotropy field is 500 times stronger in the c-direction ($H_A^{II}$) than in the easy ab plane ($H_A^{I}$)[67,71], the value of $H_A^{II} = 2.5 \times 10^3$ Oe is only 0.1% of the exchange field ($H_E = 2.5 \times 10^6$ Oe) and thus sufficiently small to justify the choice of the Heisenberg model in preference to the planar one. A possible change in the nature of the anisotropy (from Ising to XY) slightly above $T_N$ has been considered[72]. Below $T_N$, the magnetization of the sublattice shows a 3-D behavior with $\beta = 0.35$ in the $1 \times 10^{-3} < 1\text{-}T/T_N < 0.1$ range and a decrease with a $T^2$ law for the low-temperature region[68].

A considerable reduction of the moment has been detected by NMR experiments, for instance $M_{exp.} = 0.55\ \mu_B$; $M_{theor.} = g\mu_B S \simeq 1.1\ \mu_B$[63,67]. These results will be discussed in Sect. 3.6.

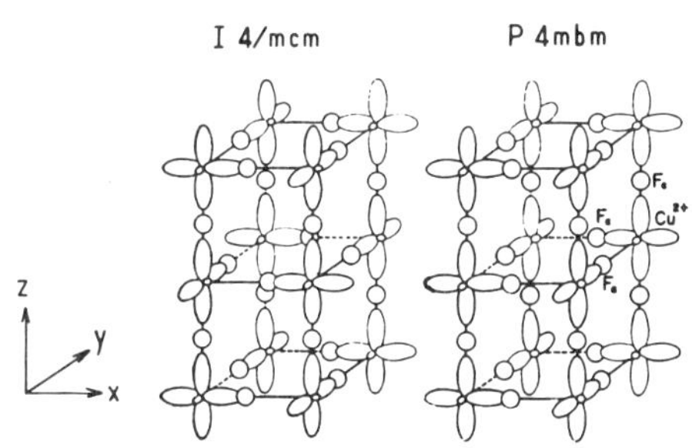


| Variety | I4/mcm | P4mbm |
|---|---|---|
| Lattice parameters | $a_0$=4.141 Å<br>$c_0$=3.923$_7$ Å | $a_0$ = 4.139$_5$ Å<br>$c_0$ = 3.930$_5$ Å |
| $T_{\chi max.}$ | 243 K | 243 K |
| $T_N$ | 38 K | 22 K |
| J/k | -190 K | -190 K |
| \|J'\|/\|J\| | 1.6x10$^{-2}$ (63)<br>2.7x10$^{-2}$ (68) | 0.6x10$^{-2}$ (63) |
| $g_{\vec{a}}$ (from ESR) | 2.28 | 2.25 |
| $g_{\vec{c}}$ | 2.17 | 2.17 |
| Magnetic moment per $Cu^{2+}$ ion | 0.54$\mu_B$ | 0.46$\mu_B$ |

**Fig. 9.** Orbital arrangements and parameters of the two varieties of $KCuF_3$

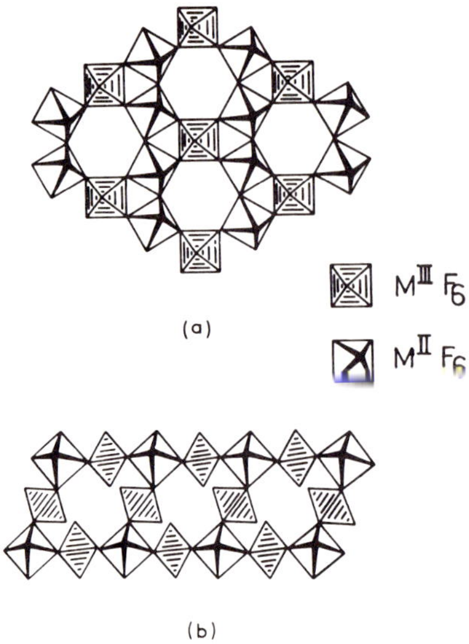


**Fig. 10 a, b.** Structure of weberite

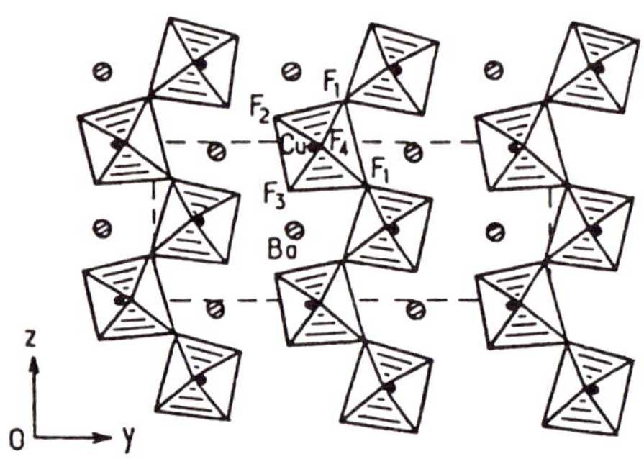


**Fig. 11.** Crystal structure of $BaCuF_4$

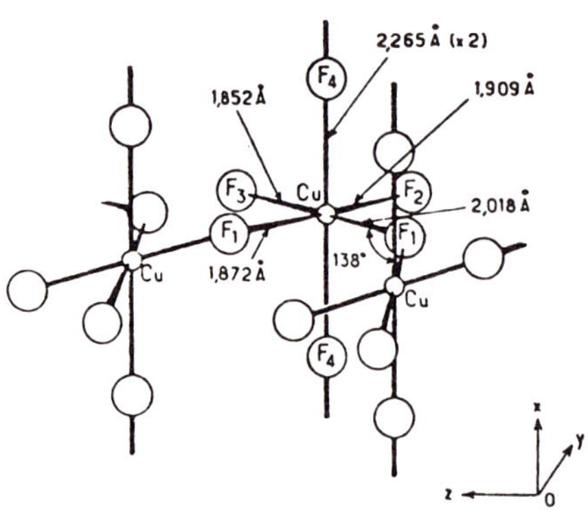


**Fig. 12.** Ferrodistortive ordering in $BaCuF_4$

#### 3.1.3.3 Weberites

A lowering in the dimensionality of the magnetic interactions can also be achieved via chemical substitutions, as for instance in the weberite series $Na_2M^{II}M^{III}F_7$. The structure of weberites, which is related to that of pyrochlore, can also be described as consisting of chains of octahedra containing the divalent $M^{II}$ cation which are linked by $(M^{III}F_6)^{3-}$ octahedra; the shared elements are corners (Fig. 10). When both $M^{II}$ and $M^{III}$ are transition elements, a 3-D ferrimagnetism occurs due to a $M^{II}$-F-$M^{III}$ superexchange mechanism ($T_C(Na_2NiFeF_7)$ = 90 K)[73]. When the chosen trivalent cation is diamagnetic, the paramagnetic $(M^{II}F_6)^{4-}$ groups form isolated chains. $Na_2NiAlF_7$ and $NaFeAlF_7$, for instance, display 3-D antiferromagnetic properties below 18 and 19 K, respectively, and a flattening of $\chi^{-1} = f(T)$ curves is observed up to 80 K[74].

### *3.1.4 1-D Antiferromagnetism in a Layered Fluoride: $BaCuF_4$*

This is again a particular case where the nature of the Jahn-Teller ordering has a marked influence on magnetic properties. In the series $A^{II}CuF_4$ (A = Ca, Sr, Ba) two kinds of structures are found. For A = Ca, Sr, the compounds crystallize in the $KBrF_4$ structure which exhibits a square coplanar coordination of the $Cu^{2+}$ ions with antiferrodistortive order, thus giving rise to paramagnetism down to 4.2 K; $BaCuF_4$ crystallizes in the $BaMnF_4$ structure consisting of puckered layers of corner-connected octahedra (Fig. 11). The Jahn-Teller ordering is of the ferrodistortive type, i.e. the long Cu-F bonds are all parallel to the $\vec{a}$-axis of the structure (Fig. 12). The only possible interaction pathway is Cu-$F_I$-Cu with an angle of 138°. No interaction is possible along the $\vec{a}$-axis (Cu-$F_4$-Cu interaction) as it involves two filled $d_{z^2}$ orbitals. Only very weak interactions may occur between the layers ($\vec{b}$ axis). This produces one-dimensional antiferromagnetism with $J/k \simeq -135$ K[75].

## *3.2 Magnetism in Layer-Type Fluorides*

### *3.2.1 Ferromagnetism in Layer-Type Fluorides*

As discussed in the preceding chapter, ferromagnetism may be induced by an ordering of the orbitals of Jahn-Teller ions. When the structural arrangement has 3-D character, the ferromagnetic layers are coupled antiparallel and antiferromagnetism is observed in the 3-D domain.

If the structure is anisotropic and consists of layers containing a transition element, the interaction between a partially filled and an empty orbital leads to ferromagnetic behavior even in the 3-D state. This is the case with compounds of formulae $A^I_2M^{II}F_4$ and $A^IM^{III}F_4$ which are derived from $K_2NiF_4$ and $TlAlF_4$ types, respectively. The involved transition ions are $Cr^{II}$, $Mn^{III}$, $Cu^{II}$. It should be noted that the $Ba_2ZnF_6$-type structure is also favorable for the establishment of 2-D ferromagnetism[76]; a positive value of $\theta_p$ has been found for $Ba_2CuF_6$ ($\theta_p = +15$ K)[77].

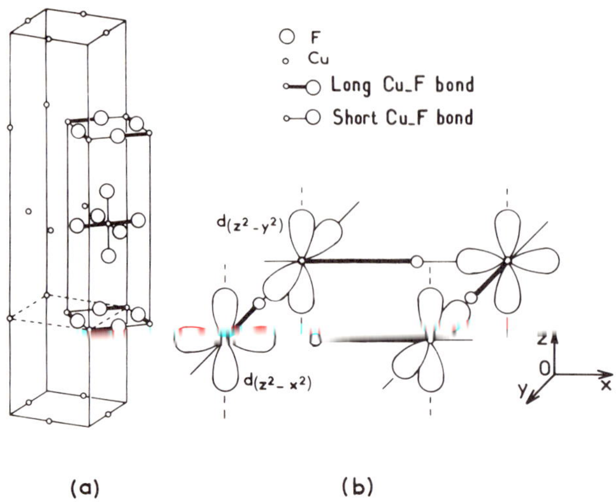


**Fig. 13a, b.** Structure of $K_2CuF_4$ (**a**) and arrangement of orbitals in the xOy plane (**b**)

### 3.2.1.1 Ferromagnetic Fluorides Derived from the $K_2NiF_4$ Type

The $A_2CuF_4$ series (A = K, Rb, Cs, $NH_4$, Tl) has been extensively studied by means of a large number of physical techniques such as magnetization, NMR, ESR, neutron diffraction, neutron scattering, spin-wave, specific heat, Faraday rotation, and optical measurements[78–88]. $A_2^IAgF_4$ compounds (A = K, Rb, Cs) show identical crystallographic features[89, 90].

It should be pointed out that, although the $A_2CrCl_4$ phases have been shown to be ferromagnetic[91], homologous fluorides $A_2CrF_4$ cannot be synthetized.

*The Structure of $K_2CuF_4$.* Light green $A_2CuF_4$ compounds are prepared by solid-state reaction and large transparent single crystals have been obtained by melting-zone techniques in a fluorinating atmosphere[92]. The structure was initially determined by Knox by analogy with $K_2NiF_4$[93]. Later, Babel pointed out that a complete indexing of the pattern required the existence of a larger unit cell[94]. The relationship between the cells is described in Fig. 13. The $(CuF_6)^{4-}$ octahedra show a tetragonal distortion (orthorhombic component superimposed) with the elongation axis being alternatively directed along the Ox and Oy axes[95].

Single crystal and neutron diffraction measurements have recently revealed that $K_2CuF_4$ possesses a multidomain structure due to displacements of F atoms[96, 97]. The symmetry of the compound has been deduced to be orthorhombic ($D_{2h}^{18}$-Bbcm, with $a \simeq b \simeq 5.87$ Å, $c = 12.75$ Å)[97] rather than tetragonal. This result is not supported by recent EPR works[274]. $F^-$ displacement is similar to that found in $KCuF_3$ and is attributed to the two possible orientations of $Cu^{II}$ $e_g$ orbitals.

*The Ferromagnetic Properties.* Ferromagnetism in $A_2CuF_4$ compounds initially found by Yamada[98] is consistent with the model of Khomskii and Kugel based on an electronic

**Table 3.** Magnetic data of $A_2CuF_4$ ferromagnetic compounds

| Compound | $T_C$ (K) | $\theta_p$ (K) | J/k (K) | J'/k (K) |
|---|---|---|---|---|
| $K_2CuF_4$ | 6.25 ± 0.03 | 16[98] | 11.2 high temp. expans. series[98]<br>8.8 specific heat[98]<br>11.4 spin wave[85] | 0.024 Oguchi method[80]<br>0.034 parallel pumping[101] |
| $Rb_2CuF_4$ | 17.5[100] | 13[79] | | |
| $Cs_2CuF_4$ | 9.8[86] | 14[86] | 8.3 high temp. expans. series[86] | |

ordering of $d_{z^2-x^2}$ and $d_{z^2-y^2}$ orbitals[99]. This model which takes into account the crystallographic results has been confirmed by NMR studies in the case of $Rb_2CuF_4$[100].

The interactions are of the ferromagnetic-Heisenberg type with 1% XY-square anisotropy arising from both anisotropic exchange and dipole-dipole interactions[99]. The xOz plane is the easy one and the spins are weakly bound in the Ox direction[80, 99].

The intralayer exchange constant J/k has been obtained from high-temperature expansion series and from the linear dependence of the spin-wave contribution on the heat capacity (Table 3). The ferromagnetic coupling between two layers has been shown for $K_2CuF_4$ to be $J'/k \simeq 0.03$ K[101]. The corresponding $|J'|/|J| \simeq 3 \times 10^{-3}$ is larger than those generally found in 2-D antiferromagnetic systems and thus a 3-D critical behavior may be expected. This hypothesis is also supported by the temperature dependence of the magnetization in the low-temperature range $0.12 < kT/J < 0.36$ which shows a $T^{3/2}$-variation law predicted by the spin-wave theory for a 3-D ferromagnet. The value of the critical exponent $\beta$, which also gives an estimate of the dimensionality of the interactions, has been calculated for $K_2CuF_4$ in the temperature range $0.85 < T/T_C < 0.99$ : $\beta = 0.22$[80, 99]. However this value, which is close to those found for other 2-D fluorides (antiferromagnetic $AFeF_4$ series, Sect. 3.2.2) does not seem to be conclusive ($\beta$(2-D) = 0.125; $\beta$(3-D Heisenberg) = 0.36). As a matter of fact, taking for $T_C$ the value determined by neutron diffraction measurements ($T_C = 6.32$ K, instead of 6.25 K) the fitting of the data to a power law does not yield a straight line and a unique value of $\beta$ cannot be obtained.

The system $K_2Cu_{1-x}Mn_xF_4$ has been studied using optical and magnetic measurements to obtain more information on the exchange constant[102, 103]. Therefore, $Mn^{2+}$ was

**Table 4.** Magnetic and ligand field parameters (at 4.5 K) ($4\delta_1$ and $3\delta_2$ are the splittings of the $^5E_g$ ground state and the $^5T_{2g}$ excited state, respectively) [from [107]]

| Compound | $T_C$ (K) | $\theta_p$ (K) | $4\delta_1$ (cm$^{-1}$) | $3\delta_2$ (cm$^{-1}$) |
|---|---|---|---|---|
| $KMnF_4$ | 6 | −45 | 15200 | 4300 |
| $RbMnF_4$ | <4.5 | −14 | ~15500 | ~4000 |
| $CsMnF_4$ | 21<br>18.9[108] | 27 | 15300 | 3900 |
| $NH_4MnF_4$ | 10 | −12 | ~14700 | ~3800 |

used as a probe at low concentrations (x < 100 ppm)[103]; the antiferromagnetic exchange constant for $Cu^{2+}$-$Mn^{2+}$ pairs has been deduced from the thermal variation of the optical transition of the exchange dipole, J/k = − 52 $cm^{-1}$ [103]. The thermal variation of J/k (15% from 290 to 100 K) is in good agreement with the value calculated from the modulation of the exchange by phonons[104]; such a variation has to be taken into account for J/k values derived from high-temperature data. The critical concentration for a ferromagnetic → paramagnetic transition has been estimated to be x ≃ 0.5 for the $K_2Cu_{1-x}Zn_xF_4$ system[105].

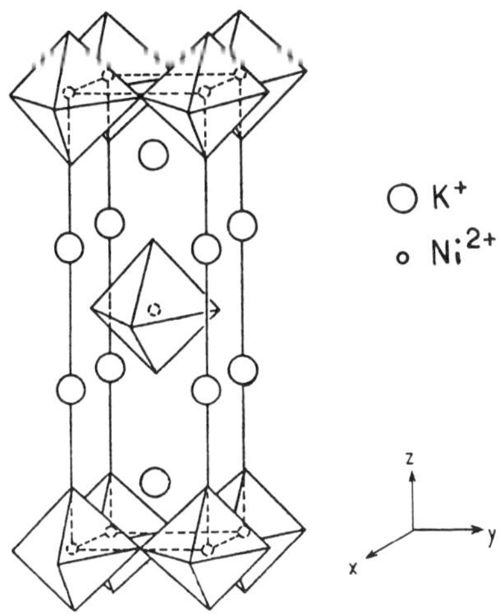


**Fig. 14.** Structure of $K_2NiF_4$

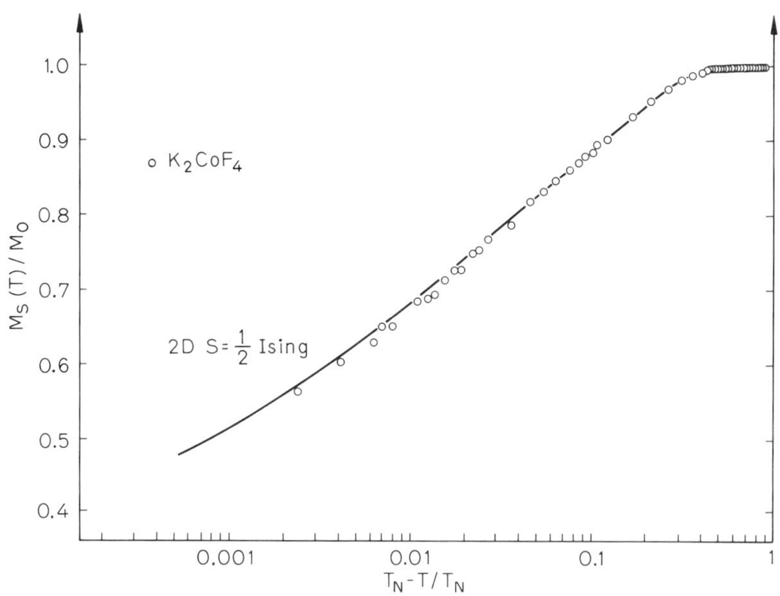


**Fig. 15.** Reduced sublattice magnetization as a function of the reduced temperature for $K_2CoF_4$. Full curve is the theoretical function of the 2-D Ising model

$Ag^{II}$ compounds such as $Rb_2AgF_4$ and $Cs_2AgF_4$ are ferromagnetic below 30 K, and $K_2AgF_4$ possesses a weak ferromagnetic component below $T_N \simeq 60$ K[89].

### 3.2.1.2 $A^IM^{III}F_4$ Ferromagnetic Fluorides

$A^IMn^{III}F_4$ compounds derived from the $TlAlF_4$-type structure and containing Jahn-Teller ions show ferromagnetic ordering[106, 107]. Magnetic and ligand field data are collected in Table 4.

The thoroughly investigated $CsMnF_4$ is isostructural with $CsFeF_4$ (tetragonal P4/nmm space group; a = 7.94 Å, c = 6.34 Å[108]). This space group would imply the presence of compressed $(MnF_6)^{3-}$ octahedra although this is very unlikely[39]. Instead of a ferrodistortive ordering of compressed octahedra, Massa and Steiner[108] proposed an antiferrodistortive order of elongated octahedra with an average phenomenon which could be due to three mechanisms:
- a planar dynamic Jahn-Teller effect,
- a disorder of the ordered layers along the perpendicular direction,
- a twinning of 90° of the two possible ordered states.

## *3.2.2 2-D Antiferromagnetic Fluorides*

### 3.2.2.1 Ising Systems

In this case, fluorides containing $Co^{2+}$ and $Fe^{2+}$ ions are the only known 2-D antiferromagnets which can be described by the Ising model.

*Fluorides with a $K_2NiF_4$ Structure.* $K_2CoF_4$ and $Rb_2CoF_4$ possess a $K_2NiF_4$ structure illustrated in Fig. 14. Their magnetic properties were first reported by Breed et al.[109]. The magnetic exchange between $Co^{2+}$ ions is highly anisotropic and the magnetic properties can be conveniently interpreted by means of a 2-D Ising model ($T_N = 107.8$ K, $J/k = -97$ K for $K_2CoF_4$; $T_N = 101$ K, $J/k = -91$ K for $Rb_2CoF_4$; in both cases $J_\perp/J_\parallel \simeq 0.3$). Neutron elastic and quasi-elastic scattering experiments performed on $K_2CoF_4$ show that the critical exponents ($\beta$, $\nu$, $\gamma$, $\eta$) coincide precisely with those obtained from Onsager's exact solution of the 2-D Ising model (Fig. 15)[110]. Recently, specific heat measurements have confirmed the above results[111].

$K_2FeF_4$ and $Rb_2FeF_4$ also crystallize in the $K_2NiF_4$ structure and have been studied by Wertheim et al.[112]. These authors concluded that the spins lie perpendicular to the Oz axis of the structure. By neutron diffraction Birgeneau et al.[113] studied the sublattice magnetization. They found a change in the critical exponent due to a transition from 2-D to 3-D character ($T_N = 56$ K). In fact, $Rb_2FeF_4$ is intermediate between the Ising and the planar Heisenberg model (Table 5).

*Fluorides with a $BaMnF_4$ Structure.* $BaFeF_4$ and $BaCoF_4$ exhibit a $BaMnF_4$-type structure previously described in Sect. 3.1.4. It consists of puckered sheets of $(CoF_6)^{4-}$ or $(FeF_6)^{4-}$ octahedra sharing four corners (Fig. 12). These sheets are separated by $Ba^{2+}$ ions[114].

According to Eibschutz et al. the susceptibility curves of $BaFeF_4$[115] show anisotropic behavior (Fig. 16) in the three directions of the crystal and indicate that the moments are parallel to the $\vec{b}$ axis. J/k has been found to be − 6.4 and − 7 K, using the Heisenberg model and the Ising evaluation for S = 2, respectively. The magnetic structure, whose unit cell is twice the crystallographic cell in the $\vec{b}$ and $\vec{c}$ direction has been determined by neutron diffraction measurements. The Mössbauer resonance indicates an internal field of 174 kOe at 4.2 K and the temperature dependence of the sublattice magnetization is expressed as follows:

$$\mathrm{Log}\,\frac{M(T)}{M(0)} = D\,\mathrm{Log}\left(1 - \frac{T}{T_N}\right)^{\beta}$$

(where D = 1.18 and $\beta$ = 0.158), both indicating an Ising tendency ($D_{theor.}$ = 1.22 and $\beta_{theor.}$ = 0.125). In fact, $BaFeF_4$ is intermediate between the two models and this seems to be the general case of $Fe^{2+}$ fluorides.

$BaCoF_4$ has the same structure as $BaFeF_4$ and its magnetic behavior is highly anisotropic. A broad maximum of $\chi$ is observed around 100 K for $H//\vec{a}$ and the three curves ($//\vec{a}$, $//\vec{b}$, $//\vec{c}$) show a maximum of the slope around 70 K. The latter temperature was shown by neutron diffraction to be the Néel temperature[116]. The exchange integral was evaluated to be around − 100 K, a value which is of the same order of magnitude as that for $K_2CoF_4$ or $Rb_2CoF_4$.

### 3.2.2.2 Heisenberg Isotropic Systems

*Fluorides with a $K_2NiF_4$ Structure.* Extensive studies have been carried out on the magnetic properties of compounds exhibiting a $K_2NiF_4$ structure. We will not list the results of all the measurements made, but only tabulate the relevant data of some compounds to facilitate their comparison.

In the case of $K_2NiF_4$, Lines compiled all the previous studies and accurately determined the exchange integral of this 2-D quadratic antiferromagnet by the application of the expansion series[117]. The obtained exchange constant (J/k ≃ − 50 K) appears to result from different sources and can be separated into a ferromagnetic interplane exchange, J′/k ≃ + 5 K, and negative intraplane exchange, J″/k ≃ − 60 K. In fact, differ-

**Table 5.** Magnetic parameters of $K_2NiF_4$-type compounds

| Compound | S | Type of interaction | $T_N$(K) | J/k(K) | Ref. |
|---|---|---|---|---|---|
| $K_2CoF_4$ | S′ = ½ | Ising | 107 | − 97 | 109 |
| $Rb_2CoF_4$ | | | 101 | − 91 | 109 |
| $K_2NiF_4$ | | | 97 | − 50 | 117 |
| $Rb_2NiF_4$ | 1 | Heisenberg | 91 | − 42 | 118 |
| $Tl_2NiF_4$ | | | 100 | − 45 | 119 |
| $K_2FeF_4$ | 2 | Ising + Planar | | | 112 |
| $Rb_2FeF_4$ | | Heisenberg | 56 | − 6.5 | 112 |
| $K_2MnF_4$ | 5/2 | Heisenberg | 42 | − 4.1 | 120 |
| $Rb_2MnF_4$ | | | 38 | − 3.3 | 121 |

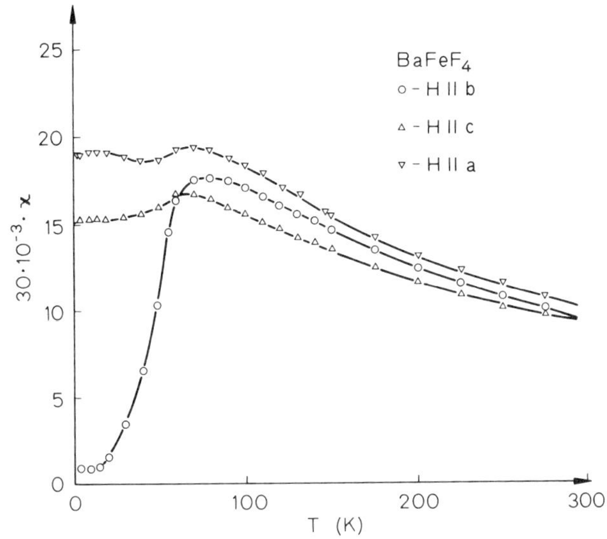


**Fig. 16.** Single-crystal susceptibility data of $BaFeF_4$

ing values of J/k were obtained by various authors due to the variety of measurement techniques[118]. Using neutron diffraction, spin-reduction of about 15% is observed and J'/J is shown to be less than $2 \times 10^{-4}$. NMR line shift and line width results indicate two different fluorine atoms (i.e. one with two Ni neighbors and the other with 4 Ni neighbors[119]. The critical behavior of the line width of these two $F^-$ atoms coincides with the predictions of Stanley and Kaplan[28] stating that there is a transition at about 100 K.

Isostructural $Tl_2NiF_4$ and $Rb_2NiF_4$ reveal a behavior similar to that of $K_2NiF_4$ (Table 5).

For S = 5/2, we may consider $K_2MnF_4$ and $Rb_2MnF_4$ which were extensively studied by Breed[120]. Once again, the $kT_N/J$ ratio is in good agreement with the predictions of Stanley and Kaplan (J/k = − 4.1 K for $K_2MnF_4$ and − 3.34 K for $Rb_2MnF_4$). A spin-flop phenomenon is observed in both compounds with critical fields around 55 kOe (cf. Sect. 3.5). An extensive review on the exchange interactions occurring in $A_2MF_4$ compounds has been published by De Jongh and Block[121].

*Fluorides with a $BaMnF_4$ Structure.* Two fluorides belong to this family: $BaMnF_4$ and $BaNiF_4$. $BaMnF_4$[114] has been studied by Holmes et al.[122]. They found an exchange constant of − 2.72 K and a spin-flop phenomenon for $H_c$ = 10.4 kOe. The anisotropy is described by the law $H_c^2 = 2 H_A H_E (E - \chi_{\parallel}/\chi_{\perp})^{-1}$ where $\chi_{\parallel}/\chi_{\perp} = 0.047$. The Néel temperature was determined by AFMR and ESR as $T_N$ = 25 K. The magnetic structure was determined by Cox et al. who showed the moments to be directed along the $\vec{b}$-axis[123]. $BaNiF_4$ susceptibility curves exhibit a broad maximum around 150 K and an antiferromagnetic order below 70 K. The high temperature fitting of the curve yields J/k = − 32 K. Neutron diffraction experiments reveal that the moments are directed along the $\vec{b}$-axis as for $BaMnF_4$[124].

Compounds belonging to the $BaMF_4$ series display both ferroelectric (M = Co, Ni, Mg, Zn) and pyroelectric properties (M = Fe, Mn) but have estimated Curie temperatures above their melting points[125, 126].

*$NaM^{III}F_4$ Compounds (M = Cr, Fe).* $NaMF_4$ compounds are isostructural with $NaNbO_2F_2$ whose structure was determined by Anderson and Galy[127–129]. Their structure (Fig. 17) consists of puckered sheets of $MF_6$ octahedra identical with those found for the $BaMnF_4$ structure (compare Fig. 17 with Fig. 11).

$NaFeF_4$ was studied by Dance et al. who applied susceptibility, neutron diffraction and Mössbauer resonance measurements[130]. J/k is equal to $-11.5$ K and the Néel temperature $T_N = 111.5$ K. The variation of the hyperfine field with temperature for $0.6 < T/T_N < 1$ gives $\beta = 0.25$, indicating the presence of 2-D and 3-D correlations.

The magnetic curve of $NaCrF_4$ obtained by Knoke and Babel[131] includes a broad maximum around 15 K and a possible 3-D order at about 8 K. If one assumes 2-D behavior above $T_N$, calculation yields $J/k \simeq -2$ K.

*Fluorides with the General Formula $AFe^{III}F_4$ ($A = K, Rb, Cs, NH_4$).* The main features of these compounds are layers of $FeF_6$ octahedra sharing four corners but some differences exist in the stacking of the layers between $KFeF_4$ and the other $AFeF_4$ compounds (Rb, Cs, $NH_4$).

The orthorhombic structure of $KFeF_4$ determined by Heger et al.[132] consists of quasi-quadratic layers; if one considers two such successive layers, the upper one is shifted by the translation a/2 $\langle 1\ 1\ 0 \rangle$ out of line of the lower one (Fig. 18). This arrangement is the same as in the $K_2NiF_4$ structure for two successive layers. The 2-D behaviour was studied by Heger et al. They found a Néel temperature of 131 K. The value of the broad maximum of the susceptibility curve leads to $J/k = -12.4$ K. Neutron diffraction data indicate a three-dimensional magnetic structure with the moments being parallel to the $\vec{c}$-axis. Mössbauer investigations by Eibschutz et al.[133] yield a value of the critical exponent of $\beta = 0.185$.

The structures of compounds of the $AFeF_4$ series (Rb, Cs, $NH_4$) are derived from the $TlAlF_4$ structure[134–136] which consists of planar sheets of $FeF_6$ octahedra separated by the alkali ions, the sheets being situated exactly on top of one another (Fig. 19). Magnetic

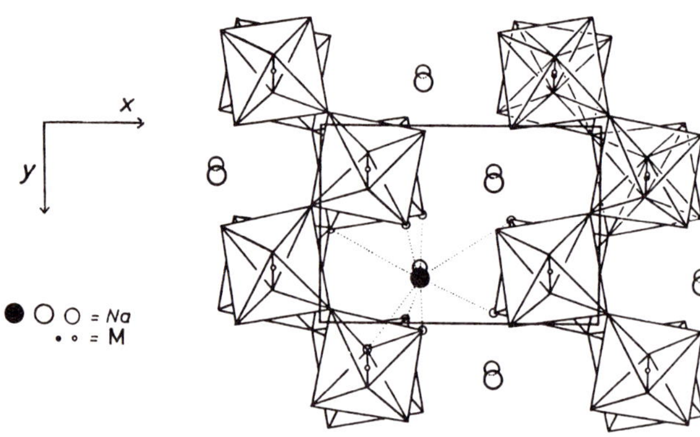


**Fig. 17.** Structure of $NaCrF_4$

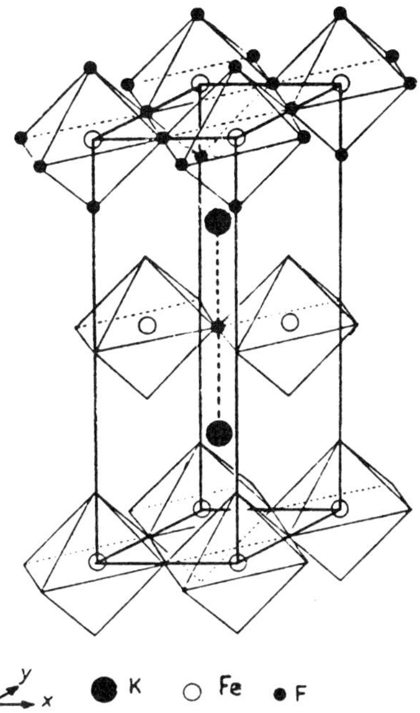


**Fig. 18.** Structure of $KFeF_4$

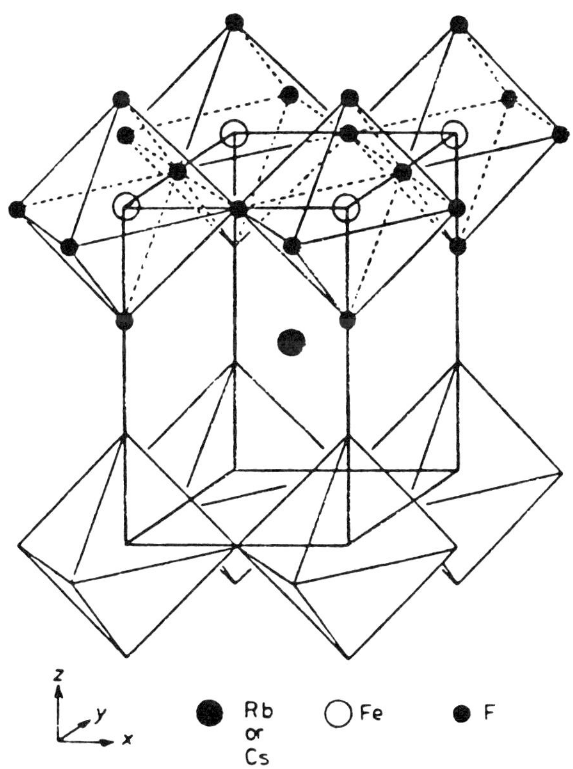


**Fig. 19.** Structure of $RbFeF_4$

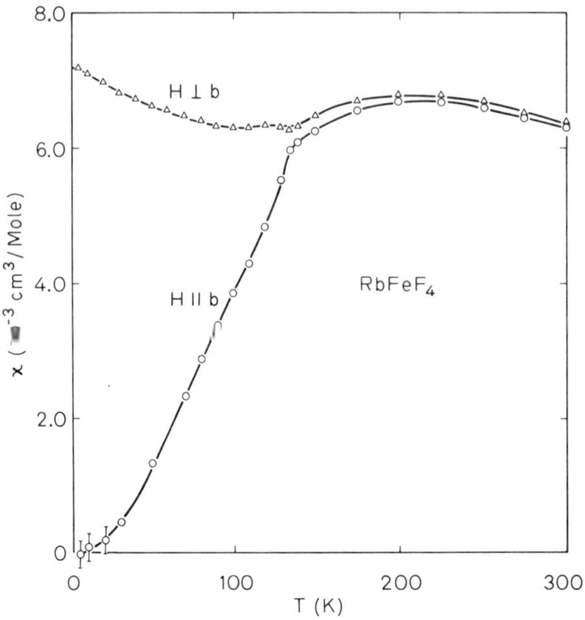


**Fig. 20.** Single-crystal susceptibility data for $RbFeF_4$

susceptibility, Mössbauer resonance and neutron diffraction measurements have been performed on $RbFeF_4$[137–139] giving $J/k = -12$ K and $T_N = 133$ K (Fig. 20). At low temperature, the results of Mössbauer resonance measurements agree with the simple two-dimensional spin-wave theory. For $0.4 < T/T_N < 1$, the $(1 - T/T_N)^\beta$ law fits the curve for $\beta = 0.245$. On single crystal, the Mössbauer effect indicates that the hyperfine field forms an angle of 16.5° with the electric field gradient tensor[138]. The hyperfine field is directed along the $\vec{c}$-axis. Neutron diffraction measurements indicate that the moments are parallel to $\vec{c}$[133]. $CsFeF_4$ behaves exactly like $RbFeF_4$ and Eibschutz et al.[139] obtained $T_N = 160$ K and $J/k = -11.2$ K from the value of the temperature of maximum susceptibility.

The Mössbauer results for $0.006 < 1 - T/T_N < 0.69$ indicate a $\beta$ value of 0.278 which coincides better with the 3-D than with the 2-D model[140]. A possible explanation of this could once again involve the presence of three-dimensional correlations even above $T_N$. The magnetic 3-D structure is the same as for $RbFeF_4$, i.e. with the spins parallel to the $\vec{c}$-axis.

**Table 6.** Magnetic parameters of $AFe^{III}F_4$ compounds

| A | Interlayer distance (Å) | $T_N$ (K) | J/k (K) | Ref. |
|---|---|---|---|---|
| Na | 6.24 | 111.5 | – 11.5 | 130 |
| K | 6.44 | 137 | – 12.4 | 132 |
| Rb | 6.25 | 133 | – 12 | 137 |
| $NH_4$ | 6.36 | 135 | – 13 | 141 |
| Cs | 6.56 | 160 | – 14.2 | 139 |

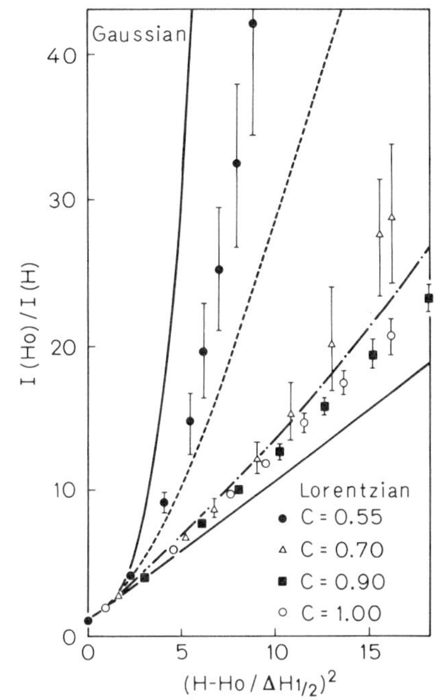


**Fig. 21.** EPR line shape of $K_2Mg_{1-c}Mn_cF_4$ at $\theta = 0°$. Dashed and dashed-dot curves are theoretical curves for 1-D and 2-D systems. Full curves are Gaussian and Lorentzian theoretical curves[143)]

More recently, $NH_4FeF_4$ was studied by susceptibility and Mössbauer resonance measurements[141)]. From an expansion series, a value of − 13 K was obtained for J/k and $T_N$ = 135 K from Mössbauer data. The law of variation of the hyperfine field with temperature has the form

$$\text{Log}\, H_{(T)}/H_{(0)} = D\, \text{Log}\, (1 - T/T_N)^{\beta}$$

with D = 1.06 and $\beta = 0.26$, in good agreement with the other fluorides. It should be noted that $T_N$ and J/k increase with the distance between the layers (Rb → $NH_4$ → Cs) which is somewhat unusual (Table 6). A possible explanation is the presence of stacking defects which may be more numerous in the cesium compound.

*Percolation Processes in $A_2MF_4$ Series.* Random two-dimensional antiferromagnets have recently been studied; besides magnetic measurements, EPR has proved to be a powerful tool in the elucidation of percolation processes.

Breed et al.[142)] studied the variation of $T_N$ and of the perpendicular susceptibility with x in the system $K_2Mg_{1-x}Mn_xF_4$. The percolation threshold is shown to be 0.50 in this system compared to 0.56 for the Ising system $K_2Mg_{1-x}Co_xF_4$. The systems $K_2Mg_{1-x}Mn_xF_4$ and $Rb_2Mg_{1-x}Mn_xF_4$ have been studied by EPR. The minimum of the line width is observed at the "magic angle" $\theta \simeq 55°$, $\theta$ being the angle between the $\vec{c}$-axis and the direction of the magnetic field. When the concentration varies, the line shape, which was almost Lorentzian for x = 1, becomes practically Gaussian for x = 0.55 (below the percolation threshold) (Fig. 21)[143, 144)]. Concerning the $Rb_2Mg_{1-x}Mn_xF_4$ system, particular attention has been paid to the temperature dependence of the line width near the percolation threshold x = 0.59. At high temperatures, the line width is dominated by the $(3 \cos^2\theta - 1)^2$ variation, due to dipolar interactions in 2-D systems. For $T < 30$ K, antiferromagnetic correlation effects govern the line width. At 1.6 K a nine-line hyperfine structure appears suggesting the critical slowdown of the spin dynamics at the scale of EPR. The hyperfine structure is believed to be due to isolated $Mn^{2+}$ ions interacting with their next-nearest neighbors locked into clusters[145)].

*Double-Layer Fluorides.* The structure of these fluorides is derived from the $K_2NiF_4$ structure and consists of sheets of two adjoining quadratic layers (Fig. 22); $K_3Mn_2F_7$ and $Rb_3Mn_2F_7$ are known to crystallize in this structure. In this case, the number of magnetic interacting nearest neighbors is five. Susceptibility experiments on powders and single crystals have been performed for both $K_3Mn_2F_7$ and $Rb_3Mn_2F_7$[146, 147)]. At low temperatures, the renormalized spin-wave theory yields J/k = −3.88 K and in the paramagnetic region the application of the high-temperature expansion series gives J/k = −4.04 K for $K_3Mn_2F_7$. The Néel temperature is 58 K and the broad maximum above $T_N$ in less pronounced, indicating that short-range correlations are less predominant (J/k = −3.56 K for $Rb_3Mn_2F_7$).

Magneto-optical properties of $K_3Ni_2F_7$ have been studied by Ferguson et al., who have reported intermediate results between those of $KNiF_3$ and $K_2NiF_4$[147bis)]. A large minimum of the inverse susceptibility curve of $K_3Fe_2F_7$ is observed around 110 K. Below 40 K, a weak ferromagnetism appears[244)].

*$Ba_2NiF_6$.* The structure of $Ba_2NiF_6$ is also characterized by layers of corner-connected $NiF_6$ octahedra separated by alternate layers of barium and fluorine atoms (Fig. 23)[77)]. As the magnetic layers are relatively far apart compared to those in $K_2NiF_4$, $Ba_2NiF_6$ has been suggested to be a stronger 2-D antiferromagnet than $K_2NiF_4$. The susceptibility curve shows a broad maximum and a rather important anisotropy. $T_N$ is evaluated to be

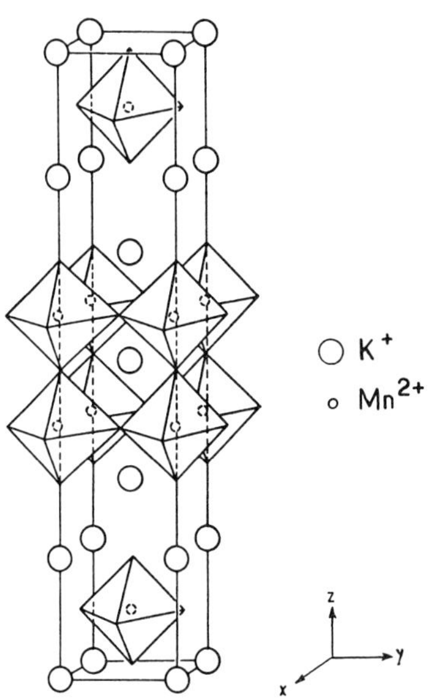


**Fig. 22.** Structure of $K_3Mn_2F_7$

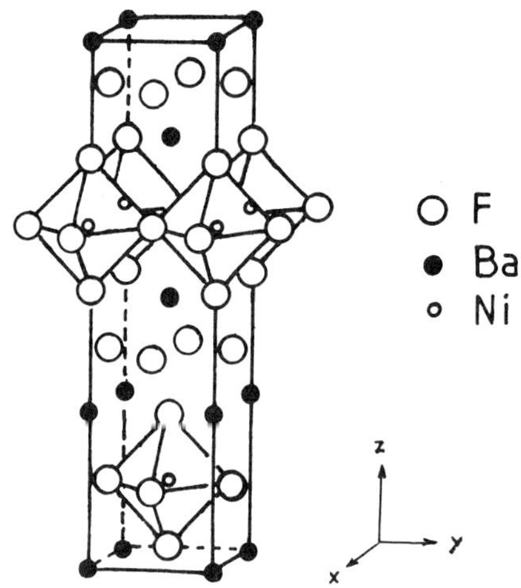


**Fig. 23.** Structure of $Ba_2NiF_6$

93 K and the exchange constant has been determined from the value of $T_{\chi\,max}$ as $J/k = -87.5$ K. The anisotropy field is higher than in $K_2NiF_4$; moreover, $H_A/H_E = 0.03$ compared with $H_A/H_E = 0.002$ for $K_2NiF_4$[148].

*$CuF_2$. 2 $H_2O$.* This hydrate has a structure consisting of antiferromagnetic layers of $CuF_2O_2$ groups parallel to the (101) plane[149]. Indirect interactions are made possible by long Cu-F bonds. The fitting of the susceptibility curves via expansion series yields $J/k = -13$ K. There is a sharp peak in the specific heat curve[150], indicating that $T_N \simeq 11$ K. The ratio J′/J is equal to $1.5 \times 10^{-2}$.

## 3.3 Magnetism in Chain Fluorides

### 3.3.1 1-D Ferromagnetism in $CsNiF_3$

The structures of $AMF_3$ hexagonal polytypes of the perovskite can be deduced from the various arrangements of the cationic and anionic layers[151, 152]. These phases are characterized by the number of layers (L) constituting the stacking sequence. Besides the cubic perovskite, the 6 L, 9 L, or 12 L structural types consist of a 3-D arrangement of $MF_6$ octahedra linked by corners and faces. The two-layer (or 2 H) structure-H for hexagonal – is unique in that it consists of infinite chains of $(MF_6)$ octahedra sharing two faces and separated by monovalent cations[6, 153].

The 2 H structure is illustrated by the compounds $BaNiO_3$, $CsNiCl_3$ and $CsVX_3$ (X = Cl, Br, I)[154–156]. The only fluoride showing these crystallographic features is $CsNiF_3$, first prepared by Babel[153]. This compound crystallizes in the hexagonal $P6_3/mmc$ space group with the lattice constants a = 6.236 Å, c = 5.225 Å. While the six Ni-F distances are equivalent, there exist two different F-F spacings leading to a trigonal distortion of the octahedra which are elongated in the $\vec{c}$-direction[153]. There are two significantly different Ni-Ni distances: in a chain Ni-Ni = c/2 = 2.61 Å and between two

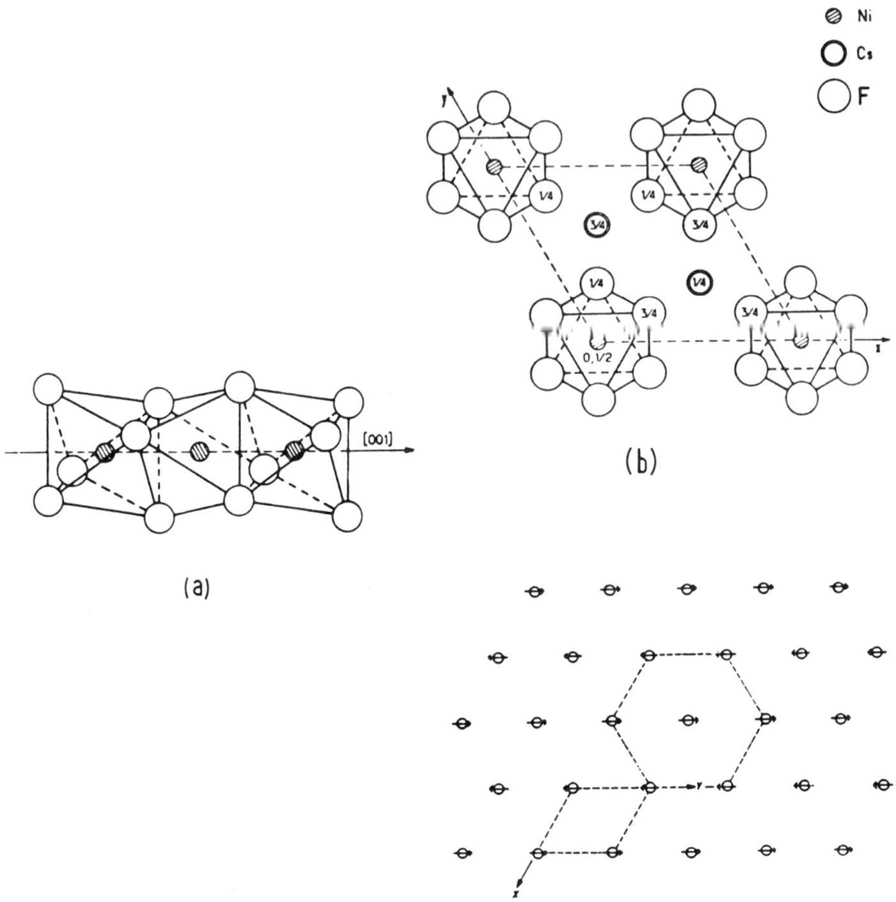


**Fig. 24 a–c.** Structure of $CsNiF_3$ (**a** and **b**) and 3-D magnetic structure below $T_N$ (**c**)

nearest chains Ni-Ni = a = 6.24 Å, this being responsible for the strong 1-D character (Fig. 24).

$CsNiF_3$ has been widely studied by many physical methods in the temperature range between 1.2 and 300 K. It is an interesting compound since it is one of the few 1-D materials consisting of ferromagnetic chains. An extensive review article has been published by Steiner, Villain and Windsor[8] so that a further amplification of the topic is unnecessary.

### 3.3.1.1 High-Temperature Behavior: 1-D Ferromagnetism

Neutron diffraction and susceptibility measurements have shown that the main interaction along the chain is ferromagnetic and that this interaction occurs over a large temperature range[157–159]. Single-ion anisotropy favors an orientation of the spin within the plane perpendicular to the chain direction. While the difference in anisotropy is large between the $\vec{c}$- and $\vec{a}$-directions, it cannot be detected in the ab plane and the magnetic properties of $CsNiF_3$ can be treated using the easy-plane Heisenberg model[8)].

At high temperatures ($70 < T < 300$ K), the magnetic susceptibility follows a Curie-Weiss law with $\theta_p = +22$ K. Applying the high-temperature expansion series, the exchange constant is obtained[160, 161)] (Fig. 25).

In Table 7 are compiled the values of the intrachain constant J/k and the single-ion anisotropy parameter D/k. Although there are some discrepancies, recent measurements[167, 168)] seem to be in agreement with a value of D close to that of J.

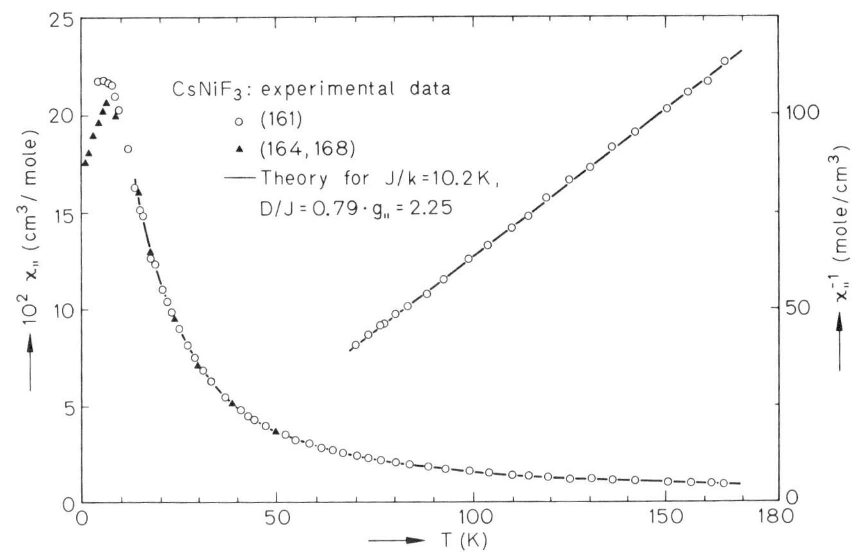


**Fig. 25.** Single-crystal susceptibility data of $CsNiF_3$

**Table 7.** Experimental J/k and D/k parameters of $CsNiF_3$ [from [168)]]

| J/k (K) | D/k (K) | Experimental technique | Ref. |
|---|---|---|---|
| $8.3 \pm 0.8$ | | susceptibility (powder) | 159 |
| $11.8 \pm 0.3$ | $4.5 \pm 2$ | spin-wave dispersion relation (neutron scattering) | 162 |
| 12 | | magnetization | 163 |
| $7.9 \pm 0.4$ | $8.1 \pm 2$ | susceptibility (crystal) | 164 |
| | $4.7 \pm 0.2$ | infrared ferromagnetic resonance | 165 |
| $9.9 \pm 0.2$ | $7.5 \pm 0.2$ | susceptibility (high temperature expansion series) | 160 |
| $6.5 \pm 1.5$ | 1.5 | susceptibility | 166 |
| $11.5 \pm 0.05$ | $8.9 \pm 0.2$ | spin waves | 167 |
| $10 \pm 0.5$ | $8.5 \pm 0.5$ | susceptibility (crystal) | 168 |

### 3.3.1.2 Low-Temperature Behavior: 3-D Antiferromagnetism

Using neutron diffraction and specific heat measurements, the 3-D transition temperature has been found to be 2.65 K[159, 169]. The magnetic moments lie in the ab plane; the 3-D antiferromagnetic structure consists of ferromagnetic planes parallel to the c-axis and coupled antiferromagnetically (Fig. 24). Under a low applied field, the 3-D structure is destroyed and the involved processes correspond to variations (increase or decrease) of the domains (see Sect. 3.5)[170]. The magnetic moment per unit $Ni^{2+}$ ion is $M_{0K} = 2.25 \pm 0.03\ \mu_B$[158, 169].

The influence of the interchain interaction has been treated on a theoretical basis. If chains are considered as a succession of magnetic units of length $1/\varkappa$ and spin $S/\varkappa$, the interaction between these is approximately $J'S^2/\varkappa$, and the 3-D ordering temperature corresponds to $k_B T_{ord.} \simeq |J'|\, S^2/\varkappa$. At low temperatures, for an Heisenberg chain, $\varkappa \simeq k_B T/2\, JS^2$ and therefore $k_B T_{ord.} \simeq S^2 \sqrt{(|J|\ |J'|)}$. Results obtained from this relationship are in good agreement with Green's calculations[171, 172]. For $T_N = 2.63$ K and $S = 1$, the exchange constants are $J/k = 12$ K, $J'/k \simeq 0.14$ K and $J'/J \simeq 10^{-2}$.

With regard to 1-D systems, it can be clearly stated that, in contrast to 2-D systems where a 2-D phase transition has been considered by Stanley and Kaplan among others[28, 78], the 1-D short-range order ⇆ 3-D long-range order transition only shows 3-D properties[7].

### *3.3.2 1-D Heisenberg Antiferromagnetic Chains*

Whereas $CsNiF_3$ exhibits chains of octahedra sharing faces, $A^{II}M^{III}F_5$ and $A^{I}_2M^{III}F_5$ series are characterized by $MF_6$ octahedra connected by corners – or eventually edges – to form infinite chains. Although $A^{I}M^{IV}F_5$ phases have been synthesized[173–176], their crystallographic and physical properties have not yet been clearly described.

### 3.3.2.1 Structures of $A^{II}M^{III}F_5$ and $A^{I}_2M^{III}F_5$ Compounds

*Structural Features of $AMF_5$ Compounds.* The arrangement of the $(MF_5)_n^{2n-}$ infinite chains has been correlated with the ratio $r_{A^{II}}/r_{M^{III}}$ by Von der Mühll and Ravez[177]. The different types are compiled in Fig. 26 and their main structural features are given in Fig. 27.

– $r_{A^{II}} \leqslant 1$ Å

The crystallographic network of $A^{II}M^{III}F_5$ compounds is composed of $MF_6$ octahedra linked by opposite (trans-)corners[178]. Five structural types have been found: $MnAlF_5$[179], $Cr_2F_5$[180], $MnCrF_5$[181], $CaFeF_5$[182], and $CaCrF_5$[183]. Strong relationships have been established between these phases and structural mechanisms proposed[177, 178, 183]. Concerning the first two structures ($MnAlF_5$ and $Cr_2F_5$), the $(MF_5)_n^{2n-}$ chains are almost linear and give rise to an octahedral surrounding of the $A^{II}$ cation (distorted for $Cr^{II}$ ions). The structure can also be described in terms of chains of $AF_6$ octahedra sharing edges.

The other three structures are composed of zig-zag $(MF_5)_n^{2n-}$ chains, the anionic surrounding of the $A^{II}$ cation being a pentagonal bipyramid. These polyhedra are also

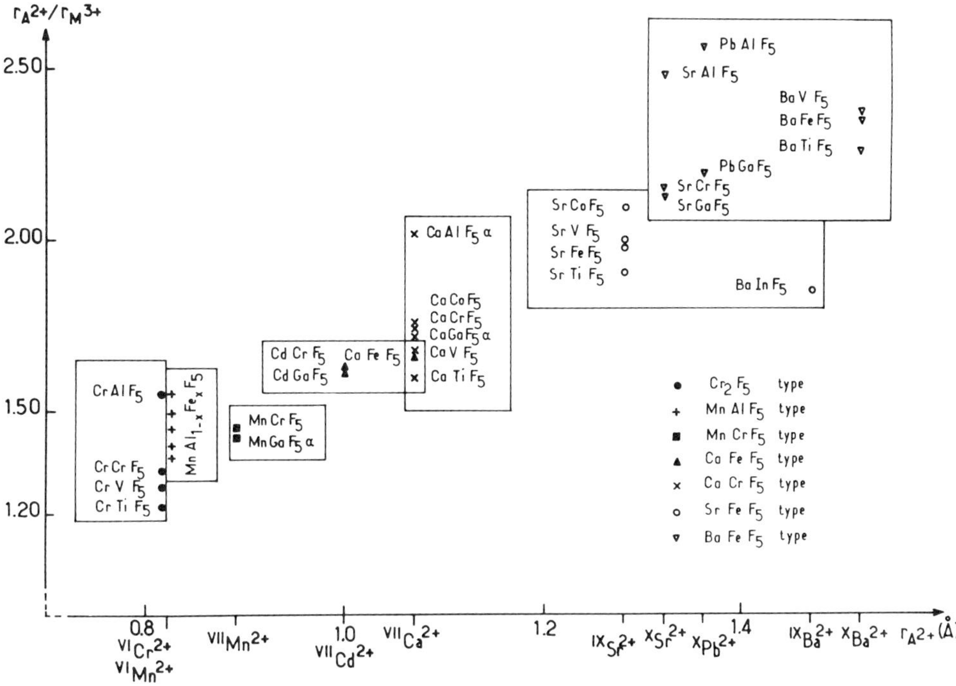


**Fig. 26.** $AMF_5$ structural types

connected by edges, thus forming infinite chains. $MnCrF_5$ is related to the $CaCrF_5$ structural type and the difference between $CaCrF_5$ and $CaFeF_5$ can be explained on the basis of the position of $A^{II}$ elements, half of the Ca file being shifted by c/2.

– $r_{A^{II}} > 1$ Å

For $r_{A^{II}} > 1$ Å and $r_{A^{II}}/r_{M^{III}} < 1.75$, the $SrFeF_5$ structure is composed of helicoïdal $(FeF_5)_n^{2n-}$ chains with corner-shared octahedra in the cis-position[184].

For $r_{A^{II}} > 1$ Å and $r_{A^{II}}/r_{M^{III}} \geqslant 1.75$, i.e. for large divalent cations associated with small trivalent ones, the $BaFeF_5$ structure shows two distinct types of chains: a simple one and a branched one[185]. It should be pointed out that for a ratio $r_{A^{II}}/r_{M^{III}} \geqslant 2.8$ ($\alpha$ and $\beta$ $BaAlF_5$, $BaBF_5$)[186, 187], or for larger $M^{III}$ cations ($BaTlF_5$)[188], the structures are no more composed of chains.

The structural features observed both with increasing $r_{A^{II}}$ and $r_{A^{II}}/r_{M^{III}}$ ratio has been correlated with an increase in the coordination number of the divalent cation (Fig. 27). The latter varies from six in $MnAlF_5$ and $Cr_2F_5$ to seven in $MnCrF_5$, $CaFeF_5$ and $CaCrF_5$; in $SrFeF_5$ its coordination number is nine whereas it is nine and eleven in $BaFeF_5$.

*Different Structures of $A^I_2M^{III}F_5$ Compounds.* The crystallographic data of $A_2MF_5$ compounds are compiled in Table 8. It should ne noted that the phases $Li_2MF_5$ and $Na_2MF_5$ are unusual and have only been reported for M = Cr, Mn without precise structural information[6, 189].

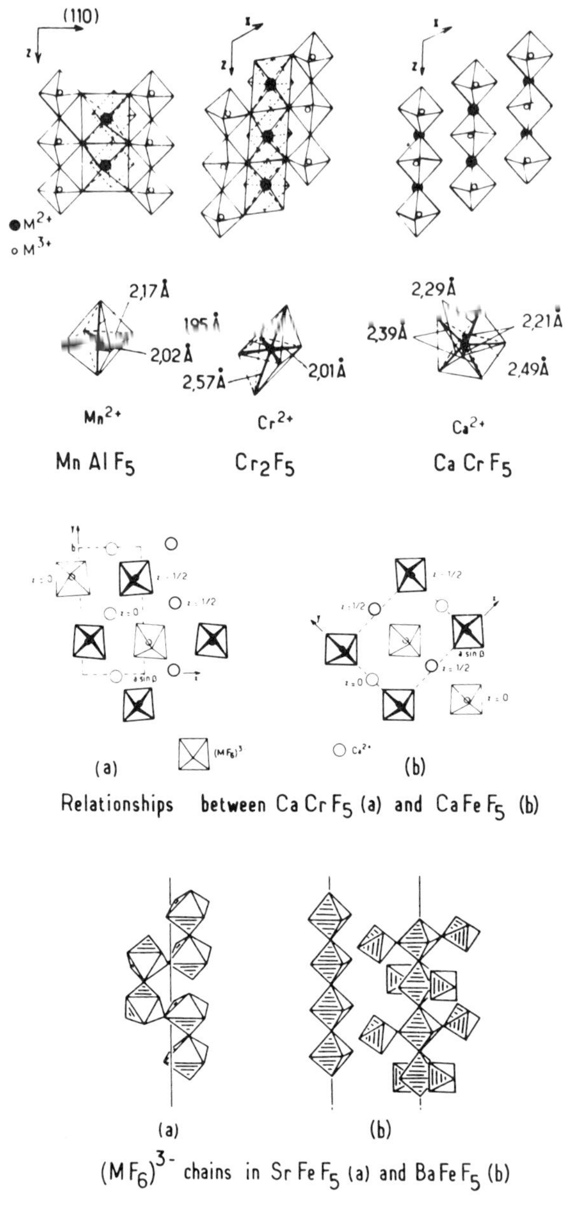


**Fig. 27.** Structural features of some $AMF_5$ fluorides

Ten years ago, all the $A_2MF_5$ fluorides were assigned to the $Tl_2AlF_5$ type (Brosset)[190]. In fact, three main structures are presently known for these compounds: $K_2FeF_5$[191], $Rb_2CrF_5$[192] which is isostructural with $K_2VO_2F_3$[193], and of course $Tl_2AlF_5$[190] (Fig. 28). For larger trivalent cations, another chain structure has been found by Bochkova et al., namely the $K_2SmF_5$ structure[194, 195]. The coordination number of $Sm^{III}$ is seven, and the $(SmF_5)_n^{2n-}$ chains are separated by potassium ions with the coordination number eight; magnetic investigations on these phases have not been carried out.

**Table 8.** Cristallographic data of $A_2MF_5$ fluorides (chain direction underlined; —: no structural data available)

| A \ M | Al | V | Cr | Mn | Fe | Ga |
|---|---|---|---|---|---|---|
| K | Pna $2_1$<br>$\underline{a}$ = 19.6 Å<br>b = 12.6 Å<br>c = 7.1 Å<br>(Z = 16)[204] | —[196, 197] | Pna $2_1$<br>$\underline{a}$ = 19.60 Å<br>b = 12.84 Å<br>c = 7.37 Å<br>(Z = 16)[198] | | Pna $2_1$<br>$\underline{a}$ = 20.39 Å<br>b = 12.84 Å<br>c = 7.40 Å<br>(Z = 16)[191] | Pna $2_1$<br>$\underline{a}$ = 20.29 Å<br>b = 12.83 Å<br>c = 7.34 Å<br>(Z = 16)[194, 205] |
| Rb | | —[197] | Pnma<br>a = 7.515 Å<br>$\underline{b}$ = 5.724 Å<br>c = 11.985 Å<br>(Z = 4)[192] | P 4/m, P 4/mmm<br>a = 6.10 Å<br>$\underline{c}$ = 4.14 Å<br>(Z = 1)[199] | Pnma<br>a = 7.53 Å<br>$\underline{b}$ = 5.78 Å<br>c = 11.985 Å<br>(Z = 4)[201–203] | Pnma<br>a = 7.49 Å<br>$\underline{b}$ = 5.77 Å<br>c = 11.96 Å<br>(Z = 4)[194, 205] |
| Cs | | —[197] | | | Pnma<br>a = 7.84 Å<br>$\underline{b}$ = 5.95 Å<br>c = 12.57 Å<br>(Z = 4)[203] | |
| Tl | C $2_1$22<br>$\underline{a}$ = 7.46 Å<br>b = 10.06 Å<br>c = 8.24 Å<br>(Z = 4)[190, 154] | | | | | |
| $NH_4$ | | —[197] | | Pnma<br>a = 6.20 Å<br>$\underline{b}$ = 7.94 Å<br>c = 10.72 Å<br>(Z = 4)[200] | | |

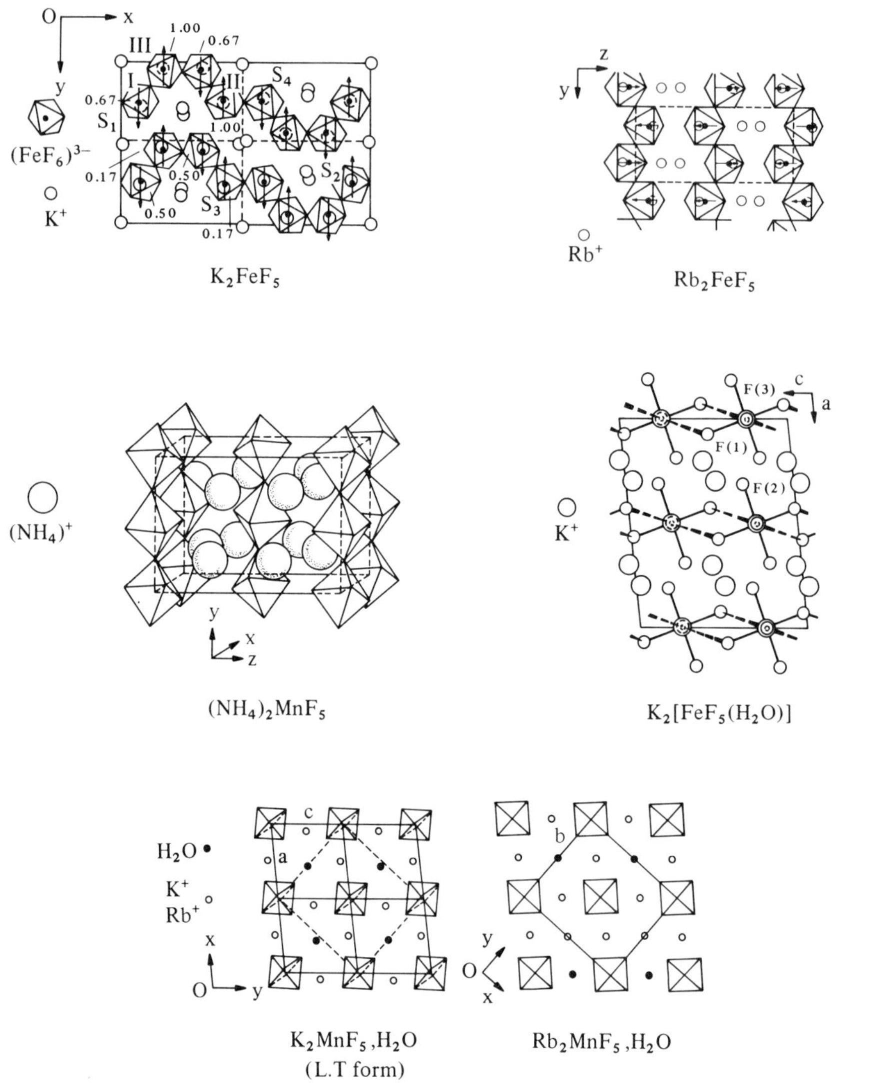


**Fig. 28.** Crystal structures of some $A_2MF_5$ and $A_2MF_5 \cdot H_2O$ compounds

$K_2MF_5$ compounds crystallize in the $K_2FeF_5$ structure of orthorhombic space group $Pna2_1$ (Vlasse et al.)[191]. The trivalent cation is octahedrally surrounded by fluorine atoms, the $MF_6$ octahedra being connected by cis-corners to form infinite zig-zag chains parallel to the $\vec{a}$-axis (Fig. 28). In $K_2FeF_5$, the octahedra are characterized by two different Fe-F bonds: $4\,Fe\text{-}F_{terminal} = 1.88$ Å and $2\,Fe\text{-}F_{bonding} = 2.02$ Å. The Fe-F-Fe angles lie between 162 and 173°. Assuming the chains to be connected with one another, the $K^+$ ions have a coordination number of nine and ten.

For larger monovalent cations, two crystallographic arrangements are observed. The fluorides $Rb_2MF_5$ (M = Cr, Fe, Ga) exhibit the orthorhombic space group Pnma

(Z = 4). Their structure consists of linear chains of $MF_6$ units linked by cis-corners. Here also, the M-$F_{bond.}$ distances are longer than the M-$F_{term.}$ distances. The M-F-M angles are close to 180° and the coordination number of $Rb^+$ ions is nine and ten (Fig. 28). $Cs_2FeF_5$ is derived from this structure.

Although the space groups of $Rb_2MnF_5$[199], $(NH_4)_2MnF_5$[200], and $Tl_2AlF_5$[190] (Table 8) are different, these compounds nevertheless have some common features: the $MF_6$ octahedra are bridged by trans fluorine atoms, the M-F-M angles varying from 143° ($(NH_4)_2MnF_5$) to 180° ($Rb_2MnF_5$). In $A_2MnF_5$ compounds, the octahedra are elongated along the chain direction due to the $Mn^{3+}$ Jahn-Teller ion.

*Hydrates.* Two hydrates $AMF_5 \cdot nH_2O$ containing mixed-valency Fe are known: $Fe_2F_5 \cdot 2H_2O$ and $Fe_2F_5 \cdot H_2O$[206, 207]. The water molecules participate in the coordination polyhedra of $Fe^{II}$ ions, thus forming $(Fe^{II}F_4(H_2O)_2)^{2-}$ octahedra. In the dihydrate, $Fe^{III}F_6$ groups are linked by opposite corners to from infinite chains held together by isolated $(FeF_4(H_2O)_2)$ units. A dehydration mechanism has been proposed by Ferey et al.[208]. In an intermediate step, one water molecule in every two is eliminated and the $Fe^{II}$ isolated polyhedra are transformed into infinite chains sharing the $(Fe^{III}F_6)$ octahedral chains. Nevertheless, these phases are typically three-dimensional and $Fe_2F_5 \cdot 2H_2O$ is ferrimagnetic below 48 K[209].

A larger group of 1-D hydrates can be found within the $A_2MF_5 \cdot nH_2O$ series. Although $H_2O$ molecules participate in the structure of these compounds in two different manners, the structural features can be explained in terms of chains.

The structures of $K_2[FeF_5(H_2O)]$[212], $Rb_2[CrF_5(H_2O)]$[210, 211], $(NH_4)_2[GaF_5(H_2O)]$, and of the oxide fluoride $Cs_2[VOF_4(H_2O)]$[213] can be considered as consisting of $[MX_5(H_2O)]^{2-}$ units separated by the monovalent ions. These structures, which are also found in many homologous chlorides and oxide chlorides, are more or less derived from the $K_2PtCl_6$ type. Due to strong hydrogen bonds between the anions, the structural features can be described in terms of infinite chains.

$K_2AlF_5 \cdot H_2O$[204, 151], $A^I_2Mn^{III}F_5 \cdot H_2O$ (with A = K, Rb, Cs, Tl)[214–217] and $A^I_2Fe^{III}F_5 \cdot H_2O$ (K = Rb; Cs)[218] are typical 1-D compounds. They consist of infinite $(MF_5)_n^{2n-}$ chains with $MF_6$ octahedra sharing two trans-corners. The structures of the Mn compounds have been elucidated by Edwards[214] and Bukovec[215, 216] while their dehydration mechanisms were studied by Günter, Matthieu and Oswald[199]. In $Mn^{III}$ and $Fe^{III}$

**Table 9.** Magnetic data of antiferromagnetic $AMF_5$ fluorides

| Compound | $T_N$ (K) | $\theta_p$ (K) | $C_M$ exp. | $C_M$ calc. | J/k (K) | Ref. |
|---|---|---|---|---|---|---|
| $MnAlF_5$ | 2.35 | −6 | 4.43 | 4.38 | – | 220, 221 |
| $CrAlF_5$ | < 4.2 | −3 | 2.57 | 2.55 | – | 222 |
| $MnCrF_5$ | 6 | −16 | 6.16 | 6.25 | – | 223 |
| $CaCrF_5$ | < 4.2 | −28 | 1.86 | 1.87 | −3.9 | 224 |
| $CaFeF_5$ | 21 | −202 | 4.18 | 4.38 | −10.8 | 224 |
| $SrCrF_5$ | < 4.2 | −18 | 1.53 | 1.87 | −6 | 228 |
| $BaVF_5$ | 20 | −45 | 0.98 | 1.00 | −21 | 228 |
| $BaFeF_5$ | 35 | −78 | 3.46 | 4.38 | −11 | 228 |

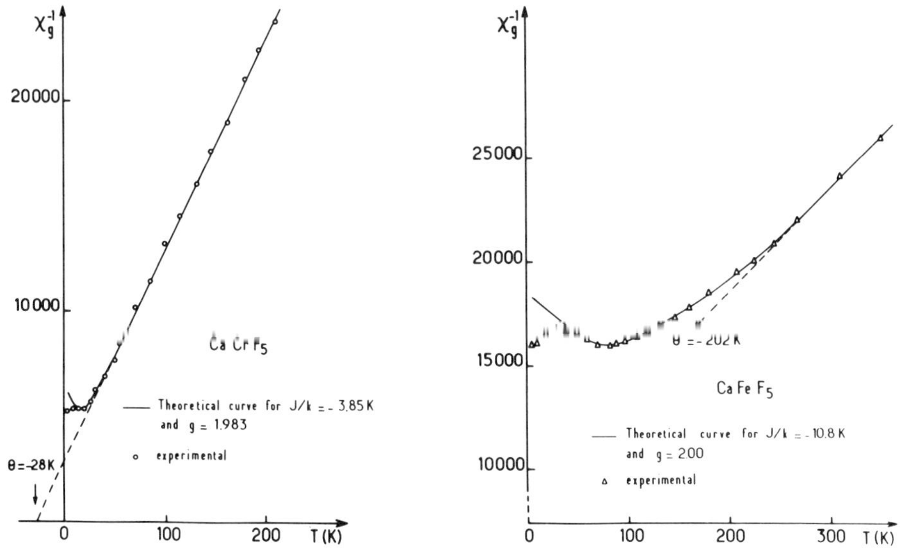


**Fig. 29.** Powder susceptibility data of $CaCrF_5$ and $CaFeF_5$

fluorohydrates, X-ray diffraction and scanning electron microscope studies have suggested a chain-controlled topotactic mechanism. The final dehydration product is the anhydrous $A_2MF_5$ fluoride and the reactions are reversible. It should be pointed out that $AMnF_5 \cdot H_2O$ (A = Ba, Sr) compounds prepared recently have the same features[219].

### 3.3.2.2 Magnetic Behavior of $AMF_5$, $A_2MF_5$ and $A_2MF_5, H_2O$ Compounds

*$AMF_5$ Stoichiometry.* Above 4 K, the magnetic susceptibility of $MnAlF_5$ follows a Curie-Weiss law with $\theta_p = -6$ K (Table 9)[220]. A 3-D ordering has been observed at $T_N$ = 2.35 K, the magnetic moments being directed along the $\vec{b}$-axis perpendicular to the $MnF_6$ octahedral chains. Although the Mn-Mn distances are very different within a chain (3.35 Å) and between two nearest chains (Mn-F-Al-F-Mn $\simeq$ 7.7 Å) – since $(MnF_6)$ chains are separated by $(AlF_6)$ ones – the two exchange constants involved are identical[221]. This behavior may be attributed to the competition between short-distance coupling with a non-favorable angle (Mn-F-Mn = 90°) and long-range interactions with angles of about 135° ($CrAlF_5$, which exhibits a similar structural arrangement, is paramagnetic down to 4.2 K[222]).

In $MnCrF_5$, although both cations are paramagnetic, the coordination of Mn atoms tends to destroy the long-distance interaction pathways. Neutron diffraction experiments have shown that below $T_N$ = 6 K the two chain sublattices are coupled independently. These results were confirmed by pulsed magnetic field measurements; they showed that antiferromagnetic interactions between $CrF_6$ octahedra are stronger than those occurring within the chains composed of $MnF_7$ units[223]. 1-D magnetism of $(CrF_5)_n^{2n-}$ chains could be considered above 6 K.

The magnetic properties of $CaCrF_5$ and $CaFeF_5$ have been investigated by Dance et al.[224]. In Fig. 29 $\chi^{-1} = f(T)$ curves are given, and in Table 9 the high-temperature parameters are listed. The curves have been fitted by application of high-temperature expansion series[21] with $S = 3/2$, $g = 1.983$ and $S = 5/2$, $g = 2.00$ for $CaCrF_5$ and $CaFeF_5$, respectively. For $CaCrF_5$, the value of the exchange constant is similar to those determined by ESR measurements of $Cr^{3+}$ or $V^{2+}$ pairs ($d^3$ configuration) coupled antiferromagnetically: $3.5\ K \leqslant J/k \leqslant 6.5\ K$[225]. Mössbauer resonance studies carried out on a 5% $^{57}FeF_3$-doped sample do not show 3-D ordering as low as 4.2 K. A spin-flop phenomenon is observed in the 1-D domain. The 3-D transition temperature of $CaFeF_5$ has been determined by Mössbauer resonance technique to be $T_N = 21$ K. Between 25 and 21 K, besides the quadrupole doublet, a sextuplet with broadened lines appears. This superposition can be explained on the basis of a slow paramagnetic relaxation process which often occurs in 1-D compounds[226–227]. The average value of the hyperfine field is 440 kOe at 4 K; this low value is related to a zero-point spin-reduction (see Sect. 3.6). The inter- to intrachain ratio has been evaluated by means of the Oguchi method: $J'/J \simeq 10^{-2}$[62], this value being higher than those observed for other chains systems.

The magnetic properties of several phases exhibiting a $BaFeF_5$ structure have been studied by Georges et al. using magnetization and susceptibility data (Table 9)[228]. At low temperature, $BaVF_5$ and $BaFeF_5$ possess a weak ferromagnetic component $\sigma_0$, and Néel temperatures have been determined from $\sigma_0$ thermal variations. Nevertheless, the large value of the extrapolated magnetization, observed for $BaFeF_5$ ($\sigma_0(0K) = 0.21\ \mu_B$) cannot be explained on the basis of the Dzialoshinskii mechanism[229]. The exchange constants within the chains have been evaluated using the molecular-field model and the relation $J/k = 3\,\theta_p/S(S+1)$. Due to the complexity of the structures, an interpretation of the low-temperature data has not yet been reported.

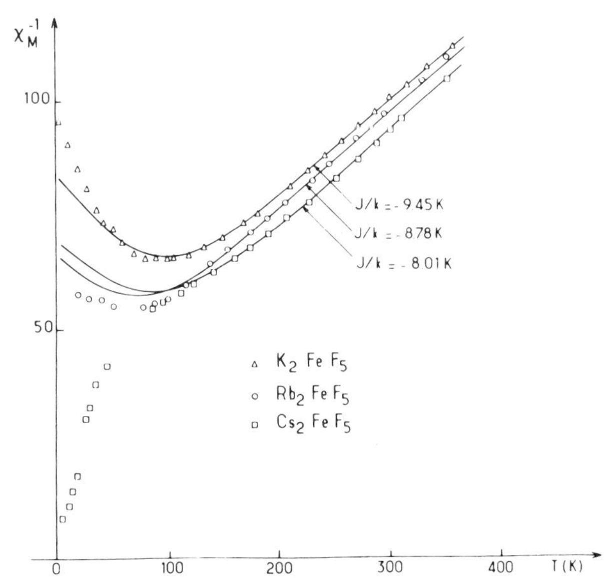


**Fig. 30.** Powder susceptibility data of $A_2MF_5$ fluorides (A = K, Rb, Cs)

**Table 10.** Magnetic parameters of $A_2MF_5$ compounds

| Compound | $T_N$ (K) | $\theta_p$ (K) | J/k (K) | J′/J | $H_{hf}$ (kOe) | Ref. |
|---|---|---|---|---|---|---|
| $K_2FeF_5$ | 11.2 | −125 | −9.45 | $3.8 \times 10^{-3}$ | 410 | 230, 203 |
| $Rb_2FeF_5$ | 9.3 | −125 | −8.78 | $2.4 \times 10^{-3}$ | 420 | 231, 202 |
| $Cs_2FeF_5$ | 5 | −135 | −8.01 | $1.1 \times 10^{-3}$ | 340 | 203 |
| $N_2H_6FeF_5$ | 9 | −150 | −10.2 | $2 \times 10^{-3}$ | 430 | 233 |
| $K_2VF_5$ | | − 75 | −11.6 | | | 197 |
| $Rb_2VF_5$ | | | −11.6 | | | 197 |
| $Cs_2VF_5$ | | | −11.6 | | | 197 |
| $(NH_4)_2VF_5$ | | | −5 | | | 197 |
| $Li_2MnF_5$ | <4.2 | | −5.6 | | | 189 |
| $Na_2MnF_5$ | <4.2 | | −9.2 | | | 189 |
| $(NH_4)_2MnF_5$ | 5.8 | | −10.6 | | | 189 |

*$A_2MF_5$ Stoichiometry.* Among the $A_2MF_5$ compounds, $A_2FeF_5$ series (A = K, Rb, Cs) has been studied extensively by Gupta, Dickson and Johnson by Mössbauer resonance measurement[230–232] and by Tressaud et al. using magnetic and neutron diffraction experiments[202, 203].

The space groups Pna $2_1$ for $K_2FeF_5$ and Pnma for $Rb_2FeF_5$ obtained by X-ray studies have been confirmed by neutron diffraction measurements. Although $Cs_2FeF_5$ can be indexed with the cell of $Rb_2FeF_5$, the presence of additional reflexions suggests a space group with lower symmetry. The high-temperature part of $\chi^{-1} = f(T)$ curves allows the calculation of the intrachain exchange constants using the method of the high-temperature expansion series[21]. A comparison between theoretical curves and experimental values is given in Fig. 30 and the magnetic data are compiled in Table 10. The lowering of $|J|/k$ observed from K to Cs compounds can be explained by the increase in the Fe-F-Fe distances within a chain.

By neutron diffraction measurements, magnetic spectra obtained below the ordering temperature can be indexed in the nuclear cell. In every case the principal mode of the $Fe^{3+}$ ions exhibits an A configuration $(S_1 - S_2 - S_3 + S_4)$[203]. The magnetic moments of $K_2FeF_5$ are parallel to the $\vec{b}$-axis. Components $G_x$ and $C_z$ can be coupled with the $A_y$ mode, leading to a slight deviation of the spins from the $\vec{b}$-axis. For $Rb_2FeF_5$, the strong intensity of the magnetic line (100) and the presence of both (010) and (001) lines clearly show the non-collinearity of the spins. A refinement of the magnetic model reveals that the principal mode is $A_z$ and that it can be associated with the $G_x$ mode. The variation of the resulting moment with temperature shows that $Rb_2FeF_5$ does not reach complete magnetic saturation even at 2 K: $M_{extrap.} = 3.4 \pm 0.2\ \mu_B$, $T_N = 8.0 \pm 0.5$ K. A diffuse diffraction maximum is observed in the diffraction range $10° < \theta_{Bragg} < 18°$ and shows an intensity maximum at 10 K, slightly above $T_N$. This effect, which is due to short-range interactions, decreases at lower temperatures when long-range ordering is achieved and also at higher temperatures, but it is still present at 50 K. Below 5 K, a similar arrangement of the moments is found in $Cs_2FeF_5$. A spin-flop behavior has been observed on single crystals of $K_2FeF_5$ and $Rb_2FeF_5$ at 3.7 and 6.5 T, respectively, using both Mössbauer resonance and magnetic measurements under high applied fields[202, 230, 232]. The ratio J′/J of inter- to intrachain exchange constants has been evaluated by the Oguchi method[21] and also by a comparison of the experimental values of the hyperfine field with the calculated values using the spin-wave theory[230, 231]. These results show that $A_2FeF_5$ compounds are good approximations of the 1-D Heisenberg magnetic model.

The magnetic properties of $M_2VF_5$ phases (M = K, Rb, Cs) have been studied by Padalko et al.[197]. The intrachain exchange constants have been calculated by means of the Bonner and Fisher method applied to 1-D Heisenberg spin lattices (Table 10) using the formula:

$$\mu^2_{\text{eff.}} = S(S+1)\, g^2 \mu_B^2 \quad \exp\left(\frac{S(S+1)}{3/4} \frac{J}{kT}\right)$$

If no data on the structures and on the magnetic behavior below 77 K are available, it seems useless to establish further hypotheses.

Similar high-temperature calculations have been carried out on $Mn^{III}$ compounds $A_2MnF_5$ (A = Li, Na, $NH_4$) and the obtained chain constants are compiled in Table 10[189]. The magnetic anisotropy of $(NH_4)_2MnF_5$ has been studied on the basis of the classical spin model with isotropic coupling in the axial crystal field. The anisotropy constant is $D/k = -3.3$ K[235]. The exchange constants calculated with the Heisenberg model are collected in Table 10. Emori et al. explained the decrease of $|J|/k$ from $(NH_4)$ to Li salts on the basis of crystallographic data[189].

Not only their structural properties but also electrical measurements of $A_2MF_5$ compounds have shown that these may be ionic conductors. In $K_2AlF_5$, for instance, the mobile species have been shown to be $K^+$ and $F^-$ [236]. At 300 °C, the conductivity is about $10^{-3}\Omega^{-1} \cdot cm^{-1}$; this high value can be related to the tunnel structural type[237].

*Hydrates.* $A_2Mn^{III}F_5 \cdot H_2O$ compounds (A = K, Rb, Cs) are suitable examples of Heisenberg chain antiferromagnets. The 1-D behavior is enhanced by the ferrodistortive ordering of $d^1_{z^2}$ orbitals, antiparallel along the $\vec{c}$ direction. In the ab plane weak $d^0_{x^2-y^2}$ interactions occur between the chains.

In addition to the above fluoride hydrates, 1-D antiferromagnetism has also been found in $FeF_3 \cdot 3\,H_2O$ and was explained by an exchange of coupled pairs of $Fe^{3+}$ ions[238]. In fact, the crystal structure is composed of linear chains of $Fe^{3+}$ ions coupled antiferromagnetically. The transition ion is located in octahedra sharing corners with axial $F^-$ ions. The Fe-Fe distance within a chain is 3.88 Å whereas in the perpendicular direction the spacing between chains is 5.66 Å[239]. The exchange integral $J/k = -20$ K is found by ESR measurements for this compound[240]. This integral varies slightly with temperature. It should be mentioned that a close value of $J/k = -23$ K was found on the basis of the hypothesis of exchange coupled pairs[238].

### 3.3.2.3 Columnar Structures

Antiferromagnetic fluorides showing columnar structures have been recently found for $Cr^{III}$ compounds. $ACr^{III}F_4$ phases (A = K, Rb, Cs, Tl) are characterized by infinite columns of formula $(Cr_3F_{12})^{3n-}_n$ whose sections are formed of three octahedra linked by corners[241–243]. The columns are separated by monovalent ions. A flattening of the $\chi^{-1} = f(T)$ curves is observed at a value of $T_{\chi_{max}}$ higher than $T_N$. For $CsCrF_4$ ($T_N = 40$ K), a

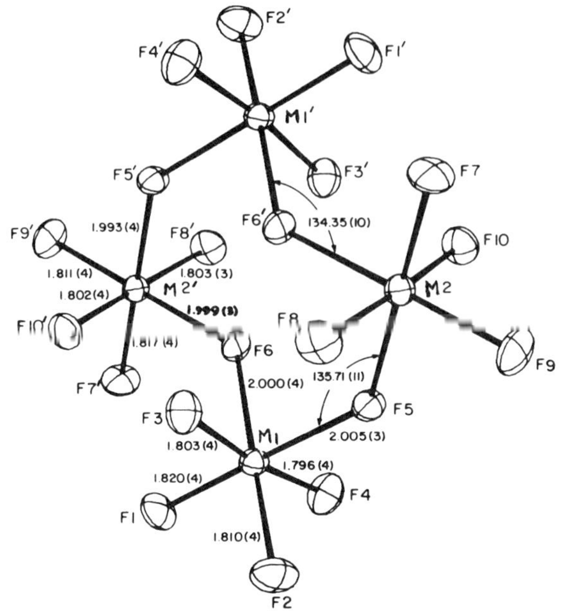


Fig. 31. Tetrameric unit in $MF_5$ (M = Ru, Rh, Os, Ir, Pt) pentafluorides

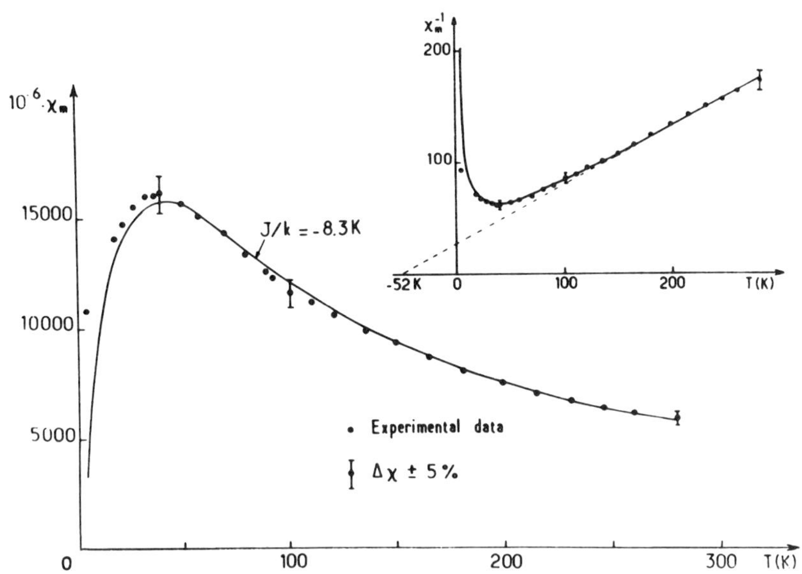


Fig. 32. Powder susceptibility data of $RuF_5$

deviation from the Curie-Weiss law is observed up to 180 K[241]. This behavior implies 1-D antiferromagnetic correlations.

## 3.4 Magnetic Clusters in Fluorides

For a solid-state chemist, it is possible to synthesize compounds with polyhedra containing transition-metal ions separated by diamagnetic elements. By cationic substitution, the following arrangements can be obtained:

– isolated ($MF_6$) octahedra as, for instance, in elpasolites ($K_2NaFeF_6$), weberites ($Na_2MgFeF_7$) and $K_2GeF_6$ ($M_2TiF_6$ compounds),

– $M_2F_9$ units of two octahedra linked by a face as in the ordered 6 H compound $RbCa_{1/3}Ni_{2/3}F_3$[244] and in $Cs_3Fe_2F_9$ structures where the arrangement corresponds to a $CsNiF_3$ structure with ordered vacancies (1 over 3) within the infinite $(MF_3)_n^{n-}$ chain[245],

– $M_3F_{12}$ units of three octahedra sharing faces as in the ordered 12 R compound $CsCd_{1/4}Ni_{3/4}F_3$[244].

These isolated paramagnetic units are incorporated into a 3-D lattice and, due to more favorable angles, long-distance superexchange interactions $M_{para}$-F-$M_{dia}$-F-$M_{para}$ may be of the same magnitude as $M_{para}$-F-$M_{para}$ interactions. Therefore, these systems are not always suitable models of low-dimensional magnetism (see, for example, 3-D properties of Mn chains in $MnAlF_5$ in Sect. 3.3.2.2).

The only type of fluoride in which the magnetic polyhedra are linked by Van der Waals interactions are $MF_5$ pentafluorides containing 4d- and 5d-transition metals. It should be noted that the oxide tetrafluorides of $Mo^{VI}$ and $T_C^{VI}$ contain $M_3O_3F_{12}$ trimers consisting of three octahedra sharing cis-fluorine atoms[246].

Amongst the pentafluorides, the magnetic data of $RuF_5$ and $OsF_5$ have been interpreted by Darriet et al.[247]. The structure of these phases consists of closely packed $M_4F_{20}$ tetrameric units. In each tetramer, octahedral $MF_6$ groups share two corners in cis position (Fig. 31)[248]. One of the main characteristics of the structure is the M-F-M angle of 135°, the distortion of the cluster leading to an anionic hexagonal close packing. As found in other fluoride series, the bridging F atoms of $MF_6$ units are more distant from the metal atom (M-$F_b \simeq 2.0$ Å) than the terminal ones (M-$F_t \simeq 1.8$ Å).

In Fig. 32 the variation of the magnetic susceptibility with temperature is illustrated for the Ru compound. The 3-D ordering temperature and the antiferromagnetic structure have been determined by neutron diffraction measurements ($T_N = 5$ K for $RuF_5$; $T_N = 6$ K for $OsF_5$). Inside the cluster the M-F-M interactions are antiparallel. Since $T_N$ is about six times lower than $T_{\chi_{max}}$ (40 K), theoretical calculations have been made to confirm the value of $T_{\chi_{max.}}$ on the basis of a tetranuclear cluster model. In the first step, only interactions between nearest neighbors have been considered in the $M_4F_{20}$ rhombus. The experimental data have been fitted to the complete Van Vleck expression by the least squares method. A good agreement is obtained for $J/k = -8.3$ K in $RuF_5$ (Fig. 32), showing that above $T_N$ the magnetic properties of these pentafluorides can be described in terms of a cluster-type behavior[249].

## 3.5 Field-Induced Transitions in Fluorides

### 3.5.1 Spin-Flop Mechanism

Most of the results discussed up to now have dealt with the properties of magnetic systems either under zero-field conditions or in a domain in which the magnetization varies linearly with the external field. When applying a high magnetic field to antiferromagnetic systems, two kinds of transitions can arise: spin-flop transition or metamagnetism. The difference between these two phenomena is related to the ratio of the exchange field ($H_E$) to the anisotropy field ($H_A$). For a smaller $H_A/H_E$ ratio, a spin-flop transition occurs which can be considered as an antiferromagnetic rearrangement; at higher $H_A/H_E$ ratio, metamagnetism leads directly from an antiparallel to a parallel configuration. Only the former mechanism has been observed in fluorides since the anisotropy fields involved are generally small compared with $H_E$.

Spin-flop behavior was predicted in 1936 by Néel on the basis of the molecular field theory applied to a two-sublattice model of uniaxial antiferromagnets[1, 250]. This model was further developed by Nagamiya, Yosida and Kubo[3] and the first experimental proof forwarded by Gorter[251]. The effect can be summarized as follows: by applying a magnetic field (H) parallel to the axis of the magnetization of an antiferromagnet, a competition occurs between the external and internal exchange fields. For systems with weak anisotropy, when H reaches a critical value, the sets of antiparallel moments of the two sublattices flip from the direction of the antiferromagnetic axis to a direction perpendicular to the applied field.

For systems showing complete anisotropic exchange, i.e. compounds described by the Ising model, the spin-flop phenomenon does not occur. At $T = 0$ K, for a square-planar Ising antiferromagnet such as $Cs_3CoBr_5$[252], the moments become parallel to the field direction at a critical value $H_c$, which is equivalent to the magnitude of the exchange field: $g\mu_B H_c = 2z\,|J|\,S$. The critical field decreases with increasing temperature and vanishes at $T = T_N$.

In contrast to the Ising model, spin-flop is observed in Heisenberg systems over a large range of applied fields and temperatures.

The establishment of an easy (or antiferromagnetic) axis along which the moments are located is due to the presence of an anisotropy energy, K. This effect can arise from single-ion or exchange mechanisms. The free energy of the system is proportional to $-(\frac{1}{2}).\ \chi_{\parallel} \cdot H^2$ if the applied field is parallel to the easy axis, and proportional to $(K - \frac{1}{2}\ \chi_{\perp} H^2)$ if the applied field is perpendicular to this axis. In antiferromagnets $\chi_{\perp}$ is larger than $\chi_{\parallel}$ and if H is parallel to the easy axis, the moments tend to align perpendicularly, thus gaining a magnetic energy of $\frac{1}{2}(\chi_{\perp} - \chi_{\parallel})H^2$. For small H values, the anisotropy is larger than the field term and the system is unchanged. However, when $-\frac{1}{2}\chi_{\parallel}H^2 = K - \frac{1}{2}\chi_{\perp}H^2$, the applied field is sufficient to balance the anisotropy effect and the moments flip over to the perpendicular direction $H_{SF} = [2K/(\chi_{\perp} - \chi_{\parallel})]^{1/2}$.

When increasing the value of H, the moments gradually rotate towards the applied field direction and when $H = H_c$, they are parallel to the field. In the molecular field model, the spin-flop is a first-order transition and the spin flop $\leftrightarrows$ paramagnetic transition a second-order process. Between $H_{SF}$ and $H_c$ magnetization increases linearly with H: $M/M_S = H(2H_E - H_A)$ where $M_S$ is the saturation magnetization of a sublattice ($M_S = \frac{1}{2}\ Ng\mu_B S$). At $T = 0$ K, the critical field can be written as $H_{SF} = [2H_E \cdot H_A/(1 - \alpha)]^{1/2}$

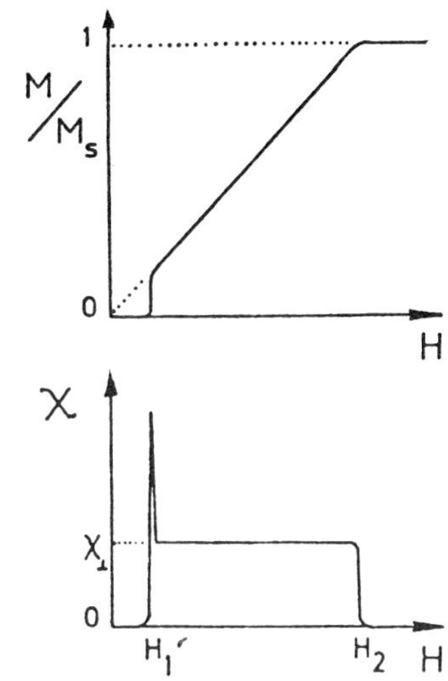


**Fig. 33.** Plot of isothermal magnetization and differential susceptibility of a weakly anisotropic antiferromagnet versus the applied external field ($M_s$ = saturation magnetization of a sublattice)

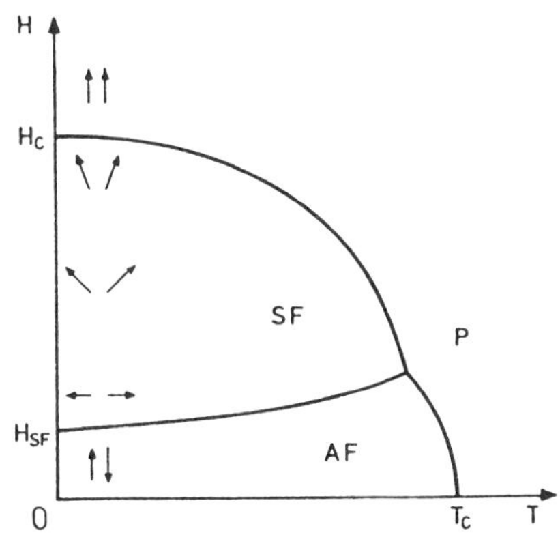


**Fig. 34.** Theoretical magnetic phase diagram of a weakly anisotropic antiferromagnet

where $\alpha = \chi_{\parallel}/\chi_{\perp}$ and $H_C = 2H_E - H_A$. According to the molecular field theory, the sum of the internal fields can be expressed by $\chi_{\perp} \simeq M_S/(H_E + H_A)$.

The spin-flop behavior is limited to cases where $H_E \gg H_A$. Since the anisotropy field increases, the difference between $H_{SF}$ and $H_c$ decreases and the two are equivalent for $H_A = H_E$. The spin-flop domain vanishes and the metamagnetic situation is identical to that involving the Ising model.

The field dependence of magnetization and of the susceptibility of a weakly anisotropic antiferromagnet – according to the molecular field theory – is described in Fig. 33 and the H = f(T) diagram in Fig. 34.

Apart from high applied field measurements, another powerful technique for the determination of the interaction fields is antiferromagnetic resonance (AFMR)[253–255]. Below the ordering temperature, the resonance frequency depends strongly on both $H_E$ and $H_A$ according to the relation $\omega/\gamma \simeq (2H_E \cdot H_A)^{1/2}$ for $H_E \gg H_A$; $\omega$ is the applied frequency and $\gamma = ge/2mc$ is the magnetochemical ratio.

The spin-flop behavior of $MnF_2$ has been extensively studied because this fluoride appears to be a good example of an uniaxial antiferromagnet[254–259]. The greatest part of its anisotropy can be accounted for by magnetic dipole interactions[254]. Decoupling of the direction of the magnetic moment and the easy axis have been determined by pulsed field and AFMR techniques by Jacobs[255] and Foner[257]. The strength of the spin-flop field is $H_{SF} = 93$ kOe. A complete agreement between experimental results and spin-wave theory was obtained.

### *3.5.2 Transitions in Layer and Chain Fluorides*

In 2-D and 1-D systems even in non-uniaxial materials, the Néel model can still be used since the anisotropic field in the plane where the moments can rotate is small compared with the exchange field in the perpendicular direction. After first considering the particular case of $CsNiF_3$, we will described several 2-D and 1-D Heisenberg fluorides showing spin-flop behavior.

*$CsNiF_3$.* In $CsNiF_3$, when a magnetic field is applied along the scattering vector, the variation of the (½, 0, 0) reflection leads to a temperature dependence of the critical field[170]. No rotation of the spins within the domain results but only an increase in the size of the preferential magnetic domains. These experiments show that, if ferromagnetic ordering in $CsNiF_3$ is due to exchange forces, single-ion anisotropy and also dipole forces occur in the ab plane. The small value of the critical field suggests the contribution of these dipole forces to be important.

*Layer Fluorides.* In 2-D systems containing $Mn^{II}$, spin-flop has been investigated for instance in $BaMnF_4$ ($H_{SF} = 10.4$ kOe)[122], $K_2MnF_4$ and $Rb_2MnF_4$ ($H_{SF} = 55$ and 51 kOe, respectively)[121]. The anisotropy parameter $\alpha = H_A/H_E$ of $A_2MnF_4$ phases is in good agreement both with the dipolar anisotropies which are calculated taking into account the spin reduction effect and AFMR measurements[7].

*$CaCrF_5$.* A discontinuity is observed between 2.5 and 12 K in the M = f(H) diagram for $CaCrF_5$ at about 30 kOe, presumably corresponding to an antiparallel rearrangement of the $Cr^{III}$ moments within the chains[224]. This rearrangement occurs very probably in the 1-D domain as no 3-D ordering has been found down to 4.2 K.

*$A_2FeF_5$ (A = K, Rb, Cs).* Since large single crystals can be prepared by the flux method[260], $A_2FeF_5$ compounds have been more thoroughly investigated. When H is applied parallel to the antiferromagnetic axis of $K_2FeF_5$, i.e. along the $\vec{b}$-direction (Fig. 28), Mössbauer measurements (Gupta et al.[230, 231]) show that the spins reorientate at 37 kOe thus becoming antiparallel perpendicular to the applied field. With an exchange field of $H_E \simeq 700$ kOe calculated from the exchange constant $J/k = -9.45$ K[203], the anisotropic field becomes $H_A \simeq 1$ kOe and $H_A/H_E \simeq 1.4 \times 10^{-3}$ [230, 231, 234].

For $Rb_2FeF_5$, agreement is also observed between magnetization and Mössbauer resonance data under a high applied field[202, 203, 232].

Antiferromagnetic reorientation takes place at $H_{SF} = 65$ kOe as shown in Fig. 35. On the basis of magnetic results, interaction fields are: $H_E \simeq 1000$ kOe, $H_A \simeq 1.5$ kOe, $H_A/H_E \simeq 1.5 \times 10^{-3}$ [202]. This phenomenon is still detected at about 30 K, i.e. at a temperature three times higher than $T_N$.

Recently, Stahlbush and Scott calculated the field dependence of the magnetic susceptibility of a chain system in the 1-D domain[261, 262]. The calculations are based on the classical Heisenberg chain model, including anisotropic and applied fields and employing the transfer integral technique. Variations of $\chi = f(T)$ have been calculated with a single-ion anisotropy parameter $|D/k| = 0.03$ K. The high-field region $[\mu H > (DJ)^{1/2}]$ is analogous to the 3-D spin-flop domain.

## 3.6 Zero-Point Spin-Reduction in Fluorides

The spin-reduction property was predicted in antiferromagnetic substances in 1952 by Anderson[263] but conclusive examples have only been obtained during the last decade, many of these being low-dimensional fluorides. De Jongh and Miedema[7], Gupta et al.[230–232] have given a detailed description of the phenomenon for 2-D and 1-D fluorides respectively.

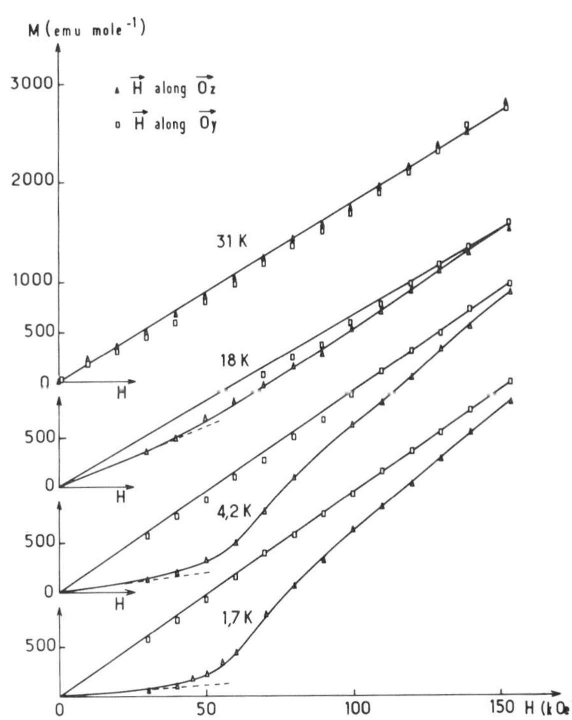


**Fig. 35.** Single-crystal magnetization data of $Rb_2FeF_5$

### *3.6.1 Spin-Reduction in Low-Dimensional Systems*

The perfectly aligned Néel state is not an eigenstate of the antiferromagnetic exchange Hamiltonian, except in the case where its anisotropy is infinite. Consequently, even at T = 0 K, the spins will be subject to deviations.

The presence of anisotropy is a stabilization factor for the two-sublattice ground state. For $\alpha = H_A/H_E \rightarrow \infty$, the energy is equal to the Ising value $E_{(\alpha\rightarrow\infty)} = -2z\,|J|S^2N$. For intermediate values of $\alpha$ the relation $E_{(\alpha)} = E_{(\alpha\rightarrow\infty)}\,(1 + e(\alpha)/zS)$ holds, where $e(\alpha)$ is a function of the anisotropy which varies from about 1/4z for $\alpha = 0$ to zero for $\alpha \rightarrow \infty$. At the same time, the magnetic moment is reduced by an amount $\Delta S_{(\alpha)}$ which also depends on the anisotropy factor $M = g\mu_B(S - \Delta S_{(\alpha)})$. $\Delta S_{(\alpha)}$ is equal to zero for $\alpha \rightarrow \infty$ and $e_{(\alpha)}$ and $\Delta S_{(\alpha)}$ have been calculated using spin-wave theory[264, 265].

The first attempts to confirm experimentally zero-point spin-reduction have been carried out on 3-D compounds. The results have been disappointing: besides a low value of $\Delta S$ [for S = 5/2, $\Delta S \simeq 3\%$], additional phenomena interfered with the results. For instance, the covalency factor reduces the effective moment, due to a delocalization of a small amount of the moments onto the ligands.

A spin-reduction property cannot be obtained for all the antiferromagnetic compounds. Thus, if ferromagnetic layers are weakly coupled antiferromagnetically, the zero-point spin-reduction will be negligible because the main interaction is the ferromagnetic one. The most impressive examples of spin reduction have been found in 2-D and 1-D Heisenberg antiferromagnetic fluorides.

### *3.6.2 2-D Heisenberg Antiferromagnetic Fluorides*

In layer-type compounds, the experimental values of the spin reduction are in good agreement with calculations based on the two-dimensional spin-wave theory. The compounds $A_2Mn^{II}F_4$ (A = K, Rb) and $A_2MF_4$ (M = Mg, Zn, Cd) doped with $Mn^{II}$ have been intensively studied by neutron diffraction and magnetic resonance[266, 267]. After correction due to supertransferred hyperfine interactions, $\Delta S_{exp.}$ values for $K_2MnF_4$ and $Rb_2MnF_4$ (0.166 and 0.170) can be compared with the calculated ones (spin-wave theory) which are 0.169 and 0.167 respectively[268].

**Table 11.** Spin reduction occurring in 2-D fluorides [from[7]]

| Compound | S | $\alpha = H_A/H_E$ | $\chi_\perp(0)/\chi_\perp^0(\alpha)$ Exp. | $\chi_\perp(0)/\chi_\perp^0(\alpha)$ Theor. |
|---|---|---|---|---|
| $CsFeF_4$ | 5/2 | $7 \times 10^{-3}$ | 0.92 | 0.90 |
| $RbFeF_4$ | 5/2 | $6.5 \times 10^{-3}$ | 0.94 | 0.90 |
| $Rb_2MnF_4$ | 5/2 | $4.7 \times 10^{-3}$ | 0.92 | 0.90 |
| $K_2MnF_4$ | 5/2 | $3.9 \times 10^{-3}$ | 0.90 | 0.90 |
| $BaMnF_4$ | 5/2 | $3.1 \times 10^{-4}$ | 0.89 | 0.89 |
| $BaNiF_4$ | 1 | $2 \times 10^{-2}$ | 0.80 | 0.78 |
| $Rb_2NiF_4$ | 1 | $1 \times 10^{-2}$ | 0.77 | 0.77 |
| $K_2NiF_4$ | 1 | $2 \times 10^{-3}$ | 0.76 | 0.74 |
| $CuF_2, 2H_2O$ | 1/2 | $3.7 \times 10^{-3}$ | 0.63 | 0.50 |

De Jongh showed that in 2-D systems, spin-reduction can also be deduced from the perpendicular magnetic susceptibility[7,269]. The experimental value of $\chi_\perp$ extrapoled to 0 K is in good agreement with the theoretical value obtained by means of the spin-wave theory ($\chi_\perp$ (0). The difference between the latter value and that calculated via the molecular field approach $\chi_{\perp(MF)}$ is essentially due to the zero-point spin reduction.

$$\chi_{\perp(MF)} = Ng\mu_B S/2\,H_{exch.}$$

with $H_{exch.} = 2z\,|J|S/g\mu_B$

$$\chi_\perp(0) = \frac{\chi_{\perp(MF)}}{1 + \frac{1}{2}\alpha}\left[1 - \frac{\Delta S_{(\alpha)}}{S} - \frac{e_{(\alpha)}}{(2+\alpha)\,zS}\right]$$

with $\alpha = H_A/H_E$, where z is the number of nearest neighbors and S the spin value.

In Table 11 are compiled the results concerning 2-D Heisenberg systems: $\alpha$ is the anisotropy coefficient and the perpendicular susceptibility is calculated from the relationship:

$$\chi_\perp^0(\alpha) = Ng^2\mu_B^2/4z|J|(1 + \alpha/2)\ .$$

This model, which gives good results for high values of S, can neither be applied to $S = \frac{1}{2}$ nor to compounds with chain structures. More recently, improvements have been achieved by Ishikawa and Oguchi by including into the spin-wave theory kinematic interactions besides dynamic ones[270]: for $CuF_2 \cdot H_2O$, $S = \frac{1}{2}$, $\chi_\perp/\chi_\perp^0(exp) = 0.63$ and $\chi_\perp/\chi_\perp^0 calc. = 0.64$.

### *3.6.3 1-D Heisenberg Antiferromagnetic Fluorides*

Concerning Heisenberg chain antiferromagnets, calculations taking into account kinematic interactions also give a better agreement. More precisely, for $S = \frac{1}{2}$, the spin-reduction is no more larger than the total spin value as previously found with the spin-wave theory[270]. $\Delta S$ is a function of the 1-D character and extrapolation to $J'/J \to 0$, i.e. to the ideal 1-D system with $S = \frac{1}{2}$, leads to $\Delta S = \frac{1}{2}$. Concerning $KCuF_3$ a good agreement is obtained between the experimental and the calculated value: $\Delta S_{exp.} = 0.23$; $\Delta S_{calc.} = 0.20$[63,69].

A spin reduction of about 65% is observed in $CuBeF_4 \cdot 5\,H_2O$. This compound which is isostructural with $CuSO_4 \cdot 5\,H_2O$ and $CuSeO_4 \cdot 5\,H_2O$ contains weakly coupled antiferromagnetic Heisenberg chains[271]. This large spin reduction influences the field dependence of the sublattice magnetization[272–273].

In $A_2FeF_5$ compounds (A = K, Rb, Cs) an important spin reduction has been detected by neutron diffraction[202,203] and Mössbauer resonance measurements[230–232]. The calculations are based only on the spin-wave theory disregardering the kinematic interactions since the spin value is important. Figure 36 describes the variation of the zero-point spin reduction with the anisotropy factor $\alpha$. $\alpha$ is here $(1 - \omega_A/\omega_E)^{-2}$ where $\omega_A$ and $\omega_E$ are the Larmor frequencies corresponding to the anisotropy and exchange fields, respectively.

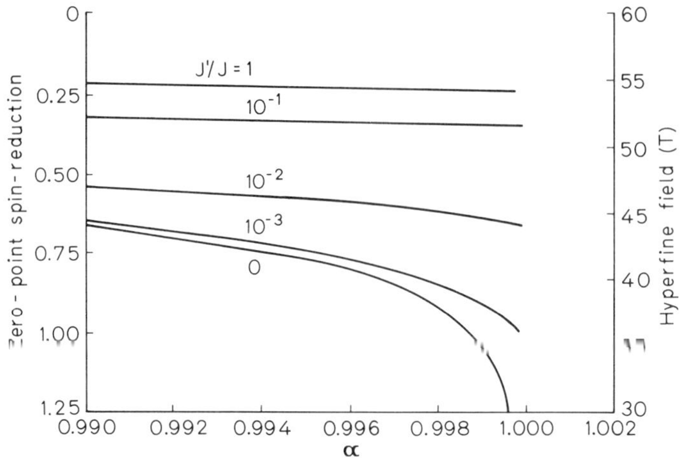


**Fig. 36.** Calculated values of the zero-point spin reduction for different values of $\alpha$ and J'/J

Hence, $\Delta S$ has a minimum but is not equal to zero for 3-D magnetic systems and increases when the dimensionality is lowered ($J'/J \rightarrow 0$). The influence of the anisotropy is more important in low-dimensional compounds, and the more isotropic the system ($\alpha \rightarrow 1$), the more is the spin reduced. For small spin values ($S = \frac{1}{2}$) addition of the kinematic interactions permits to get $\Delta S/S = 1$ for $J'/J \rightarrow 0$.

# 4 Conclusions

Table 12 compiles the important contribution of fluorides as examples for almost all theoretical models of low-dimensional magnetism. Some of these features can be summarized as follows:

– 1-D magnetism can occur in 3-D crystallographic arrangements of octahedra such as $Cu^{2+}$ perovskites. The ratio of the exchange constants corresponding to two types of Cu-Cu distances differing by only 0.3 Å can be as high as $10^2$ to $10^3$.

– Fluorides are suitable 2-D and 1-D magnetic models. In the $K_2NiF_4$ series, for instance, the antiferromagnetic behavior of fluorinated compounds can perfectly be explained by means of high-temperature expansion series which is not always the case with the homologous oxides[9)].

– In tetrameric structures, although the difference between intra- and intercluster distances is lower in fluorides than in oxides (3.73 and 4.90 Å, 3.4 and 5.5 Å in $RuF_5$ and $Na_3RuO_4$, respectively), only the former compounds display a cluster-type magnetic behavior over a large temperature range.

**Table 12.** Examples of different magnetic models

| Model | Dimensionality of interactions | Type of interaction | Examples |
|---|---|---|---|
| Ising | 2-D | AF | $A_2CoF_4$ (A = K, Rb) |
| Ising + Planar Heisenberg | 2-D | AF | $A_2FeF_4$ (A = K, Rb) $BaFeF_4$ |
| Planar Heisenberg + Ising | 2-D | AF | $BaCoF_4$ |
| Planar Heisenberg | 1-D | F | $CsNiF_3$ |
| Heisenberg | 1-D | AF | $VF_2$, $A_2FeF_5$ (A = K, Rb, Cs) $KCuF_3$, $CaMF_5$ (M = Cr, Fe) |
| | 2-D | F | $A_2CuF_4$ (A = K, Rb, Cs), $CsMnF_4$ |
| | 2-D | AF | $A_2NiF_4$ (A = K, Rb, Tl) $BaMnF_4$, $BaNiF_4$ $AFeF_4$ (A = Na, K, Rb, Cs) $A_3Mn_2F_7$ (A = K, Rb) $K_3Ni_2F_7$ $Ba_2NiF_6$ |

*Acknowledgements.* The authors are indebted to Prof. P. Hagenmuller and Dr. J. Portier for their constant interest they took in this work. Profs. D. Babel, K. Hirakawa, D. Reinen, Drs. J. Darriet, G. Le Flem, R. Sabatier, and J. L. Soubeyroux are thanked for their helpful discussions and Mr. B. Ellis for his critical reading of the english language of the manuscript. The authors also wish to thank the members of the "Fluorine Research Group" in Bordeaux, especially Prof. J. Grannec and L. Lozano.

# 5 References


1. Néel, L.: Ann. Phys. *17,* 5 (1932), *3,* 137 (1948)
2. Smart, J. S.: Magnetism, Vol. III, p. 63, New York: Academic Press 1963
3. Nagamiya, T., Yosida, K., Kubo, R.: Adv. Phys. *4,* 13 (1955)
4. Goodenough, J. B.: Magnetism and the Chemical bond, New York, London, Interscience, 1963
5. Goodenough, J. B., Longo, J. M.: Landolt-Bornstein series, Vol. III/4 a, 126, Berlin, Heidelberg, New York: Springer 1970
6. Babel, D.: Structure and Bonding, Vol. 3, p. 1, Berlin, Heidelberg, New York: Springer 1967
7. De Jongh, L. J., Miedema, A. R.: Adv. Phys. *23,* 1 (1974)
8. Steiner, M., Villain, J., Windsor, C. G.: Adv. Phys. *25* (2), 87 (1976)
9. Le Flem, G.: Rev. Int. Htes Temp. et Refract. *14,* 71 (1977); Z. Anorg. Allg. Chem. *476,* 69 (1981)
10. Ising, E.: Z. Phys. *31,* 253 (1925)
11. Onsager, L.: Phys. Rev. *65,* 117 (1944)
12. Heisenberg, W.: Z. Phys. *49,* 619 (1928)
13. Blöte, H., Huiskamp, W.: Phys. Lett. *29,* 304 (1969)
14. Bonner, J. C., Fisher, M. E.: Phys. Rev. *135,* 640 (1964)
15. Blombergen, P., Franse, J. M.: Sol. State Commun. *10,* 325 (1971)

16. Baker, G. A. et al.: Phys. Rev. *164,* 800 (1967)
17. Suzuki, M., Tsujiyama, B., Katsura, S.: J. Math. Phys. *8* (1), 124 (1967)
18. Van Dongen, E. J., Capel, H. W.: Physica *84,* 285 (1976)
19. Katsura, S.: Phys. Rev. *127,* 1508 (1962)
20. Fisher, M. E.: Am. J. Phys. *32,* 343 (1964)
21. Chi Yuan Weng: Thesis, Carnegie Mellon Univ., 1969
22. Takeuchi, S.: J. Phys. Soc. Japan *37,* 809 (1974)
23. Duffy, W., Jr., Barr, K. P.: Phys. Rev. *165,* 647 (1968)
24. Smith, T., Friedberg, S. A.: Phys. Rev. *176,* 660 (1968)
25. Fisher, M. E.: Rep. Prog. Phys. *30,* 615 (1967)
26. Griffiths, R. B.: Phys. Rev. *136,* 437 (1965)
27. Mermin, N. D., Wagner, H.: Phys. Rev. Lett. *17,* 1133 (1966)
28. Stanley, H. E., Kaplan, T. A.: Phys. Rev. Lett. *17,* 913 (1966)
29. Rushbrooke, G. S., Wood, P. J.: Mol. Phys. *1,* 257 (1958); *6,* 409 (1963)
30. Stanley, H. E.: Introduction to phase transitions and critical Phenomena, Oxford, Clarendon Press 1971
31. Schieber, M.: Experimental Magnetochemistry, Amsterdam, London, North Holland 1967
32. Bienenstock, A.: J. Appl. Phys. *37,* 1459 (1966)
33. Skalkyo, J., Jr. et al.: Phys. Rev. *2,* 1310 (1970); *2,* 4632 (1970)
34. Abragam, A.: Principles of magnetic resonance, Oxford, Clarendon Press 1969
35. Borsa, F., Rigamontti, A.: Magnetic resonance and phase transitions, New York, Academic Press 1979
36. Richards, P. M., Salomon, M. B.: Phys. Rev. *9,* 32 (1974)
37. Dietz, R. E. et al.: Phys. Rev. Lett. *26,* 1186 (1971)
38. Kokozka, G. F.: Low-dimensional cooperative phenomena, NATO Adv. Study Inst. Ser. *7,* 171 (1975)
39. Reinen, D. et al.: Z. Naturforsch. *31b,* 1574 (1976); Z. Anorg. Allg. Chem. *408,* 187 (1974); Z. Naturforsch. *24a,* 1518 (1969); Structure and Bonding *37,* 1 (1979)
40. Cousseins, J. C., Kozak, A. de: C. R. Acad. Sci., Paris *263,* 1533 (1966)
41. Scatturin, V. et al.: Acta Cryst. *14,* 19 (1961)
42. Yoneyama, S., Hirakawa, K.: J. Phys. Soc. Japan *21,* 183 (1966)
43. Hepworth, M. A., Jack, K. H.: Acta. Cryst. *10,* 345 (1957)
44. Lorin, D.: Thesis, Univ. Bordeaux 1980
45. Wollan, E. O. et al.: Phys. Rev. *112,* 1132 (1958)
46. Stout, J. W., Lau, H. Y.: J. Appl. Phys. *38,* 1472 (1967)
47. Bozorth, R. M., Nielsen, J. W.: Phys. Rev. *110,* 879 (1958)
48. Koelher, W. C., Wollan, E. O.: J. Phys. Chem. Solids *2,* 100 (1957)
49. Cable, J. W., Wilkinson, M. K., Wollan, E. O.: Phys. Rev. *118,* 950 (1960)
50. Joenk, R. J., Bozorth, R. M.: J. Appl. Phys. *36,* 1167 (1965)
51. Fischer, P. et al.: J. Phys. Chem. Solids *35,* 1683 (1974)
52. Fischer, P., Schwarzenbach, D., Rietveld, H. M.: J. Phys. Chem. Solids *32,* 543 (1971)
53. Fischer, P., Roult, G., Schwarzenbach, D.: J. Phys. Chem. Solids *32,* 1641 (1971)
54. Tressaud, A. et al.: Mat. Res. Bull. *16,* 207 (1981)
55. Stout, J. W., Boo, W. O. J.: J. Appl. Phys. *37,* 966 (1966)
56. Yoshimori, A.: J. Phys. Soc. Japan *14,* 807 (1959)
57. Nishikubo, T., Motizuki, K.: Sol. State Commun. *12,* 937 (1973)
58. Lau, H. Y. et al.: J. Appl. Phys. *40,* 1136 (1969)
59. Konishi, S., Motizuki, K.: Sol. State Commun. *27,* 1079 (1978)
60. Okazaki, A., Suemune, Y.: J. Phys. Soc. Japan *16,* 176 (1961)
61. Okazaki, A.: J. Phys. Soc. Japan *26,* 870 (1969)
62. Oguchi, T.: Prog. Theor. Phys. *13,* 148 (1955); Phys. Rev. *133,* 1098 (1964)
63. Hutchings, M. T. et al.: Phys. Rev. *188,* 919 (1969)
64. Kadota, S. et al.: J. Phys. Soc. Japan *23,* 756 (1967)
65. Hirakawa, K., Kadota, S.: J. Phys. Soc. Japan *23,* 756 (1967)
66. Hirakawa, K., Kurogi, Y.: Prog. Theor. Phys. *46,* 147 (1970)
67. Hirakawa, K., Yamada, I., Kurogi, Y.: J. Phys. *32,* 890 (1971)
68. Ikeda, H., Hirakawa, K.: J. Phys. Soc. Japan *33,* 393 (1972)

69. Kubo, H., Kaneshima, N., Hirakawa, K.: J. Phys. Soc. Japan *41,* 1165 (1976)
70. Kubo, H., Yahara, I., Hirakawa, K.: J. Phys. Soc. Japan *41,* 442 (1976)
71. Ikebe, M., Date, M.: J. Phys. Soc. Japan *30,* 93 (1971)
72. Miike, H., Hirakawa, K.: J. Phys. Soc. Japan *38,* 92 (1975)
73. Cosier, R. et al.: C. R. Acad. Sci., Paris, *271,* 142 (1970)
74. Tressaud, A. et al.: Mat. Res. Bull. *9,* 1219 (1974)
75. Dance, J. M.: Mat. Res. Bull. *16* (5), 599 (1981)
76. Reinen, D., Weitzel, M.: Z. Naturforsch. *32b,* 476 (1977)
77. Schnering, H. G. von: Z. Anorg. Allg. Chem. *353,* 1 (1967); *353,* 13 (1967)
78. Rüdorff, W., Babel, D.: Naturwiss. *49,* 230 (1962)
79. Rüdorff, W., Lincke, G., Babel, D.: Z. Anorg. Allg. Chem. *320,* 150 (1963)
80. Hirakawa, K., Ikeda, H.: J. Phys. Soc. Japan *35,* 1328 (1973); Phys. Rev. Lett. *33,* 374 (1974)
81. Le Dang Khoi, Veillet, P.: Phys. Rev. B *11,* 4128 (1975); *13,* 1919 (1976)
82. Dupas, A., Renard, J. P.: Phys. Lett. *53A,* 141 (1975)
83. Kleemann, W., Farge, Y.: J. Phys. *36,* 1293 (1975)
84. Laiho, R., Levola, T.: Sol. State Commun. *18,* 1619 (1976)
85. Funahashi, S., Moussa, F., Steiner, M.: Sol. State. Commun. *18,* 433 (1976)
86. Dance, J. M., Grannec, J., Tressaud, A.: C.R. Acad. Sci., Paris *283,* 115 (1976)
87. Yamada, I., Suzuki, H.: Sol. State Commun. *18,* 237 (1976)
88. Moussa, F. et al.: Sol. State. Commun. *27,* 141 (1978)
89. Odenthal, R. H., Paus, D., Hoppe, R.: Z. Anorg. Allg. Chem. *407,* 144 (1974)
90. Friebel, C., Reinen, D.: Z. Anorg. Allg. Chem. *413,* 51 (1975)
91. Gregson, A. K. et al.: J. Chem. Soc. Dalton Trans. *13,* 1306 (1975)
92. Hirakawa, K., Ikeda, H.: J. Phys. Soc. Japan *33,* 1483 (1972); *35,* 1608 (1973)
93. Knox, K.: J. Chem. Phys. *30,* 991 (1959)
94. Haegele, R., Babel, D.: Z. Anorg. Allg. Chem. *409,* 11 (1974)
95. Friebel, C., Reinen, D.: Z. Anorg. Allg. Chem. *407,* 193 (1974)
96. Ito, Y., Akimitsu, J.: J. Phys. Soc. Japan *40,* 1333 (1976)
97. Hidaka, M., Walker, P. J.: Sol. State Commun. *31,* 383 (1979)
98. Yamada, I.: J. Phys. Soc. Japan *28,* 1585 (1970); *30,* 896 (1971); *33,* 979 (1972); *33,* 1334 (1972)
99. Khomskii, D. I., Kugel, K. I.: Sol. State Commun. *13,* 763 (1973)
100. Gupta, L. C. et al.: J. Mag. Reson. *17,* 41 (1975)
101. Yamazaki, H.: J. Phys. Soc. Japan *34,* 270 (1973); *37,* 667 (1974)
102. Dance, J. M., Yoshizawa, H., Hirakawa, K.: Mat. Res. Bull. *13,* 1111 (1978)
103. Ferré, J., Régis, M.: Sol. State Commun. *26,* 225 (1978)
104. Zaspel, C. E., Drumheller, J. E.: Phys. Rev. *16,* 1771 (1977)
105. Okuda, Y. et al.: J. Phys. Soc. Japan *49,* 936 (1980)
106. Massa, W.: Inorg. Nucl. Chem. Lett. *13,* 253 (1977)
107. Köhler, P. et al.: Z. Anorg. Allg. Chem. *446,* 131 (1978)
108. Massa, W., Steiner, M.: J. Sol. State. Chem. *32,* 137 (1980)
109. Breed, D. J., Gilijamse, K., Miedema, A. R.: Physica *45,* 205 (1969)
110. Ikeda, H., Hirakawa, K.: Sol. State Commun. *14,* 529 (1974)
111. Ikeda, H., Hatta, I., Tanaka, M.: J. Phys. Soc. Japan *40* (2), 334 (1976)
112. Wertheim, G. K. et al.: Phys. Rev. *173,* 614 (1968)
113. Birgeneau, R. J., Guggenheim, H. J., Shirane, G.: Phys. Rev. B *1,* 2211 (1970)
114. Keve, E. T., Abrahams, S. C., Bernstein, J. L.: J. Chem. Phys. *51,* 4928 (1969)
115. Eibschutz, M., Holmes, L., Guggenheim, H. J.: J. Physique *32,* 759 (1971)
116. Eibschutz, M. et al.: Phys. Rev. *6,* 2677 (1972)
117. Lines, M. E.: Phys. Rev. *164,* 736 (1967)
118. Birgeneau, R. J., Guggenheim, H. J., Shirane, G.: Phys. Rev. Lett. *22,* 720 (1969)
119. Maarschall, E. P. et al.: Physica *41,* 473 (1969)
120. Breed, D. J.: Phys. Lett. *23,* 181 (1966); Physica *37,* 35 (1967)
121. De Jongh, L. J., Block, R.: Physica *79,* 568 (1975)
122. Holmes, L., Eibschutz, M., Guggenheim, H. J.: Sol. State Commun. *7,* 973 (1969)
123. Cox, D. E. et al.: Phys. Rev. B *19,* 5754 (1979)
124. Cox, D. E. et al.: J. Appl. Phys. *41* (3), 943 (1970)
125. Eibschutz, M. et al.: Phys. Lett. *29A,* 409 (1969)

126. Fukui, M., Hirose, T.: J. Phys. Soc. Japan *49,* 1399 (1980)
127. Anderson, S., Galy, J.: Acta Cryst. *B25,* 847 (1969)
128. Knoke, G., Verschaeren, W., Babel, D.: J. Chem. Research (5), 1979, 213
129. Dance, J. M. et al.: J. Chem. Research, S, *202* (1981); M, *2282* (1981)
130. Dance, J. M. et al.: Sol. State Commun. *19,* 1059 (1976)
131. Knoke, G., Babel, D.: Z. Naturforsch. *30b,* 454 (1975)
132. Heger, G., Geller, R., Babel, D.: Sol. State Commun. *9,* 335 (1971)
133. Eibschutz, M. et al.: 7th Ann. Conf. on Magn. and Magn. Mat., Chicago (1971); A.I.P. conference proceedings *5,* 670 (1972)
134. Brosset, C.: Z. Anorg. Allg. Chem. *235,* 139 (1937)
135. Tressaud, A., Galy, J., Portier, J.: Bull. Soc. Fr. Minér. Crist. *92,* 335 (1969)
136. Babel, D., Wall, F., Heger, G.: Z. Naturforsch. *29b,* 139 (1974)
137. Eibschutz, M., Guggenheim, H. J., [illegible]: J. Appl. Phys. *42,* 4 (1971)
138. Rush, J. D. et al.: Sol. State Commun. *18,* 1039 (1976)
139. Eibschutz, M. et al.: Sol. State Commun. *11,* 457 (1972)
140. Eibschutz, M., Davidson, G. R., Guggenheim, H. J.: Phys. Rev. *9,* 3885 (1974)
141. Ménil, F. et al.: Mat. Res. Bull. *12,* 983 (1977)
142. Breed, D. J. et al.: J. Appl. Phys. *41,* 1267 (1970)
143. Takano, H., Yokozawa, Y.: J. Phys. Soc. Japan *42,* 1059 (1977)
144. Tanimoto, M.: J. Phys. Soc. Japan *47,* 476 (1979)
145. Walsh, W. M., Jr. et al.: Phys. Rev. *20,* 4645 (1979)
146. Navarro, R. et al.: Physica B *83,* 97 (1976)
147. Arts, A. F. M. et al.: Sol. State Commun. *21,* 13 (1977)
147 bis Ferguson, J. et al.: Chem. Phys. Lett. *17* (4), 551 (1972)
148. Yamaguchi, Y., Sakuraba, T.: J. Phys. Soc. Japan *34,* 834 (1973)
149. Abrahams, S. C., Prince, E.: J. Chem. Phys. *36,* 50 (1962)
150. Clay, R. M., Stavely, L. A. K.: Proc. Low. Temp. Calorimetry Conf. (Helsinki) (Ann. Acad. Sci. Fennicae *210,* 194 (1966)
151. Wells, A. F.: Structural Inorganic Chemistry, Oxford, Clarendon Press 1975
152. Katz, L., Ward, R.: Inorg. Chem. *3,* 205 (1964)
153. Babel, D.: Z. Anorg. Allg. Chem. *369,* 117 (1969)
154. Wyckoff, R. W. G.: Crystal structures, New York, London, Interscience 1968
155. Cros, C.: Rev. Inorg. Chem. *1,* 163 (1979)
156. Niel, M. et al.: Physica B *86–88,* 702 (1977)
157. Steiner, M., Krüger, W., Babel, D.: Sol. State Commun. *9,* 227 (1971)
158. Steiner, M.: Z. Angew. Phys. *32,* 116 (1971)
159. Lebesque, J. V., Snel, J., Smit, J. J.: Sol. State Commun. *13,* 371 (1973)
160. Lebesque, J. V., Huyboom, N. F.: Commun. Phys. *1,* 33 (1976)
161. Lebesque, J. V.: Thesis, Univ. Amsterdam 1979
162. Steiner, M., Dorner, B.: Sol. State Commun. *12,* 537 (1973)
163. Mc Gurn, A. R., Montano, P. A., Scalapino, D. J.: Sol. State Commun. *15,* 1463 (1974)
164. Dupas, C., Renard, J. P.: Low-Temp. Phys. *LT14,* 5 (1975)
165. Grill, R. J., Durr, U., Weber, J.: Physica *86–88,* 673 (1977)
166. Montano, P. A.: Proc. M. M. M. Intermag. Conf., Pittsburgh 1976
167. Steiner, M., Kjems, J. K.: J. Phys. *10,* 2665 (1977)
168. Dupas, C., Renard, J. P.: J. Phys. *10,* 5057 (1977)
169. Steiner, M.: Sol. State Commun. *11,* 73 (1972)
170. Steiner, M., Dachs, H.: Sol. State Commun. *14,* 841 (1974)
171. Hennessy, M. J., Mc Elivee, G. D., Richards, P. M.: Phys. Rev. *7,* 930 (1973)
172. Richards, P. M.: Phys. Rev. Lett. *27,* 1800 (1971); *28,* 1646 (1972)
173. Sharpe, A. G., Woolf, A. A.: J. Chem. Soc. *1951,* 798
174. Hoppe, R., Liebe, W., Dähne, W.: Z. Anorg. Allg. Chem. *307,* 276 (1961)
175. Siebert, G., Hoppe, R.: Proc. Intern. Symp. Fluorine Chemistry, Durham 1971
176. Clark, H. C., Sadana, Y. N.: Canad. J. Chem. *42,* 50 (1964)
177. Von der Mühll, R., Ravez, J.: Rev. Chim. Miner. *11,* 652 (1974)
178. Dance, J. M., Tressaud, A.: C. R. Acad. Sci., Paris *277,* 379 (1973)
179. Rimsky, A., Thoret, J., Freundlich, W.: C. R. Acad. Sci., Paris *270,* 407 (1970)

180. Steinfink, H., Burns, J. H.: Acta Cryst. *17,* 823 (1964)
181. Ferey, G. et al.: Acta Cryst. *B33,* 1409 (1977)
182. Ravez, J. et al.: Bull. Soc. Chim. Fr. *41,* 1325 (1967)
183. Dumora, D., Von der Mühll, R., Ravez, J.: Mat. Res. Bull. *6,* 561 (1971)
184. Von der Mühll, R., Daut, F., Ravez, J.: J. Sol. State Chem. *8,* 206 (1973)
185. Von der Mühll, R., Anderson, S., Galy, J.: Acta Cryst. *B27,* 2345 (1971)
186. Ravez, J., Hagenmüller, P.: Bull. Soc. Chim. Fr. *1964,* 1811
187. Ravez, J.: Bull. Soc. Chim. Fr. *1969,* 1583
188. Grannec, J. et al.: Bull. Soc. Chim. Fr. *1971,* 64
189. Emori, S. et al.: Inorg. Chem. *8,* 1385 (1969)
190. Brosset, C.: Z. Anorg. Allg. Chem. *235,* 139 (1937)
191. Vlasse, M. et al.: Acta Cryst. *B33,* 3377 (1977)
192. Jacoboni, C. et al.: Acta Cryst. *B30,* 2688 (1974)
193. Ryan, R. R. et al. Acta. Cryst. *B27,* 1270 (1971)
194. Pistorius, C. W. F.: Mat. Res. Bull. *10,* 1079 (1975)
195. Bochkova, R. I. et al.: Sov. Phys. Dokl. *18,* 575 (1974)
196. Cretenet, J.: Rev. Chim. Minér. *10,* 399 (1973)
197. Padalko, V. M. et al.: Koord. Khim. *2,* 213 (1976)
198. Kozak, A. de: Rev. Chim. Minér. *8,* 301 (1971)
199. Günter, J. R., Matthieu, J. P., Oswald, H. R.: Helv. Chim. Acta. *61,* 328 (1978)
200. Sears, D. R., Hoard, J. L.: J. Chem. Phys. *50,* 1066 (1969)
201. Tressaud, A. et al.: J. Sol. State Chem. *2,* 269 (1970)
202. Tressaud, A. et al.: Sol. State Commun. *37,* 479 (1981)
203. Dance, J. M. et al.: J. Mag. Mag. Mat. *15–18,* 534 (1980)
204. Brosset, C.: Thesis, Univ. Stockholm 1942
205. Chassaing, J.: Rev. Chim. Minér. *5,* 1115 (1968)
206. Hall, W. et al.: Inorg. Chem. *16,* 1889 (1977)
207. Gallagher, K. J., Ottaway, M. R.: J. Chem. Soc., Dalton Trans. 1977, 2212
208. Ferey, G., Leblanc, M., de Pape, R.: J. Solid State Chem. *40,* 1 (1981)
209. Walton, E. G. et al.: Inorg. Chem. *16,* 2425 (1977)
210. Baumgartel, E., Teich, J.: Z. Anorg. Allg. Chem. *386,* 279 (1971)
211. Bukovec, P.: Monath. Chem. *105,* 517 (1974)
212. Edwards, A. J.: J. Chem. Soc., Dalton Trans. *1972,* 816
213. Waltersson, K.: J. Sol. State Chem. *29,* 195 (1979)
214. Edwards, A. J.: J. Chem. Soc. *1971,* 2653
215. Bukovec, P., Kaucic, V.: Acta Cryst. *B34,* 3339 (1978)
216. Kaucic, V., Bukovec, P.: Acta Cryst. *B34,* 3337 (1978)
217. Bukovec, P.: Communication at the International Symposium on Fluorine Chemistry, Avignon 1979
218. Bukovec, P.: Monath. Chem. *105,* 1299 (1974)
219. Massa, W.: Communication at the 9th European Symposium on fluorine Chemistry, Venezia 1980
220. Tressaud, A. et al.: Mat. Res. Bull. *8,* 565 (1973)
221. Wintenberger, M., Dance, J. M., Tressaud, A.: Sol. State Commun. *17,* 185 (1975)
222. Tressaud, A. et al.: Mat. Res. Bull. *8,* 1467 (1973)
223. Ferey, G., Pape, R. de, Boucher, B.: Acta Cryst. *B34,* 1084 (1978)
224. Dance, J. M. et al.: C. R. Acad. Sci., Paris *288,* 37 (1979)
225. Smith, S. R. P., Owen, J.: J. Phys. C. *4,* 1399 (1971)
226. Petrouleas, V., Simonopoulos, A., Kostikas, A.: Phys. Rev. B *12,* 4675 (1979)
227. Cheng, C., Wong, H., Reiff, N. M.: Inorg. Chem. *16,* 819 (1977)
228. Georges, R. et al.: J. Sol. State Chem. *9,* 1 (1974)
229. Dzialoshinski, I.: J. Phys. Chem. Solids. *1,* 362 (1958)
230. Gupta, G. P. et al.: J. Phys. C. *10,* L 459 (1977)
231. Gupta, G. P., Dickson, D. P. E., Johnson, C. E.: J. Phys. C *11,* 215 (1978)
232. Gupta, G. P. et al.: J. Phys. C. *11,* 3889 (1978)
233. Hanzel, D. et al.: Sol. State Commun. *22,* 215, (1977)
234. Gupta, G. P., Dickson, D. P. E., Johnson, C. E.: J. Phys. C *13,* 2071 (1980)

235. Kida, J.: J. Phys. Soc. Japan *30,* 290 (1971); *34,* 952 (1973)
236. Schoonman, J., Hellstrom, E. E., Huggins, R. A.: J. Sol. State Chem. *18,* 325 (1976)
237. Schoonman, J., Bottelberghs, P. M.: Solid Electrolytes, P. Hagenmuller, W. Van Gool (eds.), New York, Academic Press 1978
238. Jones, E. R. et al.: Sol. State Commun. *8,* 1657 (1970)
239. Teufer, G.: Acta Cryst. *17,* 1480 (1964)
240. Farach, H. A., Poole, C. P. Jr., Nicklin, R. C.: J. Mag. Reson. *23,* 221 (1976)
241. Knoke, G.: Thesis, Marburg 1977
242. Dewan, J. C., Edwards, A. J.: J. Chem. Soc., Chem. Commun. *1977,* 533
243. Babel, D., Knoke, G.: Z. Anorg. Allg. Chem. *442,* 151 (1978)
244. Dance, J. M., Tressaud, A.: unpublished results
245. Wall, F., Pausewang, G., Babel, D.: J. Less. Common. Metals *25* (3), 252 (1971)
246. Edwards, A. J., Jones, G. R., Sills, R. J. C.: J. Chem. Soc. *1970,* 2511
247. Darriet, J., Lozano, L., Tressaud, A.: Sol. State Commun. *32,* 493 (1979)
248. Morrel, B. K. et al.: Inorg. Chem. *12,* 2640 (1973)
249. Darriet, J. et al.: Mat. Res. Bull., to be published
250. Néel, L.: Ann. Phys. *5,* 232 (1936)
251. Gorter, C. J., Peski-Tinbergen, T. van: Physica *1956,* 273
252. Mess, K. W. et al.: Physica *34,* 126 (1967)
253. Keffer, F., Kittel, C.: Phys. Rev. *85,* 329 (1952)
254. Keffer, F.: Phys. Rev. *87,* 608 (1952)
255. Jacobs, I. S.: J. Appl. Phys. Suppl. *32,* 61 S (1961)
256. Johnson, F. M., Nethercot, A. H., Jr.: Phys. Rev. *114,* 705 (1959)
257. Foner, S.: Phys. Rev. *107,* 683 (1957)
258. Johnson, F. M., Nethercot, A. H., Jr.: Phys. Rev. *104,* 847 (1956)
259. De Wijn, H. W. et al.: Phys. Rev. B *8,* 299 (1973)
260. Wanklyn, B. M.: J. Mat. Science *10,* 1487 (1975)
261. Stahlbush, R. E., Scott, J. C.: Sol. State Commun. *33,* 707 (1980)
262. Stahlbush, R. E., Scott, J. C.: J. Appl. Phys. *50,* 1664 (1979)
263. Anderson, P. W.: Phys. Rev. *86,* 694 (1952)
264. Davis, H. L.: Phys. Rev. *120,* 789 (1960)
265. Lines, M. E.: J. Phys. Chem. Solids *31,* 101 (1970)
266. De Wijn, H. W. et al.: J. Appl. Phys. *42,* 1595 (1971)
267. Walstedt, R. E., Wijn, H. W. de, Guggenheim, H. J.: Phys. Rev. Lett. *25,* 1119 (1970)
268. Colpa, J. H. P., Sieverts, E. G., Linde, R. H. van der: Physica *51,* 573 (1971)
269. De Jongh, L. J.: Proc. 18th Ann. Conf. Magn. Magn. Mat., Denver, A.I.P. Conf. Proc. *10,* 561 (1972)
270. Ishikawa, T., Oguchi, T.: Prog. Theor. Phys. *54,* 1282 (1975)
271. Ingen Schenau, A. D. van et al.: Acta Cryst. *B32,* 1127 (1976)
272. Henkens, L. S. J. M. et al.: Phys. Rev. Lett. *36,* 1252 (1976)
273. Henkens, L. S. J. M. et al.: Physica *83,* 147 (1976)
274. Reinen, D., Krause, S.: Inorg. Chem. *20,* 2750 (1981)

Received July 15, 1981
D. Reinen (editor)

# Structure and Bonding in Organic Derivatives of Antimony(V)

**Vimal K. Jain, Rakesh Bohra and Ram C. Mehrotra**

Department of Chemistry, Rajasthan University, Jaipur-302 004, India

In view of a renewed interest in the structural chemistry of organic derivatives of antimony(V) during the last decade, a brief account of the same is presented highlighting points of special importance and future growth. Attention is focussed on the structural and bonding aspects. Details about their synthesis are only included when they are applicable in general.

# 1 Introduction

As evinced by the volume of publications during the last decade, there has been an increasing interest in the chemistry of antimony(V) compounds with organic ligands in general and organoantimony(V) compounds in particular. A large number of compounds of antimony(V) with organic ligands are now known but they have been mentioned in a scattered fashion in books and treatises[1–3]. The chemistry of organometallic compounds of antimony(V) has been summarized in some review articles and books[3–12], but the last one of these appeared only in 1970. Although the work published in this direction after 1970 has been abstracted in Annual Surveys of Organometallic Chemistry, there is as such no comprehensive review article dealing with both classes of compounds simultaneously, i.e., organic and organometallic compounds of antimony(V).

In this article we have attempted to summarize the recent work (particularly after 1970) focussing attention mainly on the structural and bonding aspects and including only such procedures of synthesis, etc., which have general applicability.

# 2 Alkyls, Alkenyls and Aryls

A number of pentaorganoantimony compounds are reported in the literature[13–33] and are summarized in Table 1. Pentaorganoantimony compounds have generally been prepared by the reactions of tri- or tetra-organoantimony(V) halides with the desired organolithium compounds or with Grignard reagents[20–25, 28–32]. These are covalent compounds which are monomeric in nature. The alkenyl compounds are in general hydrolytically unstable.

**Table 1.** $R_{5-n}SbR'_n$ (n = 0 – 5) Compounds and physical measurements

| Compounds<br>1 | Properties<br>2 | Physical measurements<br>3 | Ref.<br>4 |
|---|---|---|---|
| $Me_5Sb$ | b.p. 126–127,<br>m.p. −18 to −16 | IR, Raman, $^1H$ NMR | 13, 28, 31, 34,<br>35, 50 |
| $Et_5Sb$ | b.p. 64/0.4 | IR, $^1H$ NMR | 26, 28 |
| $Bu_5Sb$ | | | 29,31 |
| $(\triangleright)_5Sb$ | Yellow liquid,<br>b.p. 100/0.15 | IR, Raman, $^1H$ NMR | 43 |
| $Ph_5Sb$ | White crystals,<br>m.p. 171–172 | Dipole moment, IR<br>$^1H$ and $^{13}C$ NMR, Mass,<br>Mössbauer, X-ray | 22, 30, 38–42,<br>49, 51, 53, 55 |
| $Ph_5Sb \cdot 1/2\, C_6H_{12}$ | Monoclinic prisms,<br>m.p. 169–170 | Mössbauer, X-ray | 44, 52 |
| $(p\text{-}MeC_6H_4)_5Sb$ | Colourless crystals,<br>m.p. 189 | IR, Raman, $^1H$ and<br>$^{13}C$ NMR, X-ray | 45–49, 54 |
| $(CH_2{=}CH)_5Sb$ | Greenish oil,<br>at 110° → $(CH_2{=}CH)_3Sb$<br>$n_D^{20}$ 1.5590 | | 14, 18 |

**Table 1** (continued)

| Compounds<br>1 | Properties<br>2 | Physical measurements<br>3 | Ref.<br>4 |
|---|---|---|---|
| $(CH_2{=}CMe)_5Sb$ | Thick greenish liquid, decompose at 180, m.p. 60 | IR | 21, 24, 36, 37 |
| *cis* $(MeCH{=}CH)_5Sb$ | Colourless liquid, $n_D^{20}$ 1.5610 at 101° → *cis*$(MeCH{=}CH)_3Sb$ | IR | 16, 17, 21, 36 |
| *trans* $(MeCH{=}CH)_5Sb$ | Colourless liquid, $n_D^{20}$ 1.5490 at 160° → *trans*$(MeCH{=}CH)_3Sb$ | IR | 16, 17, 21, 36 |
| $MeSbEt_4$ | b.p. 55/0.4 | $^1H$ NMR | 13 |
| $MeSbPh_4$ | m.p. 110 | $^1H$ NMR | 15, 30 |
| 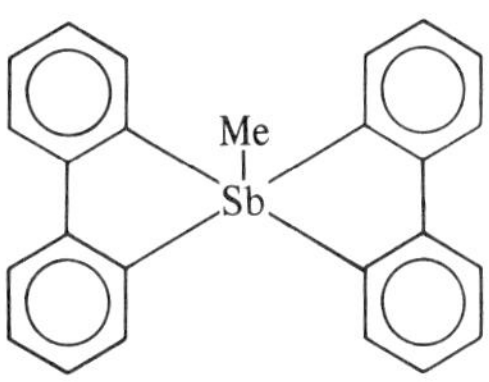 | | Mass | 51 |
| $EtSbPh_4$ | m.p. 105–106 | $^1H$ NMR | 30 |
| 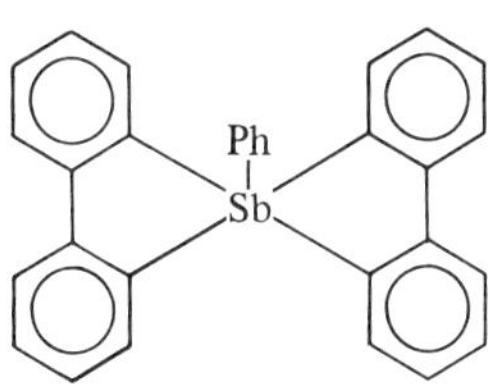 | | Mass, X-ray | 22, 37 51 |
| 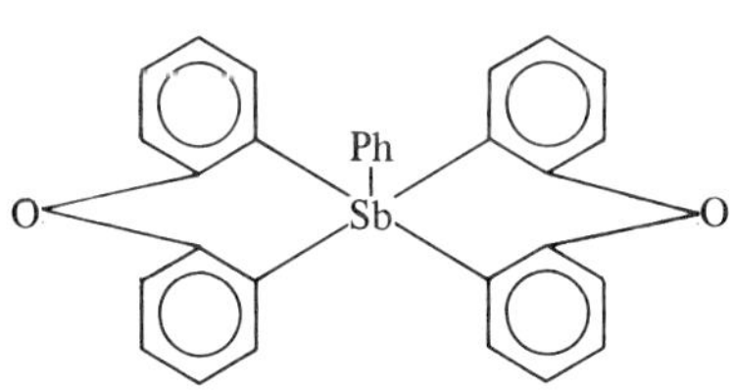 | | | 29 |
| 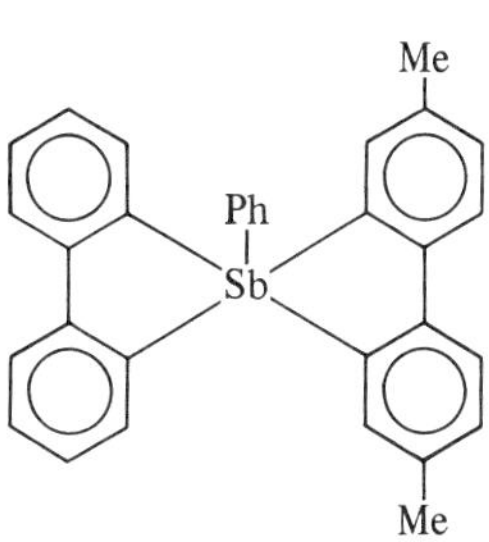 | | $^1H$ NMR | 33 |

**Table 1** (continued)

| Compounds<br>1 | Properties<br>2 | Physical measurements<br>3 | Ref.<br>4 |
|---|---|---|---|
| 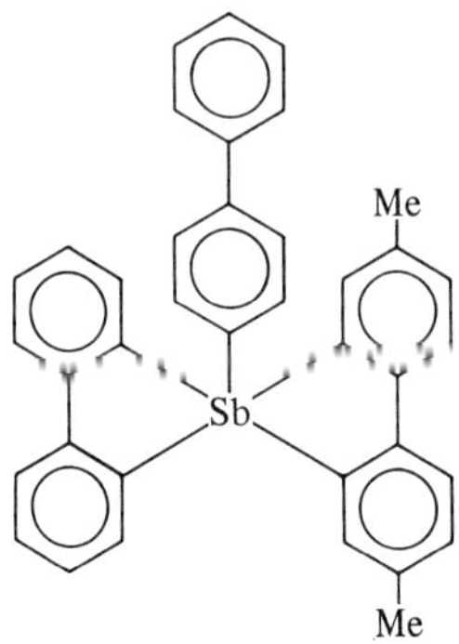 | | $^1$H NMR | 33 |
| ($p$-$CH_3C_6H_4$)$SbPh_4$ | | | 54 |
| $Me_2SbEt_3$ | b.p. 42/0.03 | IR, $^1$H NMR | 28, 32 |
| $Me_2SbBu_3$ | | | 31 |
| $Et_2Sb(CH{=}CH_2)_3$ | yellowish liquid,<br>$n_D^{20}$ 1.5470 | | 20, 23, 24 |
| $Et_2Sb(CH{=}CHMe)_3$*cis* | hydrolytically unstable<br>$n_D^{20}$ 1.5362 | IR | 20, 23, 24 |
| $Et_2Sb(CH{=}CHMe)_3$*trans* | yellow liquid,<br>$n_D^{20}$ 1.5395 | IR | 20, 23, 24 |
| $Et_2Sb(CMe{=}CH_2)_3$ | yellow hydrolytically unstable liquid,<br>$n_D^{20}$ 1.5414 | | 20, 24 |
| $Ph_2Sb(CH{=}CHMe)_3$*cis* | colourless liquid<br>$n_D^{20}$ 1.6270 | IR | 23, 24 |
| $Ph_2Sb(CH{=}CHMe)_3$*trans* | yellow substance<br>$n_D^{20}$ 1.6110 | IR | 23, 24 |
| $Ph_2Sb(CMe{=}CH_2)_3$ | yellow liquid<br>$n_D^{20}$ 1.6191 | | 24 |
| 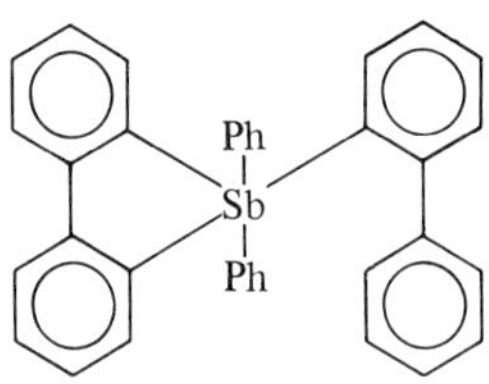 | | | 29 |
| 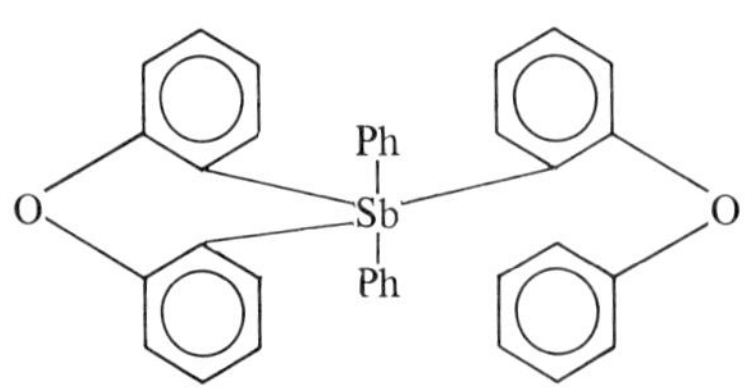 | | | 29 |

**Table 1** (continued)

| Compounds<br>1 | Properties<br>2 | Physical measurements<br>3 | Ref.<br>4 |
|---|---|---|---|
| $Me_3SbEt_2$ | b.p. 71–74/16 | $^1H$ NMR | 28 |
| $Me_3SbBu_2$ | | | 31 |
| $Me_3Sb(C{\equiv}CMe)_2$ | | IR, Raman,<br>$^1H$ NMR, X-ray | 32 |
| 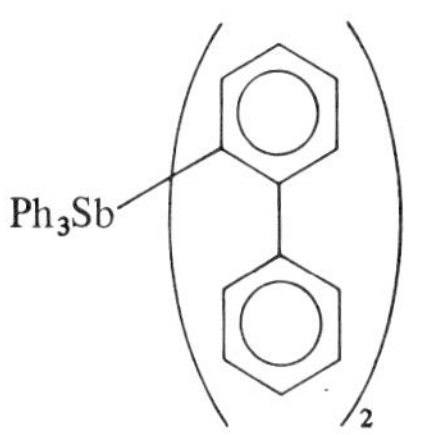 | | | 29 |
| $Me_4SbEt$ | b.p. 53-54/16 | $^1H$ NMR | 28 |
| $Me_4SbBu$ | | | 31 |

The pentaorgano compounds of group V elements ($R_5E$) with formally 10 valence electron shells have in general bonds directed to the vertices of a trigonal bipyramid. On the basis of infrared and Raman spectral data, a trigonal bipyramidal geometry has been established for pentamethylantimony[34, 35]. Infrared spectra of pentaethyl-[26], *cis* and *trans* isomers of pentapropenyl-[21, 36], diethyl tripropenyl- and diphenyl tripropenyl-[23] antimonys have also been investigated.

X-ray diffraction studies of phenyl-2,2′-biphenylene antimony[37] and $Me_3Sb(C \equiv CMe)_2$[32] have shown a trigonal bipyramidal geometry around antimony atom in these molecules. Crystals of $Me_3Sb(C \equiv CMe)_2$ belong to the monoclinic system with space group $C_2/c$ in which methyls are occupying the equatorial positions.

In place of the more common bipyramidal geometry a number of pentaorganoantimony derivatives show the alternate square pyramidal geometry also and the best known example of this class of compounds is pentaphylantimony, $Ph_5Sb$. Two dimensional[38, 39] and the three dimensional[40] X-ray diffraction analyses of $Ph_5Sb$ have shown that the molecule has a slightly distorted square pyramidal gemetry in which antimony atom is about 0.5 Å above the basal plane. The $C_{4v}$ symmetry of the molecule is distorted by alternate deviations of $C_{\overline{(axial)}}Sb\text{-}C_{(basal)}$ angles by about $\pm 4°$ from their mean value, Fig. 1. It was suggested that "crystal packing" forces in the solid state cause $Ph_5Sb$ to exhibit square pyramidal geometry. However, it has been concluded from vibrational spectra that the local $C_{4v}$ symmetry (square pyramidal geometry) is maintained by $Ph_5Sb$ in $CH_2Cl_2$ or $CH_2Br_2$ solutions also[41, 42]. The vibrational spectra of pentacyclopropylantimony have also been interpreted in terms of square pyramidal geometry[43].

In contrast to pentaphenylantimony, the crystals of pentaphenylantimony · 0.5 cyclohexane molecules have the antimony atoms in a trigonal bipyramidal geometry with virtually no distortion[44]. C-Sb-C angles differ by 2.2° from their ideal values. The axial

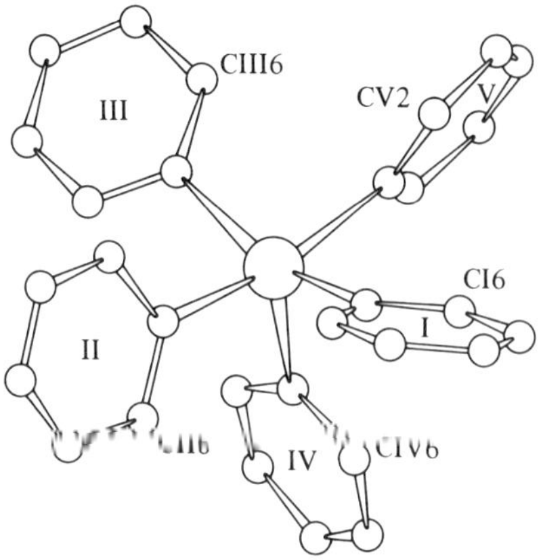


**Fig. 1.** Structure of pentaphenylantimony

bonds (av. 2.24 A°) are significantly longer than the equatorial ones (av. 2.14 A°). Cyclohexane molecule is situated halfway along the "a" axis and it makes no special contacts with other neighbouring molecules.

Again in contrast to pentaphenylantimony, the analogous penta *p*-tolylantimony, ($p$-$MeC_6H_4)_5Sb$, exhibits the more common trigonal bipyramidal geometry. The crystals are monoclinic with space group $P2_{1/c}$. The Sb-C equatorial bond length average, 2.16 A°, is slightly shorter than the axial average, 2.26 A°. Two of the C-Sb-C bond angles in the equatorial plane (113, 130°) differ considerably from 120°, but the other angles around antimony are normal[45)].

In view of the unexpected difference in the solid state structures of $Ph_5Sb$ and ($p$-tol$)_5Sb$, an attempt has been made to determine the solution stereochemistry of the latter compound also. The $^1H$ NMR spectra of ($p$-tol$)_5Sb$ have, therefore, been determined at ambient temperatures[46, 47)] and at low temperatures[45, 48)]; only one methyl proton signal (2.24 ppm in $CS_2$) has been observed even at −100°. This has been explained by assuming rapid intramolecular ligand exchange. However, in a later study[49)], broadening in this signal (the position varying from 1.9 Hz at 30° to 8.7 Hz at −129°) has been observed at −130°. The $^{13}C$ NMR spectrum of this compound shows five sharp resonances due to the non-equivalent types of carbon atoms in the *p*-tolyl ligand. The single resonances for all the five ligands has been attributed to rapid intramolecular exchange of ligand position[49)]. The $^{13}C$ NMR spectrum, therefore, virtually eliminates the possibility that the observance of a single $^1H$ NMR peak for methyl protons might be due to accidental proton magnetic equivalence of a "static" solution structure of $D_{3h}$ or $C_{4v}$ rather than to rapid intramolecular exchange. The intramolecular positional exchange barrier energy for the ($p$-tol$)_5Sb$ molecule has been estimated to be 1.46 kcal/mol from a line shape analysis (of the carbon and methyl proton resonances). In fact, there appears to be no static solution structure except at extremely low temperatures (below −130° spectra could not be taken due to experimental limitations), and the low energy barrier suggests that ligand size is probably not a dominant factor in limiting the exchange process.

$^1H$ NMR spectra of $Me_5Sb$ and $Me_nSbEt_{5-n}$ ($n = 1-5$) display only one sharp methyl proton resonance and the same is maintained at temperatures as low as −100° in $Me_5Sb$ and −80° in latter compounds[28, 50)]. This equivalence of all methyl groups even at low temperatures has been ascribed to rapid intramolecular exchange of non-equivalent axial and equatorial methyl groups.

Mass spectra of pentaphenyl-, methyl bis-2,2′-biphenylene and phenylbis-2,2′-biphenylene antimonys have been investigated by Hellwinkel et al.[51]. Mössbauer spectra of $Ph_5Sb$ and $Ph_5Sb \cdot 1/2\,C_6H_{12}$ have been determined by Long et al.[52] and IS values of 4.6 and 4.2 ± 0.1 mm/s respectively have been reported.

Phenyl bis-2,2′-biphenylene antimony reacts with butyl lithium to give pentabutylantimony which after cleavage with acid yields tetrabutylantimony cation[29]. The compounds $Ph_4SbR$ (R = Me or Et) on treatment with BuLi, PhLi or RMgX are "symmetrised" to produce $Ph_5Sb$ and other products. The formation of hexacovalent derivatives in these reactions has also been suggested according to the following Eq.[30]:

$$Ph_4SbMe + PhLi \rightleftharpoons [Ph_5SbMe]^-Li^+ \rightleftharpoons Ph_5Sb + MeLi$$

Cleavage of Sb-C bond in pentaorganoantimony compounds has also been studied[20, 26, 32]. One of the Sb-C bonds is cleaved by radical reaction process when $Ph_5Sb$[53] or $(p\text{-tol})_5Sb$[54] is treated with $CCl_4$ or $CHCl_3$ in MeOH. Photolysis and thermolysis of $Ph_5Sb$[55] have been demonstrated by $^{13}C$ tracer studies and it has been concluded that in former case free $Ph^{\cdot}$ radicals are formed whereas in latter case, an intramolecular process not involving free $Ph^{\cdot}$ radicals is involved.

The difference in the structures of $Ph_5Sb$ and $(p\text{-tol})_5Sb$ in the solid state has been the subject of considerable interest and an attempt has been made to rationalize the results on the basis of crystal packing effects. However, as indicated above, there appears to be only a small difference in the potential energies of square pyramidal and trigonal bipyramidal structures due to stereochemical non-rigidity, fluxinal nature and intramolecular exchange of axial and equatorial groups in $R_5E$ molecules. Indeed, Brock and Ibers[56] have attempted to assess the magnitude of lattice effects favouring the square pyramidal geometry in spite of the more dominant intramolecular interactions favouring the trigonal bipyramidal geometry. Later on Brock[57] has made use of non-bonded potential functions for carbon and hydrogen atoms which include Coulombic terms and calculated that when Coulombic terms have been included the results are in more harmony with experimental results of $Ph_5P$, $Ph_5As$ and $Ph_5Sb$. The difference in the molecular energies between the two $Ph_5E$ geometries is markedly dependent on the exact model used and changes as small as 0.01 A° in the C-C bond length can affect the intramolecular energy difference by as much as 1.19 kcal/mol or more.

# 3 Halides, Pseudohalides, Oxides and Related Compounds

## *3.1 Monoorganoantimony(V) Halides*

Little is known about structural aspects of mono-alkyl and arylantimony(V) chlorides[58–60], because of their unstable nature. For example, phenylantimony(V) chloride is an unstable compound, which tends to disproportionate as shown below on standing[58]:

$$2\,PhSbCl_4 \rightarrow Ph_2SbCl_3 + SbCl_3 + Cl_2\ .$$

Contrary to the unstable nature of monoorganoantimony(V) chlorides their addition compounds with oxygen donor ligands are found to be quite stable under ambient conditions. On the basis of IR and $^1H$ NMR spectral studies, the existence of six coordinated antimony atom has been suggested in these derivatives[61]. $MeSbCl_4 \cdot L$ (L = PyO or 4-MePyO) exists in *cis* and *trans* isomeric forms in solution (Fig. 2):

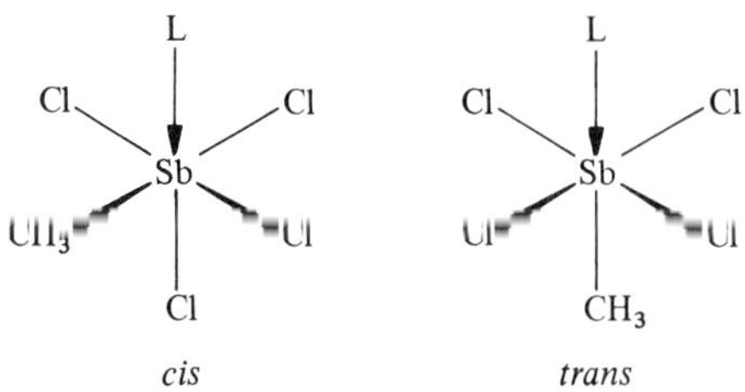


**Fig. 2.** *Cis* and *trans* isomeric forms of $MeSbCl_4L$ (L = PyO or 4-MePyO)

### 3.2 Diorganoantimony(V) Halides

Only a few dialkylantimony(V) halides are known[62–64]. Most of the diarylantimony(V) chlorides have been prepared by the reactions of antimony(III) chlorides with diazonium salts[65–69] or by some other methods[70–74] such as the halogenation of diorganoantimony-(III) halides, $R_2SbX$.

Diorganoantimony(V) halides are crystalline compounds with sharp melting points. Dialkyl derivatives are relatively less stable than their diaryl analogues and decompose on standing at ambient temperatures[62, 75, 76] in alkyl halide. Diarylantimony(V) halides can be reduced by stannous chloride, sulfur dioxide and by other reducing agents to diorganoantimony(III) halides[71, 72, 77, 78].

Kolditz and coworkers[79] found that diphenylantimony(V) chloride is a weak electrolyte in acetonitrile. There was some confusion regarding the structure of this compound. Originally Polynova and Porai-Koshits[80] on the basis of X-ray analysis described that the compound has a trigonal bipyramidal geometry. However, in a later publication[81], these workers reported the compound as a monohydrate, $Ph_2SbCl_3 \cdot H_2O$, with octahedral geometry. The trigonal bipyramid geometry of $Ph_2SbCl_3$ suggested by Gukasyan et al.[82] on the basis of Mössbauer data was contradicted by Sams et al.[83] who adduced evidence for the presence of chlorine bridges in the compound. In order to resolve the above confusion in the literature, Bordner et al.[84] reexamined the single crystal X-ray analysis of $Ph_2SbCl_3$. These workers found that the anhydrous $Ph_2SbCl_3$ exists as a dimer with chlorine bridges (Fig. 3) and the crystals belong to Pnnm space group. Bowen and coworkers[85, 86] have also supported an octahedral geometry with small deviation (8°) for anhydrous $Ph_2SbCl_3$ as well as for the other trichloro compounds [($p$-$FC_6H_4$)$PhSbCl_3$[85], ($p$-$MeC_6H_4$)$PhSbCl_3$[85] and $Me_2SbCl_3$[86, 87]] ($e^2qQ$ and $\delta$ values are very near to those reported for $Ph_2SbCl_3$).

Bone and Sowerby[88] have prepared $Ph_2SbBr_3$, $Ph_2SbBr_2Cl$ and $Ph_2SbBrCl_2$. Having prepared these three new organoantimony(V) compounds, Bone and Sowerby[89] then determined their crystal structure. Contrary to the dimeric, octahedral $Ph_2SbCl_3$, all three of the bromine containing compounds are monomeric with a distorted trigonal bipyramidal geometry having one bromine and two phenyl groups at equatorial posi-

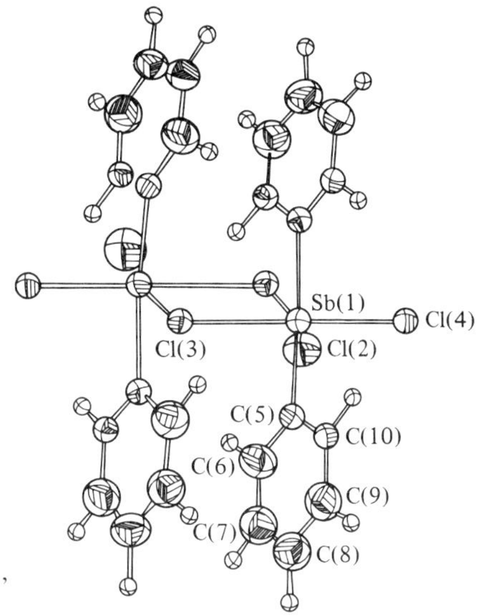


**Fig. 3.** Structure of diphenylantimony(V) chlorode, $(Ph_2SbCl_3)_2$

tions. Sb-Br equatorial distances decrease from 2.478 A° in $Ph_2SbBr_3$ to 2.462 A° in $Ph_2SbBr_2Cl$ to 2.446 A° in $Ph_2SbBrCl_2$. There is a similar but less pronounced shortening of the Sb-C bond in all three compounds. This has been attributed to increased electronegativity of atoms in axial positions in going from $Ph_2SbBr_3$ to $Ph_2SbBrCl_2$. The two bonds from antimony to the axial halogens have different lengths in all three compounds with the halogen at the greater distance being involved in weak intermolecular bonding to a neighbouring antimony atom.

An attempt to prepare $Ph_2SbBr_2F$ resulted into the formation of a crystalline compound $Ph_2SbBr_{2.5}F_{0.5}$[90]. An X-ray analysis shows that the compound is dimeric with strong fluorine bridge and formulated as $Ph_2SbBr_2F \cdot Ph_2SbBr_3$. One antimony has approximately octahedral coordination, while the second possesses a trigonal bipyramidal geometry.

The vibrational spectra of the compounds, $R_2SbCl_3$ (R = Me or Et), have been recorded in solution as well as in the solid state[64, 91–93]. These studies indicate that the compound, $Me_2SbCl_3$, possesses trigonal bipyramidal geometry in solution with two equatorial methyl groups but in the solid state, it exists as a dimer with bridged chlorine atoms and with the methyl groups *trans* to each other[93]. The Mössbauer spectral data for $Me_2SbCl_3$ have also been explained on the basis of a six coordinated structure with bridging chlorine atoms and *trans* organic groups[86, 87, 94].

The crystals of $Me_2SbCl_3$ exist in two isomeric forms. In the dimeric covalent one, $(Me_2SbCl_3)_2$, two formula units are bridged by chlorine, building a four membered ring and the crystals have been found to belong to orthorhomic system with space group *Pnma*. The second form has been suggested to have an ionic structure, $[Me_4Sb]^+[SbCl_6]^-$, in which antimony atoms have slightly distorted tetrahedral and octahedral coordination polyhedra. The crystals of this form belong to triclinic system with space group *P 1*[95].

Mössbauer spectral data for a few heterocyclic *cis*-diorganoantimony compounds of the general formulae, $[NMe_4][SbR_2Cl_4]$ and $R_2SbCl_3$ have been reported recently[96]. The large value of asymmetry parameter $\eta$ for the tetrachloro compounds has been interpreted in terms of distorted octahedral geometry and distortion has been ascribed due to formation of heterocyclic ring. However,

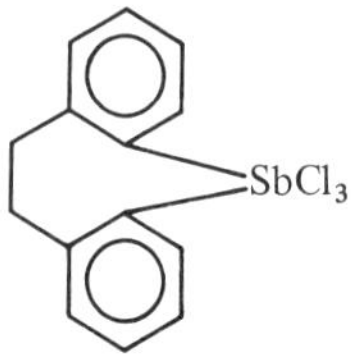

for a pentacoordinated structure has been proposed with a large C-Sb-C bond angle.

Diorganoantimony(V) halides form monomeric covalent adducts, $R_2SbX_3 \cdot L$, with various oxygen donor ligands (e.g. DMSO, HMPA, TPPO, PyO)[91, 97]. An octahedral geometry has been proposed on the basis of infrared and $^1H$ NMR spectral data. Out of the three possible geometrical isomers, the existence of the following isomer has been proposed:

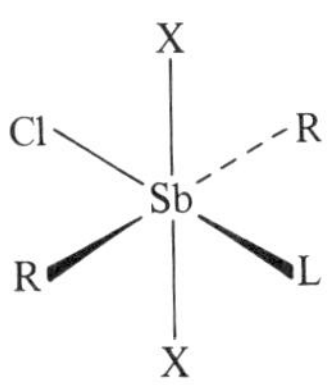


The $\nu$ Sb-O frequency shift has been correlated to the polarity of antimony oxygen bond and the acid strength of the acceptor, which increases in the following order:

$$Me_2SbCl_3 \approx Et_2SbCl_3 < Ph_2SbCl_3 < SbCl_5$$

A number of mono and diorganoantimonates, $[Y]^+[RSbCl_5]^-$ and $[Y]^+[R_2SbCl_4]^-$, have also been isolated[8, 9] and have been found to behave as 1 : 1 electrolytes.

## 3.3 *Triorganoantimony(V) Halides*

Compounds of this group are among the best known organoantimony derivatives and have been investigated in greater detail. Although these compounds can be synthesized through a number of routes[20–23, 64, 98–111], yet the most general method for the preparation of these compounds is the halogenation of triorganoantimony[62, 112–119]. Only a few mixed halogen compounds of the type $R_3SbXY$ are known and can be obtained by exchange reactions[64, 107, 116].

Triorganoantimony(V) halides are readily reduced to the corresponding antimony-(III) derivatives by various reducing agents[67, 70, 99, 114, 120, 121]. Both trialkyl and arylantimony(V) halides undergo elimination reaction, when heated above their melting points in an inert atmosphere[62, 76, 103, 113, 122]:

$$R_3SbX_2 \xrightarrow{\Delta} R_2SbX + RX$$

Conductivity studies have shown that the triorganoantimony(V) halides are essentially covalent[79, 116, 123, 124]. The solvolysis of triorganoantimony(V) halides is of considerable interest. On the basis of conductometric studies it has been found that trimethylantimony(V) halides can readily be hydrolysed in aqueous solutions[123, 125]. Nefedov et al.[126] have demonstrated that $(CH_3)_3Sb$ and $(CD_3)_3SbCl_2$ undergo isotope exchange in alcoholic solution to give $(CD_3)_3Sb$ and concluded that the exchange reaction involves principally an exchange of electrons:

$$CD_3Sb^{++} + Me_3Sb \rightleftharpoons (CD_3)_3Sb + Me_3Sb^{++}$$

Molecular weight measurements of a few triorganoantimony(V) halides show their monomeric behaviour[50, 79, 127]. Dipole moments of a few of these derivatives have also been measured and correlated with structure[128, 129]. The dimagnetic susceptibilities of tribenzylantimony(V) chloride and a number of triarylantimony(V) halides have been determined by Parab and Desai[130]. In the aryl derivatives, the question of $P_\pi - d_\pi$ resonance between the benzene ring and vacant d-orbitals of the antimony atom has been considered by Jaffe[131] and by Rao et al.[132].

For triorganoantimony(V) halides a trigonal bipyramidal geometry has been suggested on the basis of several IR and Raman investigations[114, 115, 117, 133–146]. The presence of only one Sb-C stretching frequency in the infrared spectra of trimethylantimony(V) halides indicates that these molecules belong to highly symmetric $D_{3h}$ point group[114]. The low frequency IR and Raman spectra of $R_3SbX_2$ (R = Et, $Pr^i$; X = F, Cl, Br, I) have been interpreted in terms of $C_{3v}$ symmetry of the molecule[142]. Coordinate analysis of trimethylantimony(V) halides[137, 139, 140] and of its deutrated analogues[139, 140] has also been carried out using modified Urey-Bradley force field. The (Sb-C) equatorial stretching force constant decreases in the order $Me_3Sb^{++} > Me_3SbF_2 > Me_3SbCl_2 > Me_3SbBr_2 > Me_5Sb$ indicating the strengthening of Sb-C bond as the axial group becomes more electronegative[137]. The intensity of Raman active metal halogen stretching mode for the series $Me_3SbX_2$ (X = F, Cl, Br) decreases on increasing the electronegativity of the halogen[139]. It has also been observed that the fundamental frequencies of vibration in $R_3SbX_2$ (R = Me or Ph; X = halogen) have been found to be dependent on the mass and electronegativity of the halogen and the molecular weight and moment of inertia of the molecule[141]. The force field calculations of $Me_3SbX_2$ (X = F, Cl, Br, I) give force constants, mean vibrational amplitudes bond charge, path length parameters and rotation distortion constants which agree well with experimental determinations and normal coordinate calculations[145].

The use of NMR spectroscopy has also been made to elucidate the nature of bonding in trialkyl- and triarylantimony(V) halides[50, 147–153]. In the $^1$H NMR spectra of trimethylantimony(V) halides at −32°, sharp ringing signals have been observed but in case of difluoride, the methyl protons have been found to couple with the fluorines. At ambient temperatures, the broadening of the signal and absence of coupling between fluorine nuclei and methyl protons in difluoride, have been explained by intermolecular exchange of halogen. The appearance of a single methyl proton signal and splitting at −32° by two equivalent fluorines are consistent with the trigonal bipyramidal geometry. This conclusion is supported by $^{19}$F NMR studies also[148]. NMR spectral data of $Ph_3SbF_2$ and $Ph_3SbFCl$ have also been found to be consistent with the trigonal bipyramidal geometry[50]. Halide exchange reactions in triorganoantimony(V) have also been investigated by Moreland et al.[149–152] by NMR technique. A trigonal bipyramidal geometry has also been proposed for the mixed halides, $R_3SbXY$. It has been concluded from $^{13}C$ and $^1H$ NMR spectral data that in $Ph_3SbX_2$ (X = Cl or Br) and in $Bu^n_3SbBr_2$, the antimony atom withdraws electrons from the hydrocarbon group[153] and these compounds obey the additivity rule of the $^{13}C$ NMR.

Equilibrium constants of dihalide ions for the formation of triphenylantimony(V) halides have been determined by extraction method. These constants (log $\beta$) have been determined as 19.5, 7.0, 2.5, 1.1 ($\pm$ 1.0) respectively for triphenylantimony(V) hydrxide, fluoride,chloride and iodide[154]. With the help of high energy photo electron spectra of $Ph_3SbBr_2$, $Ph_3SbO$ and $Ph_3SbS$, Madelung potentials at Sb sites have been calculated, but no correlation has been found[155] between the formal oxidation state and 3d (3/2, 5/2) binding energies. The small value of asymmetry parameter $\eta$ for triorganoantimony(V) halides in NQR spectra has been interpreted in terms of the $D_{3h}$ symmetry[156–158].

Mössbauer spectral data of a number of triorgano (as well as a few tetraorgano) antimony(V) halides at liquid nitrogen and/or helium temperatures have been determined to establish the bond hybridization and the structure of the compound[52, 82, 83, 85, 86, 159–162]. Mössbauer data for $Ph_3SbX_2$ and $Ph_4SbX$ (X = an electronegative group) have been found to be consistent with trigonal bipyramidal geometry in which X groups occupy one or both axial positions[52, 82, 83, 85, 159]. In the compounds, $Ph_3SbX_2$, an approximately linear relation between isomer shift ($\delta$) and quadrupole splitting constant ($e^2qQ$) has been observed[83]. Electron population along the Sb-R bond, $\sigma$R (R = Me or Ph) increases with the number of electronegative groups attached to antimony atom[85, 86]. The 5s character of the apical bonds in the $R_3SbX_2$ and $R_4SbX$ compounds has been found to be appreciable and varies slightly in the similar series of compounds. Further, isomer shift and quadrupole splitting variations within these series are generally consistent with electronegativity variation of the X groups[52]. It has also been observed that bridging chlorines produce a somewhat longer $\sigma$Cl and smaller value of $\langle r^{-3} \rangle p$ than the non-bridging ones[85].

As early as 1938, Wells[163] had shown a trigonal bipyramidal geometry for trimethylantimony(V) chloride, bromide and iodide by means of X-ray diffraction analysis. The antimony distances have been found significantly longer than the sum of covalent radii, which according to Wells indicated the nature of these bonds to be intermediate between covalent and ionic bonds (Fig. 4). The crystal structure of $Me_3SbF_2$ has also been determined recently (1978). Like the other trimethylantimony(V) halides, $Me_3SbF_2$ also have trigonal bipyramidal geometry with apical fluorine atoms[164]. X-ray diffraction analysis of $Ph_3SbCl_2$, have also been found to be consistent with trigonal bipyramidal coordination polyhedra around antimony atoms in which phenyl groups occupy the equatorial positions[165, 166]. An X-ray diffraction study of *trans* tris(2-chlorovinyl)antimony(V) chloride

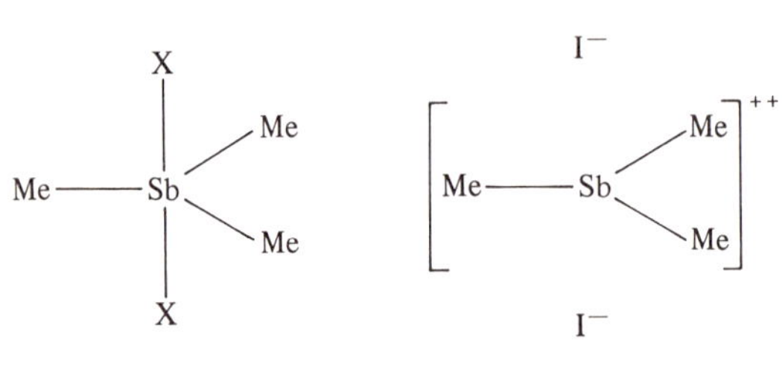


**Fig. 4.** Structure of $Me_3SbX_2$

showed that three chlorovinyl groups are coplanar with antimony atom and that both Sb-Cl bonds are nearly perpendicular (84°) to this plane[167]. The X-ray data for triorganoantimony(V) halides are summarized in Table 2.

**Table 2.** X-ray data of $R_3SbX_2$

| Compound | Crystal system | Space group | Cell parameters, etc. | | | Ref. |
|---|---|---|---|---|---|---|
| 1. $Me_3SbF_2$ | Monoclinic | $P2_c$ | a = 965.1<br>b = 785.8<br>c = 804.5 | Pm | | 164 |
| 2. $Me_3SbCl_2$ | Hexagonal | $C_6^{-}2c$ | a = 7.27<br>c = 8.44<br>Sb-Cl = 2.49 | A°<br>A° | | 163 |
| 3. $Me_3SbBr_2$ | Hexagonal | $C_6^{-}2c$ | a = 7.38<br>b = 8.90<br>Sb-Br = 2.63 | A°<br>A° | | 163 |
| 4. $Me_3SbI_2$ | Hexagonal | $C_6^{-}2c$ | a = 7.53<br>c = 9.59<br>Sb-I = 2.88 | A°<br>A° | | 163 |
| 5. $Ph_3SbCl_2$ | Orthorhomic | $P2_12_12_1$ | a = 13.44<br>b = 10.92<br>c = 12.26<br>Sb-Cl = 2.48 | A°<br>A° | | 165, 166 |
| 6. *trans* $(CHCl{=}CH)_3SbCl_2$ | | $C2/c$ | a = 20.82<br>b = 6.99<br>c = 17.23<br>Sb-Cl = 2.45 | A°<br>A° | $\beta = 101°50'$ | 108, 167, 168 |
| *cis, trans, trans* $(CHCl{=}CH)_3SbCl_2$[a] | Monoclinic | $Pc$ | a = 22.2<br>b = 16.14<br>c = 6.97 | A° | | 108 |

[a] The *cis, cis, cis* isomer described earlier was in error and the cell dimensions are = a = 15.96, b = 6.98, c = 21.92 space group *Pbc*[168])

## 3.4 Tetraorganoantimony(V) Halides

Like triorganoantimony(V) halides, derivatives of this class had also been prepared in the middle of nineteenth century[169–171]. The procedure adopted, which is still the principal method for preparing the tetraalkyl antimony halides involves the quaternization of trialkylantimony with an alkyl halide[13, 172]. Since the triarylantimony as well as diarylalkylantimony can not be quaternized by alkyl halides, other methods[13, 21, 23, 98, 173–179], e.g., the reaction of triphenylstibine, chlorobenzene and anhydrous aluminium chloride[179], have been utilized to prepare these derivatives.

Infrared and Raman spectroscopy have been extensively employed to elucidate the nature of bonding in these derivatives. For example, the Raman spectrum of tetramethylantimony chloride in aqueous solution has been found to be consistent with tetrahedral arrangement of the methyl groups around antimony atom[180]. Infrared spectrum of tetramethylantimony mercurate(II) has been investigated by Cullen et al.[181] and the formation of onium ion has been suggested. Shindo and Okawara[134] have described the infrared spectra of tetramethylantimony iodide and nitrate and suggested the *Td* symmetry for $C_4Sb$ skeleton. The infrared and Raman spectra of tetramethylantimony fluoride and hydroxide have been interpreted in terms of $C_{3v}$ symmetry and trigonal bipyramidal geometry with axial fluorine (hydroxide) groups has been suggested[182]. $^{1}H$ and $^{19}F$ NMR results indicate that in solution the five groups around antimony atom are rapidly exchanging their positions[182]. Contrary to this, an X-ray diffraction study of $Me_4SbF$ has showed that the structure is formed with fluorine bridged polymeric chains of six coordinated antimony atoms[164]. The mass spectrum of $Me_4SbF$ exhibits fluorine containing fragments in relative intensities, $Me_3SbF^+$ (100%), $Me_2SbF^+$ (15%) $MeSbF^+$ (25%) and $SbF^+$ (8%)[183].

Several investigators[132, 179, 184] have examined the ultraviolet absorption spectra of tetraphenylantimony halides. The fine structure in the 250–270 m$\mu$ region is characteristic of unperturbed benzene rings and indicates that there is comparatively little p$\pi$-d$\pi$ resonance between $\pi$ orbitals of the benzene ring and the vacant d orbitals of the antimony atom.

An ionic structure for tetraphenylantimony chloride[179], bromide[136] and perchlorate[136] has been suggested on the basis of infrared spectral data. In contrast to this, the laser-Raman spectra of a number of tetraphenylantimony compounds, $Ph_4SbX$ (X = F, Cl, Br, OH, 1/2 $SO_4$, $ClO_4$) have been measured and structural conclusions have been drawn from the low frequency region. It has been observed that $Ph_4SbClO_4$ has four coordinated antimony atoms while the other compounds are pentacoordinated (probably trigonal bipyramidal) in the solid state[185]. All these compounds have been found to be tetracoordinated in the methanolic solution except the molecular species, $Ph_4SbF$, which is pentacoordinated. The conclusions drawn from vibrational data have also been supported from Mössbauer studies[52] that the Sb-X bonds in most of the above compounds have considerable covalent character, except the ionic salt, $[Ph_4Sb]^+ClO_4^-$.

Pentacoordinated nature of antimony in tetraorganoantimony compounds has been confirmed by the X-ray analysis of $Ph_3SbMeF$[186]. The molecule has trigonal bipyramidal geometry with one axial and two equatorial phenyl groups. The methyl group occupies equatorial position while the fluorine axial.

Kok et al.[187] have used pulsed NMR spectroscopy to investigate magnetic relaxation of the $^{121}Sb$ nucleus in tetramethylantimony iodide, tetraphenylantimony sulfate, fluoride and chloride. It has been concluded that the geometry of the tetraphenylantimony cation in water or methanol deviated significantly from the tetrahedral.

## 3.5 *Pseudohalides, Oxides and Related Compounds*

Organoantimony pseudohalides of the types $R_3SbX_2$ and $(R_3SbX)_2O$ have generally been prepared by the metathetical reactions of halides with silver or sodium salts in appropriate amounts[188]. The technique which tends[189, 190] to yield hydrolysed products was finally exploited by Goel and Ridley[188] to obtain pure anhydrous triorganoantimony(V) azides, $R_3Sb(N_3)_2$ (R = Me or Ph).

Vibrational frequencies due to skeleton vibrations for both the trialkyl and triarylantimony(V) pseudohalides, $R_3SbX_2$ [R = alkyl or phenyl; X = $N_3$[188, 190, 191–193], NCO[188, 193] NCS[188, 194–196]] and $(Ph_3SbX)_2O$ [X = $N_3$, NCO, NCS][188], have generally been interpreted in terms of a $D_{3h}$ skeleton symmetry. A trigonal bipyramidal geometry having pseudohalide moieties at the axial positions has been suggested. $^1H$ NMR spectral data for the compounds of the type $R_3SbX_2$ (R = Me or Ph; X = $N_3$ or NCO) also suggest a trigonal bipyramidal geometry with three equatorial methyl or phenyl groups[193].

The conclusions derived from vibrational as well as $^1H$ NMR spectral analysis for the compounds, $R_3SbX_2$, has been supported from the X-ray diffraction studies also[197]. For example, crystals of $Ph_3Sb(NCO)_2$ belong to monoclinic system, space group $P2_{1/n}$ with Z = 4 in a unit cell. The NCO groups have been found to be N bonded to antimony atom which has a trigonal bipyramidal geometry with the nitrogen atoms at the apical positions (Fig. 5):

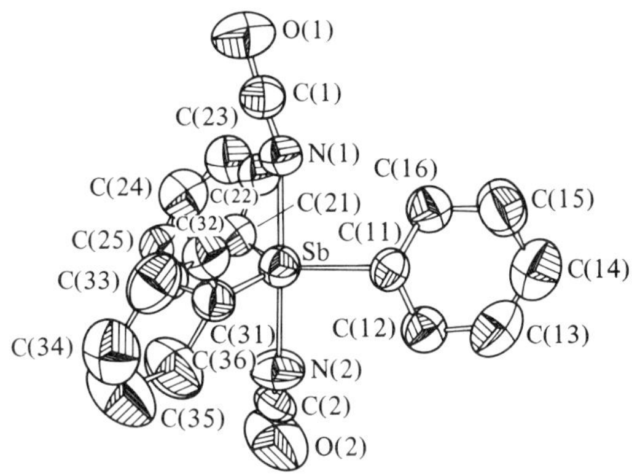


**Fig. 5.** Structure of $Ph_3Sb(NCO)_2$

X-ray diffraction analyses of $(Me_3SbN_3)_2O$[198] and $(Ph_3SbN_3)_2O$[199, 200] have been carried out and antimony atom has been found in a slightly distorted trigonal bipyramidal environment with bridging oxygen at one vertex and azide moiety at the other. The crystals of $(Ph_3SbN_3)_2O$ belong to monoclinic system with space group $C_{2/c}(C_{2h}^6)$ while $(Me_3SbN_3)_2O$ forms cubic shape crystals with space group $Pa^3$.

Infrared and Raman spectra of tetramethylantimony azide in the solid state and in dichloromethane solution indicate the presence of partly distorted tetrahedral $Me_4Sb^+$ cations[201]. A similar structure has also been proposed for $Me_4Sb(NCS)$ and $Me_4Sb(CN)$[201]. Contrary to the alkyl derivatives, the tetraphenylantimony pseudohalides, $Ph_4SbX$ (X = $N_3$, NCO, NCS), however, have covalent pentacoordinated structure[202].

Oxy bis(triorganoantimony perchlorates), which were earlier reported by Doak et al.[114, 115] as cationic derivatives $[(R_3Sb)_2O]^{++}$ on the basis of infrared spectroscopy, have been shown to be hydrated compounds, $[(R_3Sb)_2O(ClO_4)_2](H_2O)_2$[203]. The anhydrous derivatives contain pentacoordinate antimony atom in the solid state. Conductance and IR measurements of both the hydrated and anhydrous oxy bis(triphenylantimony perchlorate) suggest that these may exist as either ion pairs or non-ionic species in dichloromethane. The oxybis (trimethylantimony perchlorate) has also been found to be dissociated in nitromethane. On treating these diperchlorates with O-donor ligands, cationic complexes of the type $[(R_3SbL)_2O(ClO_4)_2$ (L = N,N-dimethylacetamide, dimethylsulfoxide, diphenylsulfoxide, pyridine N-oxide, triphenylphosphine oxide and triphenylarsine oxide) have been isolated. The conductance measurements showed that these complexes behaved as 1:2 electrolytes in nitromethane[203, 204]. Raman spectrum of an aqueous solution of $Me_3Sb(ClO_4)_2$ showed that it contains $ClO_4^-$ and planar cation $Me_3Sb^{++}$[205]. Hydrolysis of tris (pentafluorophenyl)antimony(V) perchlorate gave the oxo derivative[119]. Finally the structure of $(Me_3SbClO_4)_2O$ has been confirmed by X-ray analysis and has been found to be similar to the other oxo- compounds discussed above. The molecules of $(Me_3SbClO_4)_2O$ as well as $(Me_3SbCl)_2O$ have slightly distorted trigonal bipyramidal environment around antimony atom[198]. The data from the laser-Raman spectrum of $Ph_4SbClO_4$ in the low frequency region has been interpreted in terms of tetracoordinate polyhedra around antimony atom[185].

Derivatives, $R_3Sb(NO_3)_2$[115, 119, 134, 137, 205] and $Ph_4SbNO_3$[202] are covalent compounds in which nitrate groups behave as a unidentate moiety and occupy the axial positions. The Raman spectrum of aqueous solution of $Me_3Sb(NO_3)_2$ shows that solution contains $NO_3^-$ and planar $Me_3Sb^{++}$ cation[205], but the solid state IR and Raman spectra indicate that the anhydrous compound has monomeric pentacoordinate structure[137, 205]. Covalent compounds of the type $(R_3SbNO_3)_2O$ (R = Me or Ph) have also been prepared and characterized by infrared spectroscopy[134, 206].

Infrared spectral studies of $R_3SbX$ and $(Ph_3Sb)_2XO$ (X = $SO_4$, $SeO_4$, $CrO_4$ or $C_2O_4$)[114, 115, 207] in the solid state indicate them to have non-ionic polymeric structures with bridging anion groups and antimony atoms in a pentacoordination sphere. The data from infrared spectra, molecular weight and conductivity measurements in benzene and in nitrobenzene respectively for the compounds $(Ph_4Sb)_2X$ (X = $SO_4$, $SeO_4$ or $CrO_4$) indicate the presence of pentacoordinated structures both in solution as well as in solid state.

Compounds of the general formula $(R_3SbX)_2CH_2$ (X = halogen, pseudohalide or nitrate) have been prepared and characterized by infrared spectroscopy[208]. Three characteristic bands [(*viz.*, $\ell(CH_2)$, $\nu_{as}$(Sb-C-Sb) and $\nu_s$(Sb-C-Sb)] for Sb-$CH_2$-Sb moiety have been observed in the KBr region. A linear relationship between $\ell(CH_2)$ and $\nu_{as}$(Sb-C-Sb) has been obtained, which has been interpreted by considering changes in the s-character of antimony orbitals used for the Sb-C bonds. Cleavage of trimethylantimony(V) derivatives into $Me_4Sb^+$ and $Me_3Sb^{++}$ has been observed when these are treated with a base:

$$(BrMe_3Sb)_2CH_2 \xrightarrow{Ag_2O} Me_4Sb^+ + Me_3Sb^{++} \xrightarrow{HCl} Me_4SbCl + Me_3SbCl_2$$

In general, the trialkylantimony oxides have been considered to be molecularly associated in the neat form as well as in $CCl_4$ solution[209]. The observed monomeric nature of $Et_3SbO$ in chloroform has been attributed to an association of the type $Et_3Sb$-O-$HCCl_3$ which is not possible in $CCl_4$[209]; thus an additional band at 478 $cm^{-1}$ observed in the IR spectra in $CCl_4$ (also as a neat liquid) might be due to molecular association. Jensen and Nielsen[133] could not identify any absorption due to the Sb=O group in $Ph_3SbO$ and have considered a pseudo ionic character of the compound. However, in a later detailed study[210] a band at $744 \pm 3$ $cm^{-1}$ has been assigned to antisymmetrical stretch of Sb-O-Sb. The existing confliction in the earlier literature about the nature of the triphenylstibine oxide[9, 133, 209, 211] has also been removed by these workers[210] and on the basis of vibrational spectral studies it has been proposed that the Sb-O structural environment is similar to that of the polysiloxane backbone $(\text{-SiO-})_x$. Mass spectral studies also suggest a polymeric nature[210]. It has been found that the $\nu$ E=X stretching frequencies in $R_3EX$ (R = alkyl; E = P, As or Sb; X = O, S or Se) varies with the total mass of the molecule and the masses and electronegativities of the respective atoms[212]. The E-S bond in the drivatives $R_3ES$ shows some double bond character, which appears to be due to overlap of the field sulfur p orbitals with the empty d orbitals of the central atom E. The vibrational spectra of trialkylantimony sulfides, selinides and tellurinides have been reported[133, 209, 212, 213]. The positions of the $\nu$E-X modes for the derivatives $R_3EX$ (X = Se or Te) have been interpreted in terms of some degree of multiple bonding between the two atoms[133, 209, 212]. Crystallographic data for the compounds, $Ar_3SbX$ (Ar = Ph or *p*-tol, X = S or Se)[214], suggest that the free molecules have a trigonal tetrahedral configuration with the 3rd order symmetry axes.

Mass spectra of $Ph_3SbS$ and $Ph_3Sb(OH)_2$ have been reported by Glidwell[215]. In the mass spectra of $Ph_3Sb(OH)_2$, a molecular ion $Ph_3SbO_2H_2^+$ has been observed which indicates that the compound is a true dihydroxide rather than an oxide hydrate, as is the case of the arsenic analogue. X-ray diffraction analysis of $Ph_4Sb(OH)$ shows that this is a covalent compound with trigonal bipyramidal geometry. The oxygen atom occupies an axial position and the axial Sb-C distance is significantly longer than the mean equatorial Sb-C distance[216].

The infrared spectrum of $Me_4Sb(OH)$ in the solid state indicates a covalent trigonal bipyramidal geometry with OH group at axial position[217]. $\nu_{as}$ Sb-C vibration for $Me_3Sb(OH)_2$ has been assigned at 566 $cm^{-1}$ [114].

Tetraphenylantimony hydroxide has been found[218] to undergo decomposition at 50–70° in *p*-xylene solution under nitrogen atmosphere in dark to give benzene and $Ph_3SbO$ which has been reported to be monomeric in nature, contrary to this Venezky et al.[210] have shown its polymeric behaviour. Thermal decomposition of $Ph_4Sb(OH)$ in xylene has been studied by McEwen et al.[219, 220], who have proposed a radical mechanism involving stibonyl radical in the propagation step. The results obtained by these workers favour the following propagation sequences:

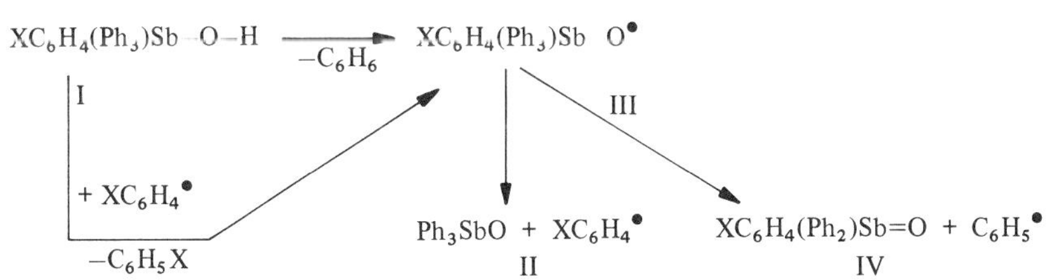


For the preparation of stibonic, $RSbO(OH)_2$, and stibinic, $R_2SbO(OH)$ acids, many methods have been developed but the reaction of trivalent antimony compound with diazonium salt appears to be the most generally applicable[8, 9]. Arylstibonic acids are rather insoluble in both polar and non-polar solvents and molecular weight measurements show the presence of associated species. On heating, these tend to decompose without showing definite melting points. It has been proposed by Schmidt[58, 221] that these acids exist in the solid state as trimers (Fig. 6).

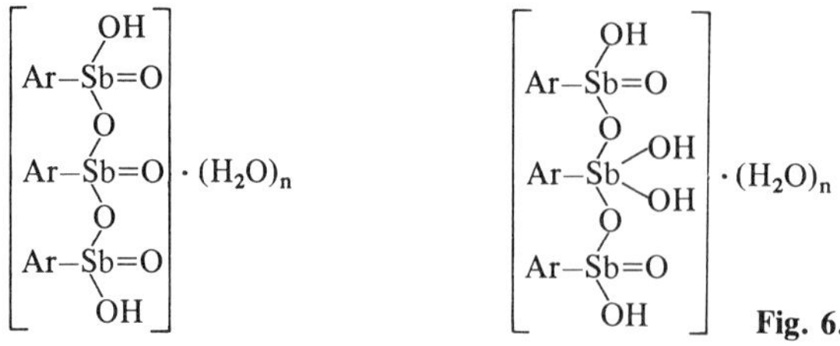

**Fig. 6.** Schmidt's structure for stibonic acids

The structure proposed by Schmidt has been supported by Fargher and Gray[222]. However, Macallum[223] on the basis of molecular weight data in phenol, rejected the Schmidt concept and suggested that the aromatic stibonic acids are monomeric in nature. The problem has subsequently been studied by many workers[224–227] confirming Schmidt's findings and indicating[227] the formation of $[RSb(OH)_5]^-$ anions in alkaline solutions. Inspite of these investigations, the structure of polymeric stibonic acids themselves can not be yet considered to be established with definiteness.

Mössbauer spectra of several arylstibonic, diarylstibinic acids, triphenyl stibine oxide and tetraphenylantimony hydroxide have been studied[228]. It has been observed that there is a linear increase of $\delta$(isomer shift) with increasing the aryl groups in the series $Ph_4SbOH$, $Ph_3SbO$, $R_2SbO_2H$ and $RSbO_3H_2$. Orbital population along the Sb-ligand bonds for various assumed geometries (octahedral or trigonal bipyramidal) have been obtained by using additive model and it has been concluded that stibonic and stibinic acids probably have trigonal bipyramidal geometry with bridging Sb-O-Sb bonds in apical positions. However, the Mössbauer spectral data of dialkylstibinic acids, $[R_2SbO(OH)]$, have indicated a tetrahedral coordination polyhedra around antimony[277].

Many tertiary stibine imides and amides having general formulae, $R_3Sb{=}NR'$ ($R'$ = aryl, sulfonyl, silyl, germyl or stannyl group)[22, 229–237], $(R_3SbCl)_2NH$[148] and $R_3Sb(NR'_2)_2$[238, 239], are known. The Sb=N bond in triorganoantimony imides can be cleaved with various ligands like alcohol, phenols, benzaldehyde, HCl etc.[236]. On the basis of spectroscopic studies a pentacoordinated geometry has been suggested for $R_3Sb(NR'_2)_2$[238, 239].

Little is known about antimony ylids[240–242]. Several of these have been postulated as intermediate and the compound I has been isolated[233].

I

The dipolar and nucleophilic character of the ylid increases and their stability decreases in the series P, As, Sb and Bi. As the dipolar character increases, the electronic spectra exhibit absorption maxima at longer wave lengths. It has been concluded by Singer et al.[243–245] that $p\pi$-$d\pi$ overlap between the heteroatom and carbon atoms varies in the order: S > P> As > Sb.

The dipole moment of triphenylstibonium tetraphenylcyclopentadiene ylid in benzene at 25° has been found to be 2.2 D, a value considerably lower than the dipole moments of the P or As analogue. This result was unexpected, since the earlier work had suggested that the dipolar character of these ylids increases in the order Sb > As > P[242].

# 4 Alkoxides, Thiolates, Glycolates and Peroxy Compounds

## 4.1 Alkoxides

Three following general methods have been employed for the preparation of antimony pentaalkoxides: (i) reaction of alcoholates with $SbCl_5$, (ii) oxidation of $(RO)_3Sb$ with bromine to form $(RO)_3SbBr_2$ and treatment of the bromo compound with alcoholates and (iii) alcoholysis of $(RO)_5Sb$[246–250].

Antimony(V) ethoxide and isopropoxide have been found to be volatile compounds; these are dimeric in cyclohexane or carbon tetrachloride solutions. By contrast, the low volatility of $Sb(OMe)_5$ has been ascribed to its polymeric nature in solid state. The antimony atom in these pentaalkoxides prefers hexacoordination state which may be acquired either by molecular association or on coordination with suitable donors such as alcohol, ammonia etc.,[246, 248, 249, 251]. Antimony pentaethoxide also reacts with sodium ethoxide to give the bimetallic alkoxide, $NaSb(OEt)_6$[249].

Preparation and properties of alkoxide halides of antimony(V) of the type $(RO)_nSbX_{5-n}$ (n = 1–4) have been described in a number of publications[248–258]. These compounds exist in dimeric form. The chloroantimony ethoxides, $(EtO)_nSbCl_{5-n}$, ionize in the solvents of high dielectric constants[259, 260]:

$$2(EtO)_nSbCl_{5-n} \leftrightarrows [(EtO)_{n-1}SbCl_{5-n}]^+[(EtO)_{n+1}SbCl_{5-n}]^-$$

In the solvents of low dielectric constant aggregates of the undissociated molecule tend to be formed. Exchange experiments with $Cl^-$ show that in solid state, $(EtO)SbCl_4$ exist as $[SbCl_4][(OEt)_2SbCl_4]$[260].

Arbuzov et al.[250, 261–263] have studied the stereochemistry of chloroantimony alkoxides, $(RO)_nSbCl_{5-n}$ (n = 1–5). Infrared and Raman spectral data of $(MeO)_nSbCl_{5-n}$[263] indicate that these are centrosymmetric bridged dimers. Raman spectra of compounds with n = 3–5 exhibit $\nu$Sb-O doublets of terminal MeO group at 530–541 and 550–570 $cm^{-1}$; vibrations of the four membered $Sb_2O_2$ ring, observed in the region 500–517 $cm^{-1}$ in the IR spectra of 1–5, are absent. The $\nu$C-O bands of the bridged and terminal OMe group in the series 1 → 5 are shifted to higher wave numbers (60 and 31 $cm^{-1}$ respectively). The stability of dimers in the series 1 → 5 increases in the following order:

$$1 < 2 < 3 < 4 \ll 5$$

At 100–120° and in MeCN solution, dimers with n = 1–3 dissociate to monomers. However, the dimeric monochloride, $(MeO)_4SbCl$, appears to dissociate only partially in MeCN[263]. A monomer dimer equilibrium for $(MeO)SbCl_4$ has been observed in benzene with predominance of the dimeric form[263].

Arbuzov and coworkers[262] have in a recent publication discussed the stereochemistry of the derivatives $(MeO)_nSbCl_{5-n}$ using NMR spectroscopy. According to them a non-degenerate structure is formed when n = 1, but for $Sb(OMe)_5$ (n = 5), only a single 25-fold degenerated structure is formed by 2 tetragonal bipyramids. For intermediate values of n = 2–4, the spectra were found to be rather complicated and these have been explained as a superposition of resonance lines from different isomeric structures, arising from the OMe groups occupying in different apical and equatorial positions in tetragonal bipyramidal moieties.

Infrared, Raman and $^1$H NMR spectra of other tetrachloroantimony alkoxides have been studied and discussed in terms of hexacoordination polyhedra around antimony atom[249, 251, 255]. The mass spectral studies of $(MeO)SbCl_4$, $(EtO)SbCl_4$, $(EtO)_{2.5}SbCl_{2.5}$ and $(EtO)_3SbCl_2$ suggest that all these compounds are built up by bioctahedra and in addition to the monomeric ions dimeric ions are present in the vapour state[254].

The hexacoordinated antimony atom in alkoxyantimony(V) chlorides has been confirmed by the X-ray diffraction analyses of $(MeO)SbCl_4$[264] and $(EtO)SbCl_4$[265]. Both compounds crystallize in the monoclinic form with space group$P2_{1/n}$. The molecular structures consist of pairs of octahedrally coordinated antimony atoms linked by oxygen bridges (Fig. 7).

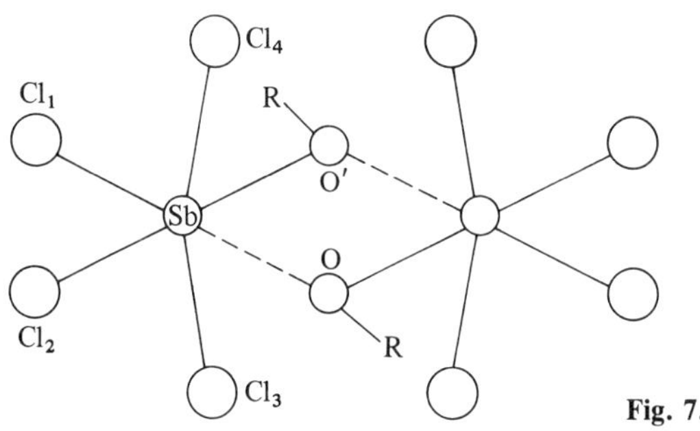


**Fig. 7.** Structure of $(RO)SbCl_4$ (R = Me or Et)

In the above compounds, therefore, antimony appears to acquire its favoured coordination number through alkoxy bridges and further coordination seems to be unlikely. However, preferential coordination by ligands which form stronger bonds than alkoxy bridges may be possible. Thus a number of adducts of monoalkoxyantimony tetrachlorides, $(RO)SbCl_4$ [R = Me[266–272], Et[266–271], $CF_3CH_2$[273, 274], $C_2H_4Cl$[266–269] or $C_2H_2Cl_3$[267, 275], with various oxygen and nitrogen donor ligands, L (e.g. pyridine N-oxide, hexamethyl phosphorimide, triphenylphosphine oxide, fromamides, ureas, dimethylsulfoxide, etc.), have been isolated. The absence of a bond in the region 480–520 cm$^{-1}$, which has been ascribed to the Sb–O–Sb moiety, involving bridging alkoxy groups, indicates the break-down of alkoxy bridges as a result of formation of acceptor-donor linkage with stronger ligands[266, –269, 273–275]. On the basis of vibrational spectral studies a *cis* octahedral configuration for these complexes has been suggested[270, 272]:

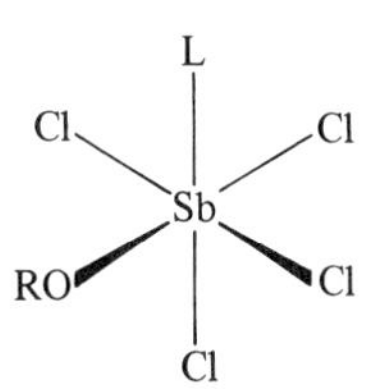


**Fig. 8.** *Cis* configuration of monoalkoxyantimony tetrachloride adducts

Contrary to the diorganoantimony(V) halides, the corresponding alkoxy derivatives, $R_2Sb(OR')_3$ (R = Me, Et, Pr, or Bu, R′ = Me or Et), have been found to be thermally stable and may be prepared conveniently by the reactions of $R_2SbBr_3$, with sodium alkoxides in *n*-pentane at − 20° to − 40°. Dimethyl- and diethylantimony(V) methoxides have been shown to be associated; the degree of association for the former derivative approximates to 2 whereas for the latter it appears from 1.4 to 1.1 with decreasing concentration range from 1.1 to 0.4 wt.% in benzene. Both the Raman and $^1H$ NMR spectral data suggest that $Me_2Sb(OMe)_3$ possesses a dimeric octahedral structure in which methyl groups occupy *trans* positions, both in the solid state and in concentrated solution[276] (Fig. 9).

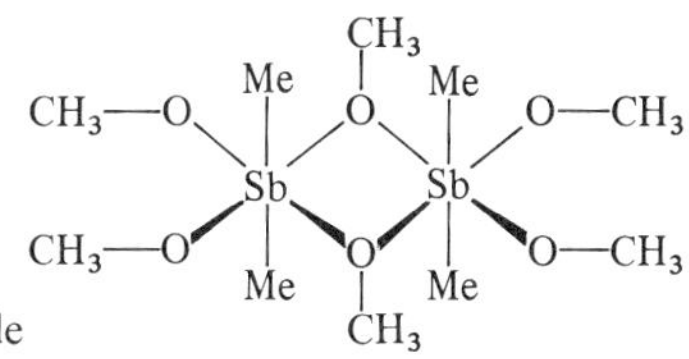


**Fig. 9.** Dimeric structure of dimethylantimony(V) methoxide

A dimeric structure proposed by Meinema and Noltes for dialkylantimony(V) methoxides has been confirmed by recent Mössbauer studies of $R_2Sb(OMe)_3$[277]. Compounds have octahedral geometry around antimony atoms with methoxo bridges.

The best method for the preparation of triorganoantimony(V) alkoxides and siloxides seems to be the reaction of triorganoantimony(V) halides with sodium alkoxides or siloxides[119, 278–284]. Triorganoantimony(V) monoalkoxy derivatives of the type, $R_3Sb(OR')X$ (X = an electronegative group) are also known but the investigations on these derivatives are very limited[79, 285–289].

Most of the triorganoantimony(V) alkoxides are monomeric, moisture sensitive and are volatile products. Both infrared and $^1H$ NMR spectral data for triorganoantimony(V) alkoxides and siloxides have been interpreted in terms of trigonal bipyramidal geometry around antimony atom[278–281].

The crystal structural studies of $Ph_3Sb(OMe)_2$ by three-dimensional single crystal X-ray diffraction techniques have confirmed[280] the above conclusions from spectroscopic data. Trigonal bipyramidal conformation around antimony atom has been observed in which three benzene rings occupy equatorial positions. Methoxy groups occupy both axial positions with Sb-O bond distances 2.039(8) and 2.027(8) A°. It has been suggested that the slight distortion of the O-Sb-O angle to 175.3 (3)° may be the result of packing requirements (Fig. 10).

Tetraorganoantimony(V) alkoxides[32, 179, 182, 217, 278–281, 290–294] and siloxides[278, 293] have generally been prepared either by reacting pentaorganoantimony with alcohol or tetraorganoantimony(V) halides with sodium salts of the corresponding ligand. IR and $^1H$ NMR spectra of these monomeric alkoxy compounds have been found to be consistent with the assumed pentacovalent structures[179, 182, 279, 281, 294]. In the siloxy derivatives Sb-O bond has considerable ionic character[278]. Since the UV spectra of tetraphenylantimony alkoxides have been found to be identical with one another and with $Ph_4SbOH$, which has been shown to have trigonal bipyramidal geometry, a pentacoordinated geometry has also been proposed for these alkoxides[179].

Trigonal bipyramidal geometry for these compounds has been confirmed by the X-ray structure determination of $Ph_4Sb(OMe)$[280]. The compound crystallizes with eight

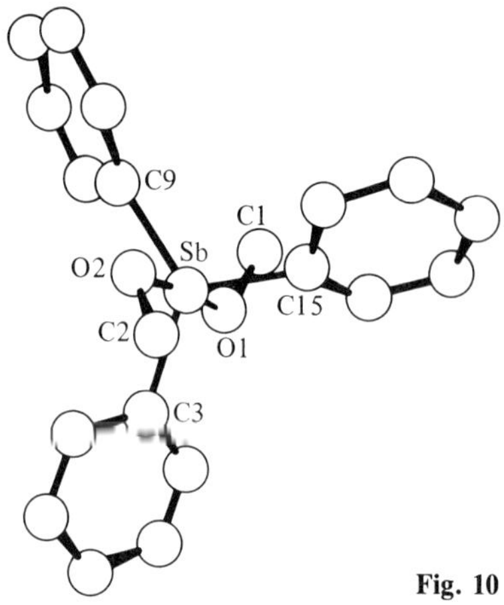


**Fig. 10** Molecular structure of triphenylantimony(V) methoxide

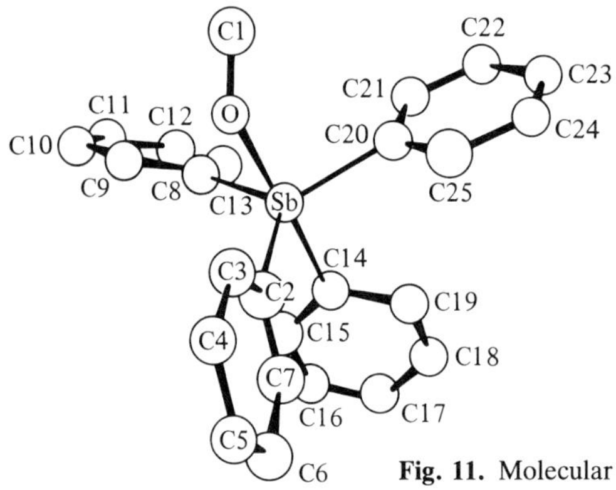


**Fig. 11.** Molecular structure of tetraphenylantimony methoxide

molecules in a space group *Pbca* of the orthorhombic system. Antimony atom has trigonal bipyramidal geometry with methoxy and one phenyl group at apical positions. The Sb atom is displaced 0.10 A° out of the equatorial plane towards the apical phenyl group which might be due to the steric interactions among the bulky benzene rings (Fig. 11).

The cleavage of the phenylantimony and Sb-O bond has been studied by Razuvaev et al.[292, 293] in the above compounds. Tetraphenylantimony alkoxides are thermally unstable and decompose to triphenylantimony, aldehyde or ketone and benzene. These reactions proceeds via the formation of simple ethers which is a basic characteristic of tetraphenylphenoxyantimony.

$$Ph_4SbOPh \longrightarrow Ph_3Sb + Ph{-}O{-}Ph$$

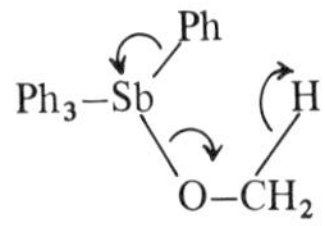


$$\longrightarrow Ph_3Sb + PhH + CHO \qquad 90\%$$

$$Ph_4SbOMe \longrightarrow Ph_3Sb + PhOMe \qquad 10\%$$

$$Ph_4SbOCHMe_2 \longrightarrow Ph_3Sb + Me_2CO + PhH$$

However, the thermal decomposition of $Ph_4SbOSiPh_3$ has been explained by a disproportionation mechanism and following products have been isolated

$$
\begin{array}{l}
2\,Ph_4SbOSiPh_3 \rightarrow Ph_3Sb(OSiPh_3)_2 + Ph_5\,Sb \\
\qquad\qquad\qquad\qquad\qquad\qquad\qquad\qquad \downarrow \\
\qquad\qquad\qquad\qquad\qquad\qquad\qquad Ph\text{-}Ph + Ph_3Sb \\
\qquad\qquad\qquad\qquad\qquad\qquad\qquad\ 46\% \qquad 45\%
\end{array}
$$

In tetraarylantimony alkoxides, cleavage of the one aryl group has been observed in the presence of sodium alkoxide and it has been postulated that these reactions proceed via an ionic reaction mechanism in which "ate" complexes are formed as intermediate[279, 281].

## 4.2 Thiolates

Due to intramolecular redox process, attempts to prepare triorganoantimony(V) thiolates by the reactions of triorganoantimony(V) halides or ethoxides with thiols alone have been found to be unsuccessful[295–297]. However, these may be prepared by the following reaction route in the presence of trimethylamine at low temperatures[296]:

$$Me_3SbCl_2 + 2\,RSH + 2\,Et_3N \rightarrow Me_3Sb(SR)_2 + 2\,Et_3N\cdot HCl$$

These compounds are monomeric colourless crystalline solids which decompose at room temperatures. The $^1H$ NMR spectral data suggest that the molecules have trigonal bipyramidal structures in which methyl groups occupy the equatorial positions. The kinetics of decomposition reactions have also been investigated by $^1H$ NMR spectra[296].

Tetraorganoantimony(V) thiolates have been prepared either by the reaction of pentaorganoantimony with thiols[298] or of tetraorganoantimony methoxide with thiols[299]. Like the triorganoantimony(V) thiolates these compounds also decomposed into triorganoantimony and thio ethers. However, the thermal decomposition of $Ph_4Sb$-$(SC_6H_4S$-*p*) (X = OMe, Me or Br) appears to proceed via a radical mechanism to yield $Ph_3Sb$, $PhSC_6H_4X$-*p* and disulfide[300]. The $^1H$ NMR spectra of monomeric $Me_4Sb(SR)$ indicate that the four methyl groups are equivalent which has been suggested to be due to ligand pseudorotation[298]. $^{19}F$ NMR spectra of $Ph_4Sb(SC_6H_4F$-*m*) and $Ph_4Sb(SC_6H_4F$-*p*) have been determined[299, 301] and shielding of $^{19}F$ chemical shifts have been observed compared to other organometallic *m*- and *p*-fluorothiophenol derivatives of heavy non-transition elements, which has been explained by greater electron-donating ability of $Ph_4SbS$ group due to the greater polarity of the metal-sulfur bond. Reactions of $Ph_4SbSC_6H_4R$ (R = H, *p*-Me, *o* or *p*-OMe, *p*-Br and *p*-$NH_2$) with allylhalides, acyl halides, halogens and $Ph_3SbCl_2$ have been studied and complete halide mercaptide exchange has been observed[302].

## 4.3 Glycolates

The crystal structure of $PhSb[(OCMe_2)_2]_2$ has been determined by X-ray diffraction analysis[303]. The central antimony atom shows slightly distorted trigonal bipyramidal

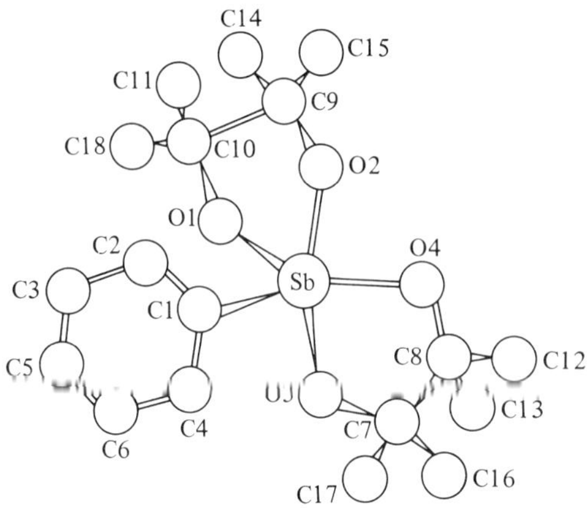


**Fig. 12.** Structure of $PhSb[(OCMe_2)_2]_2$

configuration (Fig. 12) with the phenyl group at equatorial position. To determine the solution stereochemistry of the molecule, $^1H$ and $^{13}C$ NMR spectra have been recorded and results have been explained by low temperature pseudorotation. Authors have also concluded that the methyl groups stay at *cis* and *trans* positions with respect to the phenyl group.

Triorganoantimony(V) glycolates, $R_3Sb(OXO)$, have been found to be monomeric in which diol group appears to chelate antimony through two oxygen atoms[304, 305]. These compounds are pentacoordinated, probably in trigonal bipyramidal configuration in which oxygen atoms of the ligand occupy one axial and one equatorial positions.

The interaction of equal molar amounts of triphenylantimony and tetramethyl-1,2-dioxetane in $CDCl_3$ at ambient temperatures has been found to give 77% yield of stable insertion product[306].

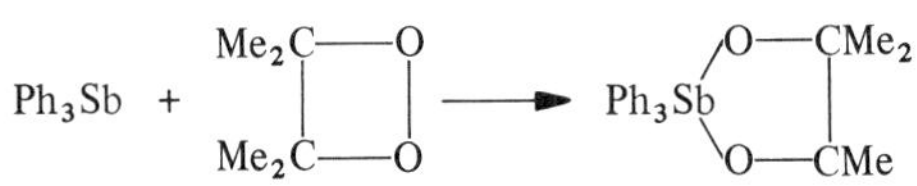


The results obtained in this study have been believed to be consistent with a biphilic insertion of the group V atom into the peroxy bond of the dioxetane.

## 4.4 Peroxy Compounds

A number of peroxy compounds of the type $R_3Sb(OOR')_2$, $R_3Sb(OOR')X$ and $R_4Sb(OOR')$ have been prepared and studied by many workers[286, 293, 307–315]. Crystals of $Ph_3Sb(OOBu^t)_2$ belong to tetragonal system with space group $P42_{1/c}$. Antimony has trigonal bipyramidal geometry with peroxy oxygens in the axial positions[316]. The structure of $[Ph_3Sb(OOBu^t)]_2O$ has also been determined by X-ray diffraction analysis[317] and the coordination polyhedra of the antimony atoms have been found to be slightly distorted trigonal bipyramidal with all the three phenyl groups in equatorial positions.

Pyrolysis and photolysis of a number of tetraphenylantimony peroxides have been studied by Razuvaev et al.[309–311] who demonstrated that among the many decomposition products, triphenylantimony appears to be the common one. The photolysis reaction appears to proceed via a radical reaction mechanism[310].

## 5 Carboxylates and Thiocarboxylates

Laber and Schmidt[318–321] have prepared antimony(V) carboxylates of the type $Cl_4SbOOCR$ by the reactions of antimony pentachloride and sodium salt of carboxylic acid. Bonding in these derivatives have been studied by vibrational spectroscopy. On the basis of vibrational spectra of $(SbCl_4)_2C_2O_4$, $D_{2h}$ symmetry with two planar 5-membered-$SbOC_2O$ rings having a common C-C bond has been proposed. Compounds of the type $(RCOO)_3SbX_2$ (R = 3–12 carbon atoms containing hydrocarbon group and X = F, Cl, Br) have also been mentioned in patent literature[322].

Reaction of antimony pentafluoride with trifluoroacetic acid anhydride yields a binuclear antimony complex[323]:

$$(CF_3CO)_2O + 2\,SbF_5 \rightarrow Sb_2(CF_3COO)F_9 + CF_3COF$$

On the basis of $^{19}F$ NMR spectrum it has been suggested that the nine fluorine atoms bonded to antimony are exchanging rapidly at temperatures above − 70° and the following structure has been proposed (Fig. 13):

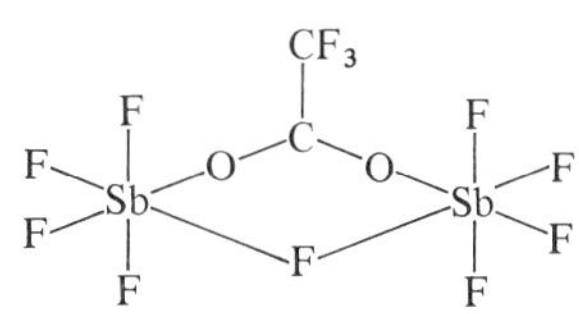

**Fig. 13.** Structure of $Sb_2(CF_3COO)F_9$

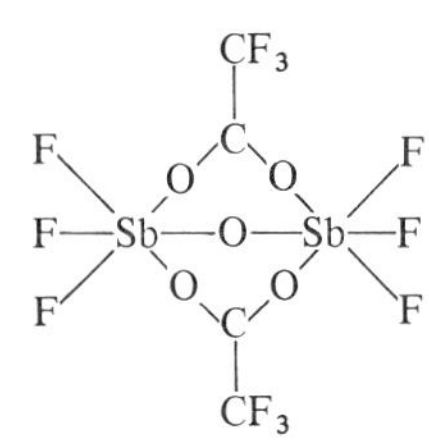

**Fig. 14.** Structure of $Sb_2O(CF_3COO)_2F_6$

This binuclear compound on sublimation under vacuum yields a triply bridged binuclear oxo compound (Fig. 14); the crystals of which belong to monoclinic system with space group $C_c$. This triply bridged binuclear molecule has $C_{2v}$ symmetry and the angles around the antimony atoms are all close to 90°. $^{19}F$ NMR spectrum of this compound has also been reported.

Meinema and Noltes[276] have prepared thermally stable monomeric diorganoantimony(V) acetates by the following general reactions:

$$R_2Sb(OMe)_3 + AcOH_{(excess)} \rightarrow R_2Sb(OAc)_3 + 3\,MeOH$$

$$Me_2SbCl_3 + 3\,AgOAc \rightarrow Me_2Sb(OAc)_3 + 2\,AgCl$$

From the Raman and IR spectral studies, these workers have concluded that $Me_2Sb(OAc)_3$ possesses an octahedral structure in which two methyl groups occupy axial positions, whereas two ester-type and one bidentate acetato group occupy the equatorial positions:

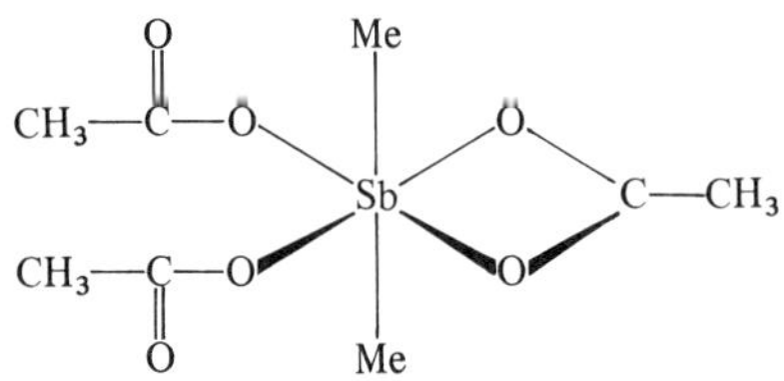

Non-rigidity of the molecule has been suggested due to the presence of only one acetato methyl proton resonance in the $^1H$ NMR spectrum at − 80°C.

Findings of Meinema and Noltes have recently been confirmed by Mössbauer spectral studies[277)] of dialkylantimony(V) acetates. The molecules are monomeric with octahedral geometry in which one acetato group is acting as bidentate moiety.

Of a variety of methods employed for the preparation of triorganoantimony(V) carboxylates[22, 64, 117, 134, 153, 324–335)], the following methods appear to be of more general importance:

$$R_3SbO + 2\,R'COOH \rightarrow R_3Sb(OOCR')_2 + H_2O$$

$$R_3SbX_2 + 2\,R'COOH + 2\,Et_3N \rightarrow R_3Sb(OOCR')_2 + 2\,Et_3N\cdot HX$$

Triorganoantimony(V) carboxylates are monomeric. Pyrolysis of trimethylantimony bis phenylbromoacetate in xylene yields trimethylantimony(V) bromide and mandelic acid polyester[325)]. However, the pyrolysis of triphenylantimony(V) acetate proceeds via a radical reaction and yields triphenylantimony and acetic acid[333)]. Trimethylantimony bis thio benzoate, on the other hand, appears to dissociate reversibly as shown below[334:]

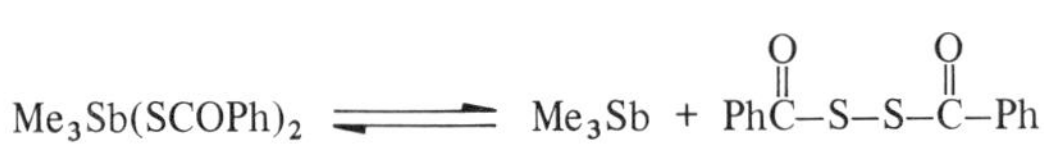

The reaction of trimethylantimony sulfide with acyl halide proceeds via a four centered intermediate yielding halo trimethylantimony thiocarboxylates[335)]:

$$Me_3SbS + RCOX \rightarrow Me_3Sb(SCOR)X$$

The above derivatives may also be obtained by the reactions of $Me_3Sb(SCOR)_2$ with $Me_3SbX_2$ in chloroform. X-ray powder patterns of these compounds indicate them to be a distinct group of compounds rather than mixtures of halide and bisthiocarboxylate.

In general, the infrared and Raman spectra of triorganoantimony(V) carboxylates have been best interpreted in terms of trigonal bipyramidal structure containing planar $R_3Sb(V)$ moiety. These spectra[115, 117, 134, 137, 139, 140, 207, 327, 332, 336] indicate that they contain in solution as well as in the solid state ester-type R′COO groups, as the symmetric and asymmetric $-CO_2-$ stretching frequencies have been observed in their spectra arising from $C_{2v}$ symmetry of $-CO_2-$ group. In some cases splitting of $\nu(C{=}O)$ band into a doublet has been observed which has been explained to be due to conformational[336] or crystal lattice effect[134]. In triorganoantimony(V) carboxylates (derived from $\alpha$-substituted acetic acid), the $\nu C{=}O$ frequency has been found to show a linear dependence upon the pK or the Taft constant, $\sigma^*$, for the parent acid[336]. Deuteration studies have confirmed the assignments made for $\nu C{=}O$ and $\nu C{-}O$ frequencies in triorganoantimony(V) acetates[139, 336]. On the basis of infrared spectral data of trimethylantimony(V) thiocarboxylates, Sb-S linkages have been suggested both in solution as well as in the solid state[334, 335].

$^1H$ and $^{13}C$ NMR spectra of triorganoantimony(V) carboxylates have also been found to be in accordance with pentacoordinate geometry[153, 283, 327, 330, 332, 336, 337]. Chemical shifts due to $Me_3Sb$ protons in trimethylantimony(V) carboxylates depend on the strength of the parent acid. A linear correlation between the chemical shift and the pK or the Taft constant, $\sigma^*$, of the parent acid has been established[336]. The $^1H$ NMR data of trimethylantimony halo thiobenzoates in $CH_2Cl_2$ indicate the following type of partial conversion:

$$2\,Me_3Sb(SCOPh)X \rightleftharpoons Me_3SbX_2 + Me_3Sb(SCOPh)_2$$

The Mössbauer spectra of triorganoantimony(V) carboxylates have been measured and interpreted in terms of trigonal bipyramidal environment around antimony atom[83, 85, 338]. On the basis of point charge calculations, planar $C_3Sb$ skeleton has been suggested, since this will produce a large axially symmetric electric field gradient (e.f.g.) as evident by the observed Mössbauer data. Mössbauer parameters of some trimethylantimony(V) carboxylates are given in Table 3.

**Table 3.** Mössbauer parameters of some trimethylantimony(V) carboxylates[338]

| Compound | IS($\delta$)[a, b] mm/s | QS($\Delta$)[b] mm/s |
|---|---|---|
| $Me_3Sb(OOCCH_3)_2$ | $-5.17 \pm 0.08$ | $-23.3 \pm 0.1$ |
| $Me_3Sb(OOCCH_2Cl)_2$ | $-5.27 \pm 0.10$ | $-25.1 \pm 0.3$ |
| $Me_3Sb(OOCCHCl_2)_2$ | $-5.30 \pm 0.20$ | $-25.2 \pm 0.5$ |
| $Me_3Sb(OOCCCl_3)_2$ | $-5.50 \pm 0.05$ | $-27.5 \pm 0.3$ |
| $Me_3Sb(OOCCH_2F)_2$ | $-5.30 \pm 0.20$ | $-24.4 \pm 0.4$ |
| $Me_3Sb(OOCCHF_2)_2$ | $-5.40 \pm 0.10$ | $-26.4 \pm 0.7$ |
| $Me_3Sb(OOCCF_3)_2$ | $-5.50 \pm 0.10$ | $-28.0 \pm 0.4$ |
| $Me_3Sb(OOCCH_2NC)_2$ | $-5.37 \pm 0.06$ | $-25.8 \pm 0.9$ |
| $Me_3Sb(OOCCH_2Br)_2$ | $-5.24 \pm 0.05$ | $-23.8 \pm 0.5$ |
| $Me_3Sb(OOCCHBr_2)_2$ | $-5.40 \pm 0.10$ | $-26.2 \pm 0.1$ |

[a] Isomer shift relative to $Ba^{121}SnO_3$

[b] All data recorded with samples at 8 °K and $Ba^{121}SnO_3$ source at 80 °K

X-ray diffraction analysis of $Ph_3Sb(OAc)_2$ suggests a distorted trigonal bipyramidal geometry around antimony atom[339)]. The distortion of the angles between phenyl groups from the expected 120° has been explained by weak intramolecular interactions between the non-bonded O atoms of the unidentate acetate groups and the central antimony atom.

Matsumura et al.[340)] have prepared triorganoantimony(V) glycolates and thioglycolates of the type $R_3Sb(XCH_2COO)$ (R = Me, Ph or cyclo $C_6H_{12}$; X = O or S). Compounds of phenyl and cyclohexylantimonys are monomeric in which -COO- group is linked to antimony atom by an ester-type linkage. Except $Ph_3Sb(SCH_2COO)$, which appears to have a similar structure in the solid as well as in solution, other compounds (R = Ph or cyclo $C_6H_{12}$) have an associated structure in the solid state with hexacoordinated antimony atom.

The compound $Me_3Sb(OCH_2COO)$ crystallizes as needles or cubes from methanol and both forms exhibit different X-ray powder patterns and IR spectra. The existence of monomer-dimer equilibrium in chloroform at room temperatures has been suggested. With higher concentrations at ambient temperatures or over a wide range of concentrations below − 15 °C, the dimeric form predominates. On the basis of $^1H$ NMR spectral studies at − 38 °C, the following structure (Fig. 15) has been proposed for this dimeric species:

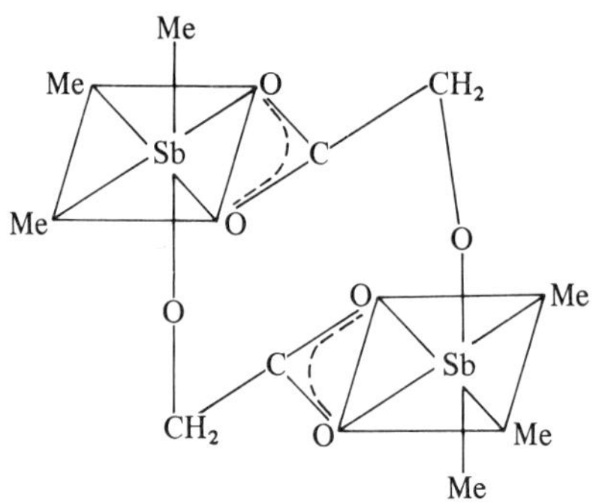


**Fig. 15.** Dimeric structure of $[Me_3Sb(OCH_2COO)]_2$

The following general methods have been employed for the preparation of tetraorganoantimony(V) carboxylates[202, 341)]:

$$Ph_4SbBr + AgOOCR' \rightarrow Ph_4SbOOCR' + AgBr$$

$$R_5Sb + R'COOH \rightarrow R_4SbOOCR' + RH$$

All the tetraorganoantimony(V) carboxylates reported till now have been found to be non-ionic and monomeric in solution as well as in vapour phase. Tetramethylantimony(V) carboxylates are volatile in nature. The vibrational spectra of tetramethylantimony(V) acetate in solution and in the liquid state indicate a trigonal bipyramidal structure for the molecule but in the solid state, antimony atom appears to be hexacoordinted with acetate group assuming a bidentate type of bonding[202, 341)] (Fig. 16).

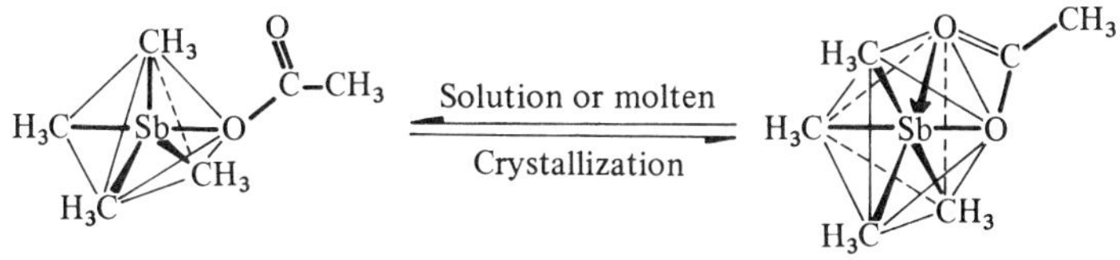


**Fig. 16.** Penta- and hexacoordinated structure of tetramethylantimony(V) acetate

The investigations on tetramethylantimony(V) formate, trifluoroacetate, trichloroacetate, propionate, pivalate and benzoate and on tetraphenylantimony(V) formate and acetate gave similar results[202, 341]. Contrary to this, tetraphenylantimony(V) trifluoracetate and trichloroacetate possess pentacoordinate structure in the solid state as well as in solution. Bis(tetraphenylantimony) oxalate possesses six-coordinated antimony both in solution and in the solid state[202]. $^1$H NMR spectrum of $Me_4SbOAc$ shows the equivalence of four methyl groups at ambient temperatures and $-45°$[341].

Contrary to the above results, a slightly distorted trigonal bipyramidal geometry has been observed for tetraphenylantimony(V) formate by means of X-ray analysis[342]. Slight distortions of the angles between the non-bonded O atom of the unidentate formate group and antimony atom has been observed.

Tetra-methyl- and phenyl-antimony(V) carboxylates react with one mole of carboxylic acid to form monomeric 1:1 adducts, $R_4SbOOCR' \cdot R''COOH$[343]. The presence of strong O-H...O hydrogen bonds in these adducts has been demonstrated by IR and $^1$H NMR data:

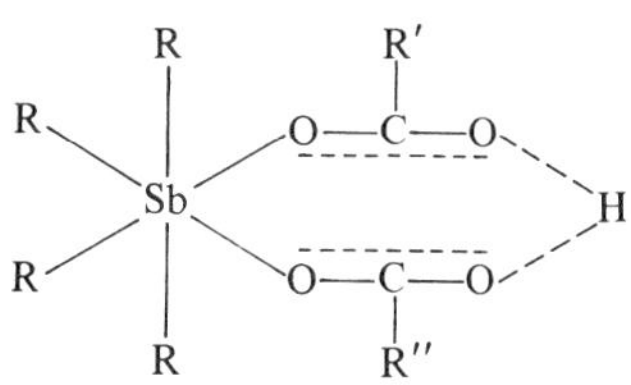

A compound, where R′ and R″ are linked, has been obtained by the reaction of pentamethylantimony with glutaric acid. However, the similar reaction has been found to be unsuccessful with oxalic, malonic or succinic acids[343].

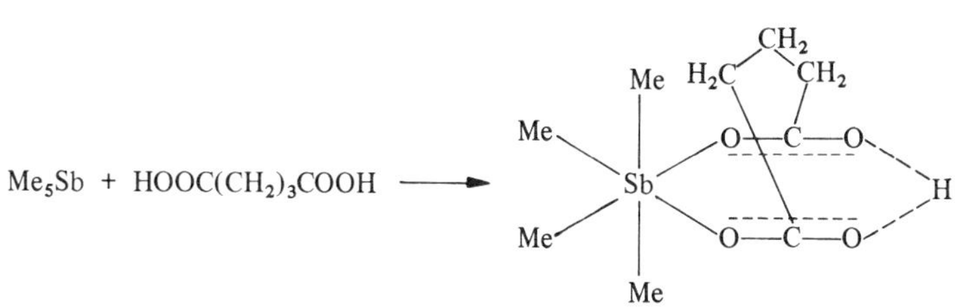

# 6 β-Diketonates

Acetylacetonatoantimony tetrachloride[344, 345] can be prepared by the reaction of antimony pentachloride with acetylacetone in $CCl_4$ or $CHCl_3$. Both IR and $^1H$ NMR spectral data indicate that acetylacetonate group is acting as a chelating ligand producing an octahedral environment around antimony atom[346–351].

Monoorganoantimony(acetylacetonato) trichlorides have been prepared by the reaction of either arylstibonic acid in HCl[347–349] or of monoorganoantimony(V) chloride with acetylacetone[351–353] at low temperatures. All the monoorganoantimony(acetylacetonato) trichlorides are monomeric and non-electrolyte in solution. The infrared spectra of these compounds show that acetylacetone moiety behaves as a bidentate ligand[347–349, 351].

On the basis of observed doublet and singlet for acac-Me protons in the $^1H$ NMR spectra, Okawara et al.[347, 348, 352, 354] have suggested an asymmetric structure (Fig. 17. I) for $PhSb(acac)Cl_3$ and a symmetric structure (Fig. 17. II) for $MeSb(acac)Cl_3$. These results have been reinvestigated by Meinema et al.[351], who observed two distinct acac-Me proton resonances in $CCl_4$ for both the above compounds as well as for the ethyl analogue and proposed an asymmetric structure for all three of them. The magnitude of the separation between methyl resonances increases in aromatic solvents[351, 352, 354] indicating an enhancement in the non-equivalence of two acac-Me groups as a result of aromatic ring current effect on asymmetrically solvated solute molecule. Dipole moment data also support the asymmetric geometry (Fig. 17. II) for these compounds[351].

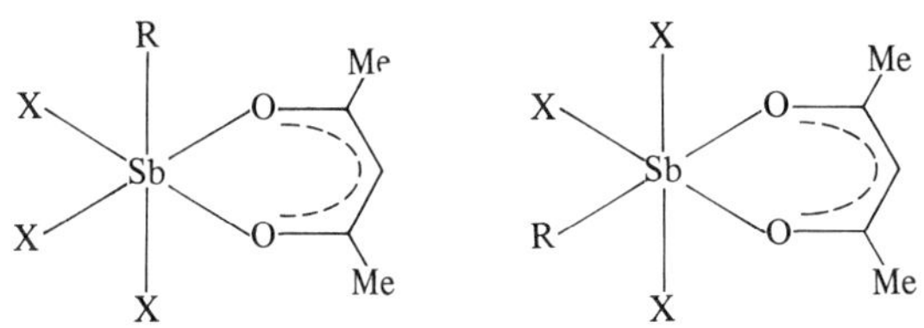


**Fig. 17.** Symmetric and asymmetric structures of $RSb(acac)Cl_3$

Results of Meinema et al.[351] also get support from the X-ray analysis of $MeSb(acac)Cl_3$[355, 356]. The crystals belong to a tetragonal system in which the methyl group occupies the basal plane. Antimony acquires a slightly distorted octahedral geometry

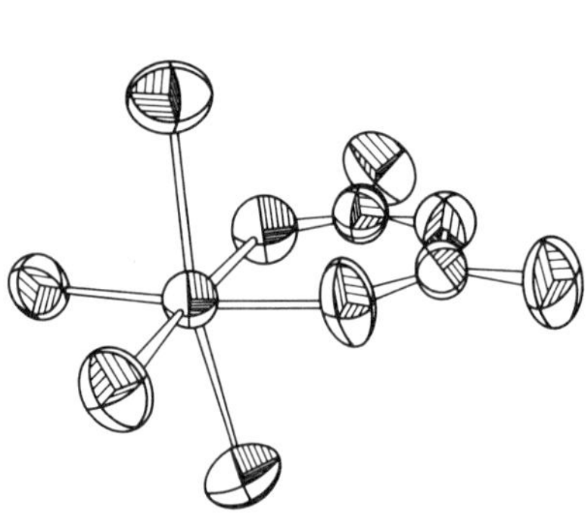

**Fig. 18.** Stereoscopic structure of $MeSb(acac)Cl_3$

(Fig. 18) with substituents on apical positions, slightly bent towards planar acetylacetonato group.

Diorganoantimony(V) $\beta$-diketonates, $R_2Sb(R'COCHCOR'')X_2$, are monomeric and non-conducting in nature. Meinema and Noltes[349)], on the basis of IR and $^1H$ NMR studies, suggested that $Ph_2Sb(acac)Cl_2$ in chloroform solution exists as a mixture of penta- and hexacoordinated isomeric forms. However, later on[351, 357)], it has been shown that $Ph_2Sb(acac)Cl_2$ exists exclusively in the hexacoordinated form in chloroform as well as in benzene solutions. Three possible geometric forms (Fig. 19. III–V) have been considered for compounds of the type $R_2Sb(acac)X_2$:

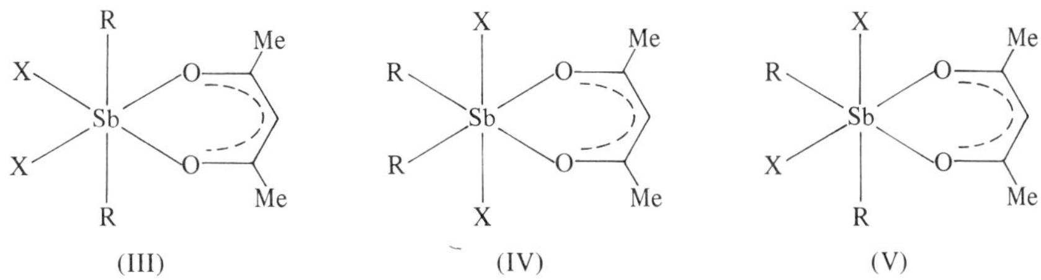

*trans*—diorgano *cis*—diargano

**Fig. 19. III-V.** Possible geometric forms of $R_2Sb(acac)X_2$

On the basis of the equivalence of the acac-Me groups and their peak separations in the $^1H$ NMR spectra in a number of solvents of the compounds (when R is an aryl group), it has been shown that these exist as a mixtures of symmetrical forms (III) and (IV); the ratio of two varied with temperature and with the polarity of the solvent[357–359)]. Compounds with R = methyl or ethyl groups and with acac or dpm ($Bu^tCOCHCOBu^t$) moieties exist exclusivly in *trans*-dialkyl configuration (Fig. 19. III)[351)].

The conclusions drawn from $^1H$ NMR spectral studies have been confirmed by X-ray diffraction studies of $Me_2Sb(acac)Cl_2$ and $Ph_2Sb(acac)Cl_2$. Crystals of $Me_2Sb(acac)Cl_2$ belong to monoclinic system with space group $P2_{1/m}$[356, 360)]. Antimony atom has been found to possess a slightly distorted octahedral geometry in which methyl groups occupy apical positions and are bent towards the planar acetylacetonato group.

The compound, $Ph_2Sb(acac)Cl_2$ crystallizes from benzene into a mixture of two kinds of crystals: (i) long prismatic (m.p. 192 °C decomp.) and (ii) polyhedron (m.p. 184.5 °C decomp.)[361)]. Crystals of both forms belong to monoclinic system with space group $P2_{1/c}$[356, 361, 362)] and have a distorted octahedral geometry in which the phenyl groups occupy the *trans* positions. Deviation form $C_{2v}$ symmetry has been ascribed mainly to rotational positions of the phenyl groups[362)]. In the higher melting form, one phenyl ring is approximately perpendicular (89°) to the bisector and the other is parallel (3°), the dihedral angle between the two ring being 92° (Fig. 20). In the lower melting form, both rings are neither perpendicular nor parallel to the bisector and each phenyl ring makes an angle of about 65 or 77° to the bisector and the dihedral angle between them is about 38°[356)].

The large and positive value of $e^2qQ$ (26.4 ± 0.4) and small value of asymmetry parameter $\eta$(O) in Mössbauer spectrum of $Ph_2Sb(acac)Cl_2$ have also been interpreted in terms of *trans*-Ph configuration[363)].

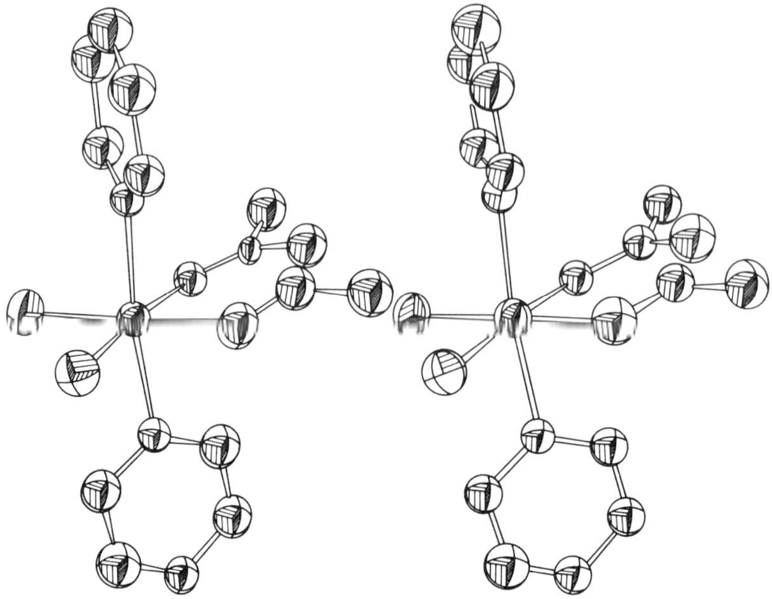

Higher–melting form

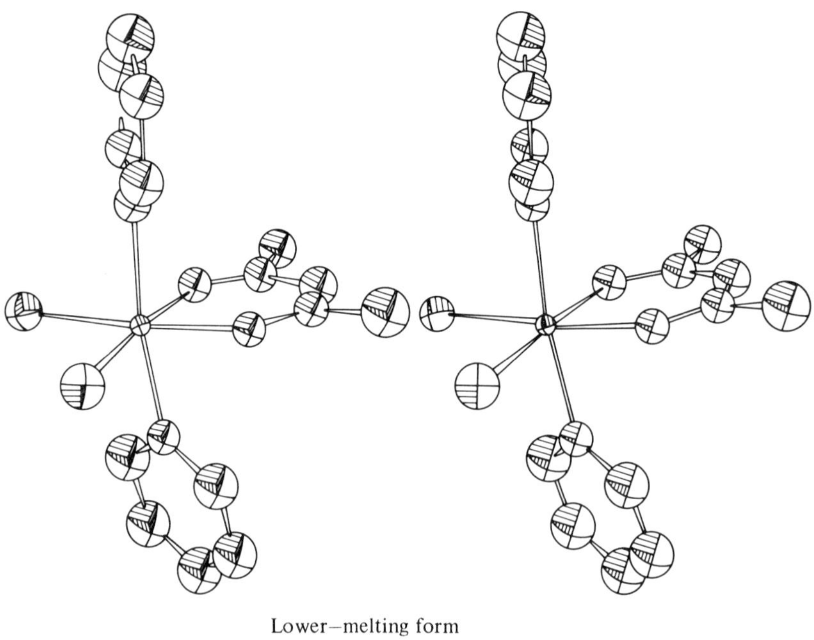

Lower–melting form

**Fig. 20.** Stereoscopic drawing of $Ph_2Sb(acac)Cl_2$

Isomerization between two geometric forms of diorganoantimony(V) $\beta$-diketonates (acac, dpm, pac) has been studied by $^1H$ NMR techniques[358, 359, 364]. Meinema and Noltes[359] have reported that $Ph_2Sb(dpm)Cl_2$ and $Ph_2Sb(pac)Cl_2$ adopt the *trans*-diphenyl configuration in the crystalline state and also in freshly prepared solutions but show geometrical isomerization in solution either on keeping or on heating. In contrast to this, the dialkyl analogues do not undergo isomerization process[364].

Meinema and Noltes[359, 364] have also studied hexacovalent heterocyclic antimony(V) $\beta$-diketonates of the type $R_2Sb(L)Cl_2$ [$R_2$ = $(CH_2)_4$, $(CH_2)_5$, $o, o'$-$C_6H_4C_6H_4$, $o, o'$-$C_6H_4CH_2C_6H_4$ and HL = acacH, dpmH or pacH ($Bu^tCOCH_2COMe$)]. Out of the possible three isomers, the structure in which the heterocyclic rings occupy *cis*-positions and chlorine, *trans* has been considered as the most plausbile one. Dipole moment data also favour this structure[364]. In freshly prepared solution, the presence of *trans*- dichloro isomer (Fig. 21. VI) only has been suggested:

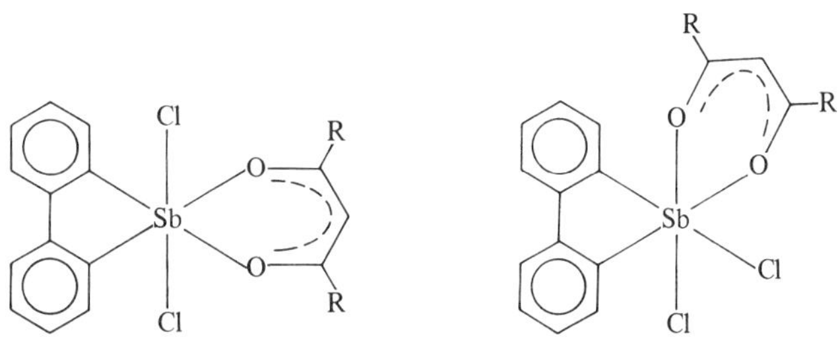

VI (*trans*) VII (*cis*)

**Fig. 21. VI, VII.** *Cis* and *trans* isomers of heterocyclicantimony(V) $\beta$-diketonates

However, isomerization process has been found to be rapid and leads to an equilibrium mixture of both the *trans* and the *cis* isomers (Fig. 21. VI, VII)[359, 364]. The isomerization process proceeds via the twist mechanism and does not appear to involve an antimony-oxygen bond repture[364].

Meinema et al.[365, 366] have measured $^1H$ NMR benzene solvent shifts for diorganoantimony(V) $\beta$ diketonates (R = Me, Et, Ph or 1/2 $o, o'$-$C_6H_4C_6H_4$). The observed shifts have been interpreted in terms of a non-specific tengential approach of the benzene molecules to the electron deficient sites of the solute.

The $\beta$-diketone exchange reactions in organoantimony(V) compounds of the type $R_2Sb(acac)Cl_2$ and $R_2Sb(dpm)Cl_2$ (R = Me, Ph or 1/2 $o, o'$-$C_6H_4C_6H_4$) have been studied by $^1H$ NMR spectroscopy[364, 367]. The compounds having *trans*-organo groups undergo fast $\beta$-diketone exchange as compared to the corresponding *cis* isomers, which are, in fact, inert to exchange with free $\beta$-diketones. The mechanism of $\beta$-diketone exchange has been considered to involve metal-oxygen bond cleavage as the rate determining step[367].

Triorganoantimony(V) $\beta$-diketonates have been prepared by the following reaction routes:

$$R_3SbX_2 + Na(R'COCHCOR'') \longrightarrow R_3SbX(R'COCHCOR'') + NaX$$
$$Ph_3SbCl(OMe) + acacH \longrightarrow Ph_3Sb(acac)Cl + MeOH$$
$$R_3Sb(OMe)_2 + R'COCH_2COR'' \longrightarrow R_3Sb(OMe)(R'COCHCOR') + MeOH$$

$$R_3SbBr_2 + R'COCH_2COR'' \xrightarrow{MeOH} R_3Sb(OMe)(acac) + 2\,NaBr + acacH$$

$$R_3SbBr_2 + 2\,Naacac \longrightarrow Ph_3Sb(acac)Cl + H_2O$$

$$(Ph_3SbCl)_2O + 2\,acacH \longrightarrow Ph_3Sb(acac)(OH)$$

Attempt to synthesize $Me_3Sb(acac)Cl$ by Meinema et al.[351] employing reaction routes of $Me_3SbCl_2$ with Naacac in methanol were unsuccessful, but a similar bromo derivative, $Me_3Sb(acac)Br$, has been prepared by us[368] by the reaction of $Me_3SbBr_2$ with Naacac in chloroform.

These compounds are monomeric and the infrared spectra show that the ligand moieties are acting as a chelating ligands[119, 351, 368, 369]. Their stereochemistry has also been elucidated with the help of $^1H$ NMR spectroscopy[351, 368–370]. A *trans*-Ph configuration (Fig. 22. IX, X) has been proposed for the compounds of the type $Ph_3Sb$ (R′COCHCOR″) X (X = Cl, Br or OH and R′ = R″ = Me or Bu$^t$)[351, 369, 370]. The compounds, $Ph_3Sb(OMe)(R'COCHCOR'')$ (R′ = Me and R″ = Me or Ph), exist as a mixture of *cis*- and *trans*-phenyl configurations in solution at ambient temperatures. However, the presence of only *cis*-Me isomer (Fig. 22. VIII) has been suggested for $Me_3Sb(OMe)$-(R′COCHCOR″) in solution.

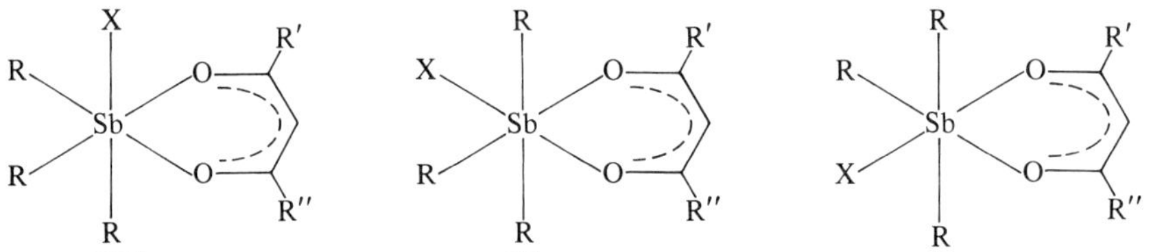

VIII IX X

*cis*–R *trans*–R

**Fig. 22. VIII–X.** Possible geometric isomers of triorganoantimony(V) $\beta$-diketonates

Tetraorganoantimony(V) $\beta$-diketonates, $R_4Sb(R'COCHCOR'')$[349, 371], are monomeric and hexacoordinated in the solid as well as in solution. On the basis of low temperature $^1H$ NMR studies the presence of two axial and two equatorial organo groups have been suggested for these complexes[351].

Ebina et al.[305, 372–375] have synthesized and studied a new class of triorganoantimony(V) compounds derived from fluorinated $\beta$-diketones and hydrated fluorinated $\beta$-diketones.

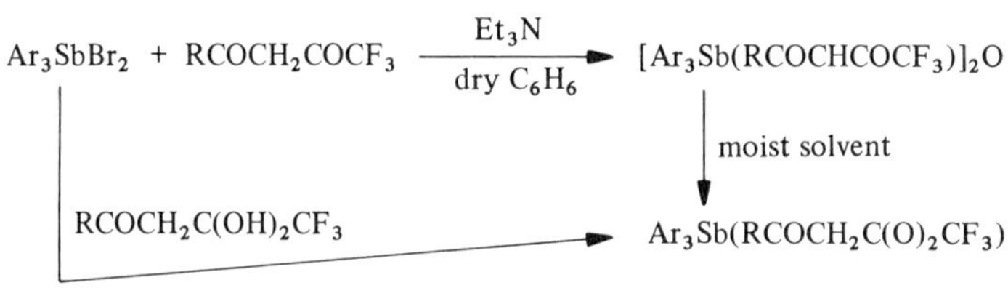


Triarylantimony(V) bromides react with fluorinated $\beta$-diketones in the presence of triethylamine to yield the oxygen bridged enol type $\beta$-diketone complex $(Ar_3SbL)_2O$, which when heated in moist organic solvent yields hydrated $\beta$-diketone complexes of

triarylantimony $Ar_3Sb(RCOCH_2C(O)_2CF_3)$. These complexes have been studied by IR, $^1H$ and $^{19}F$ NMR spectroscopy.

The crystal structure of $\mu$-oxo-bis[tris(*p*-chlorophenyl)(1,1,1-trifluoro-2,4-pentanedionato-O,O′)antimony(V)] · 1/2 chloroform has been determined[375]. The compound crystallizes in triclinic system, space group $P\bar{1}$. The binuclear Sb-complex consists of two distorted octahedra joined by an O atom which is situated at inversion centre; the Sb-O-Sb bridge is linear. This seems to be the first example of a linear Sb-O-Sb linkage. The bridged oxygen and one of the oxygens of $\beta$-diketone occupy *trans*-positions. The $\beta$-diketonate forms two distinct Sb-O bonds:

$$Sb\text{-}O(CF_3) = 2.180(4) \text{ and } Sb\text{-}O(Me) = 2.417\ (3)\ A°.$$

The crystal structures of 4,4′-dihydroxy-2-one complexes of the type, $Ar_3Sb(RCOCH_2C(O)_2CF_3)$ have also been determined[372–374]. In these complexes, hydrated $\beta$-diketones act as a terdentate ligand. The three O atoms and three carbon atoms of the aryl groups are bonded to the Sb atom in *facial* positions. The *gem*-diol oxygen atoms form an unusual four membered chelate ring (Fig. 23). The Sb-O distance for the third oxygen has been found to be unusually long, a fact which suggests that carbonyl oxygen is only weakly coordinated.

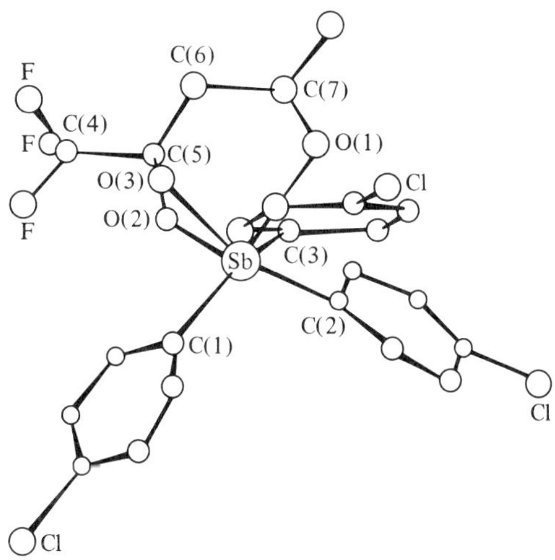


**Fig. 23.** Molecular structure of $(p\text{-}ClC_6H_4)_3Sb[CF_3C(O)_2CH_2COMe]$

# 7 Schiff Base Complexes

Diorganoantimony(V) complexes with potentially dianionic tridentate Schiff base ligands have been prepared by the exchange reactions of diorganoantimony(V) chlorides with the corresponding Schiff base complexes of trimethylantimony(V) or dimethyltin(IV)[376, 377]. These reactions proceed to completion as a result of the greater Lewis acidity of $R_2Sb(V)$ than that of $Me_3Sb(V)$ or $Me_2Sn(IV)$.

$R_2SbCl_3 + Me_3SbL(Me_2SnL) \rightarrow R_2SbClL + Me_3SbCl_2(Me_2SnCl_2)$

These are yellow orange monomeric compounds. On the basis of IR and $^1H$ NMR studies, an octahedral coordination environment for the antimony atom with *meridional* arrangement of the ONO ligand atoms and a linear C-Sb-C skeleton has been proposed[376, 377]. The $^1H$ NMR spectrum of $Ph_2Sb(Sah)Cl(SahH_2 = HOC_6H_4CH{=}NC_6H_4OH)$ has been explained in terms of a mixture of *cis-* and *trans*-Ph geometric forms, since the spectrum displays two singlets for >N=C–H proton resonances.

Large positive value for quadrupole coupling constants (21.70–26.98 mm $s^{-1}$) and negligible asymmetry parameter (O) observed in Mössbauer spectra for the complexes, $Me_2Sb(Sah)Cl$, $Me_2Sb(Bah)Cl$, $[Ph_2Sb(Sah)Cl] \cdot CCl_4$, $Ph_2Sb(Bah)Cl$ and $Ph_2Sb(Aah)Cl$ ($BahH_2$ = Ph–C(OH)=CH–C(Me) = $NC_6H_4OH$, $AahH_2$ = Me–C(OH)=CH–C(Me) = $NC_6H_4OH$) have indicated an effective octahedral geometry with *meridionally* disposed tridentate ligand[378]:

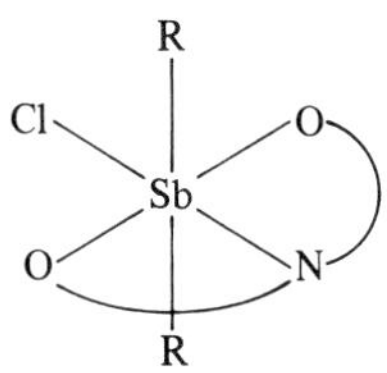


Triorganoantimony(V) complexes of monofunctional bidentate and bifunctional tridentate Schiff bases have been prepared[379–382] by the following general reactions:

$$
\begin{aligned}
R_3SbCl_2 + NaL &\longrightarrow R_3SbClL + NaCl \\
R_3SbCl_2 + Na_2L' &\longrightarrow R_3SbL' + 2\,NaCl \\
R_3Sb(OMe)_2 + HL &\longrightarrow R_3Sb(OMe)L + MeOH \\
R_3Sb(OMe)_2 + H_2L' &\longrightarrow R_3SbL' + 2\,MeOH
\end{aligned}
$$

(LH = monofunctional bidentate Schiff bases, $L'H_2$ = bifunctional tridentate Schiff bases)

Attempts to prepare $Me_3Sb(Bah)$ and $Me_3Sb(Aah)$ resulted in the formation of a viscous products of unknown structures[380]. Complexes of the type $R_3Sb(Aat)$ ($AatH_2$ = Me–C(OH)=CH–C(Me) = $NC_6H_4SH$) could not be isolated. All the compounds reported till now have been found to be monomeric in nature. On the basis of infrared and ultraviolet spectral data, a six coordinated geometry has been suggested.

$^1H$ NMR spectra of trimethylantimony(V) complexes indicate a non-rigid octahedral structure in solution[380, 382]. However, $Me_3Sb(Sah)$ and $Me_3Sb(Sat)$ ($SatH_2$ = $HOC_6H_4CH$ = $NC_6H_4SH$) have a rigid configuration, since the signal due to Me-Sb protons splits into two peaks[380]. An X-ray structure determination revealed that antimony atom in $Me_3Sb(Sah)$ is hexacoordiated and possessed a distorted octahedral geometry (Fig. 24).

The negative values of quadrupole coupling constants, $e^2qQ$ (– 14.0 to – 20.6), and nearly unity asymmetry parameter, $\eta$(0.76–1.01), in the Mössbauer spectra of

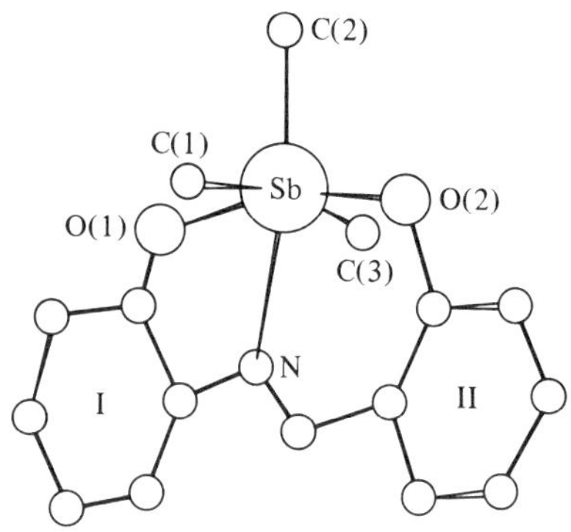


**Fig. 24.** Structure of the $Me_3Sb(Sah)$ molecule

$Me_3$ (Sah)[378], $Ph_3Sb$ (Sah)[378], $Ph_3Sb$ (Aah)[378], $Ph_3Sb$ (Sat)[378] and $Ph_3Sb$ (MeSah) ($MeSahH_2$ = $HOC_6H_4C(Me){=}NC_6H_4OH$)[383] indicate an effective octahedral geometry with *meridionally* disposed tridentate ligand and a T-shaped $R_3Sb(V)$ moiety.

# 8 Oxinates

(8-Hydroxyquinolino)antimony(V) choride has been obtained as a orange colour hygroscopic monomeric non-electrolyte solid by the reaction of antimony(V) chloride and 8-hydroxyquinoline in $CCl_4$[384]. On the basis of UV, IR and $^1$H NMR data, an hexacoordinated geometry has been suggested.

Monomeric monoorganoantimony(V)(8-hydroxyquinolino)(V) chlorides have been prepared by the reactions of monoorganoantimony(V) chlorides with oxine in absolute ethanol[385]. $^1$H NMR and UV spectral data of these compounds indicate an octahedral configuration around antimony atom. Out of the two possible geometric forms, the one with R group on equatorial position has been considered as the most plausible form:

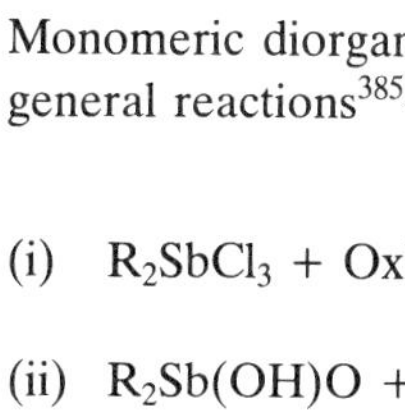


Monomeric diorganoantimony(V) compounds have been synthesized by the following general reactions[385]:

(i) $R_2SbCl_3 + OxH \xrightarrow{\text{ethanol}} R_2Sb(Ox)Cl_2 + HCl$

(ii) $R_2Sb(OH)O + OxH + 4\,N\ HCl \xrightarrow{\text{ethanol}} R_2Sb(Ox)Cl_2 + 2\,H_2O$

((i) R = Me, or Ph; (ii) R = Et or Bu)

Like the monoorganoantimony(V) compounds these diorganoantimony derivatives also depict an hexacoordinated geometry in solid state as well as in benzene, chloroform and ethanol solutions. Since the IR spectrum of $Me_2Sb(Ox)Cl_2$ possesses only one Sb-C, two Sb-Cl stretching frequencies and only one methyl proton resonances at $-100\,°C$ in the $^1$ NMR spectrum, the following *trans*-Me structure has been suggested[385)] out of the three possible geometric forms:

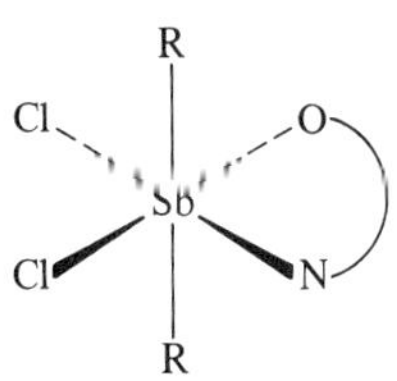

However, the observed Mössbauer spectral data, $\eta = 0.75$ and $e^2qQ = -13.3$ mm $s^{-1}$ on $Ph_2Sb(Ox)Cl_2$, indicated a *cis* distorted octahedral configuration in which C-Sb-C angle 104° has been indicated[363)].

A number of yellow crystalline monomeric triorganoantimony(V) oxinates have been prepared by the following reaction routes:

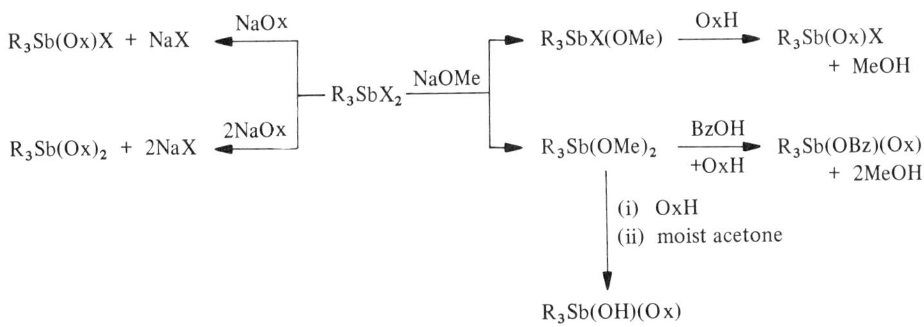

The nature of oxinate ligand (unidentate or bidentate) in triorganoantimony oxinates appear to be unresolved. UV spectra in benzene solution as well as solid state reflectance spectra support the non-chelated nature of oxine in $R_3Sb(Ox)Cl$ (R = Me or Et); the ligand appears to be weakly chelated in $Ph_3Sb(Ox)Cl$ in benzene solution and the Sb-N bond appears to be completely reptured in ethanol or chloroform solutions[385)]. The IR spectra of $R_3Sb(Ox)Cl$ (R = Me or Et) also show only one $\nu$Sb-C vibration suggesting trigonal bipyramidal geometry.

In contrast to the above, chelation (in some cases weak) of the oxine moiety in $R_3Sb(Ox)X$ (R = Me or Ph and X = Cl, Br or OH) has been suggested on the basis of UV spectra[386–389)], which appeared to vary with concentration[389)]. However, UV and IR as well as benzene induced $^1$H NMR solvent shifts on 2-H proton of 8-hydroxyquinolinato ligand suggests that in carboxylato and nitrato complexes the Sb-N interaction is probably weak[387, 389, 390)]. Benzene induced solvent shifts (0.6–0.7 ppm) in $^1$H NMR spectra of halo and hydroxo derivatives indicate the chelation of the ligands in these compounds[387, 389, 390)].

Kawasaki et al.[387, 390] have reported the rate constant and activation parameters Ea, $\Delta H \neq$, $\Delta S \neq$ and $\Delta G \neq$ by $^1H$ NMR line broadening techniques for the ligand exchange reactions in trimethylantimony(V) and methylphenylantimony(V) oxinates.

$$Me_nPh_{3-n}Sb(Ox)X + Me_nPh_{3-n}Sb^*(Ox)y \rightleftharpoons Me_nPh_{3-n}Sb(Ox)Y + Me_nPh_{3-n}Sb^*(ox)X$$

(n = 0–3, O × H = 8-hydroxyquinoline or substituted quinolines and X = an electronegative group)

The following order of ligand lability has been observed:

$$MeCOO \lesssim OC_6H_4NO_2-p < OOCCH_2Cl < OOCCHCl_2 < OOCCCl_3 < Cl < Br$$

Involvement of a bridging intermediate in these exchange reactions has been suggested.

The magnitude of red shift increases in the UV spectra as the number of phenyl groups increases in the complexes, $Me_nPh_{3-n}Sb(Ox)X$, which has been ascribed to the stronger chelation by the ligand in this order, but in exchange reactions the rate constant and activation energy decrease in the same direction. From the $^1H$ NMR data, it has been concluded that these compounds possess a chiral centre and the rate conversion between enantiomers is relatively slow[390].

Recently, a suggestion of seven coordination has been made for antimony atom in $Ph_3Sb(Ox)_2$ on the basis of UV spectral data[384]. The Mössbauer spectral data on $Ph_3Sb(Ox)Cl$ have suggested that antimony atom adopt *mer*-octahedral geometry with a chelated oxine group in the solid state[383].

Monomeric tetraorganoantimony(V) oxinates have been prepared by the reactions of tetraorganoantimony(V) alkoxides and oxine[371, 385]. From the UV spectral studies of these compounds, variation in coordination state (penta- or hexa-) of antimony has been observed by varying the R and the polarity of the solvent. Generally pentacoordinated species predominate in polar solvents like EtOH. The $^1H$ NMR spectrum of $Me_4SbOx$ in $CDCl_3$ exhibits one singlet which points to a rapid equivalence between stereochemically non-rigid penta- and hexa-coordinate species. From the splitting in the methyl signal at – 100° in toluene $d_8$, the following structure is indicated[385]:

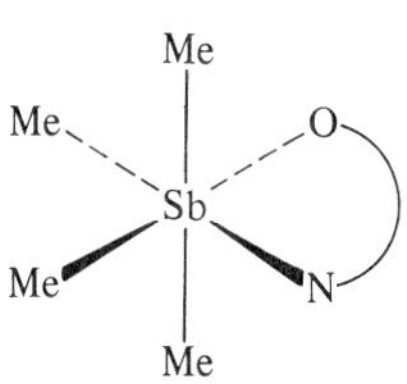


## 9 Oximates

Hygroscopic, monomeric triorganoantimony(V) oximates have been synthesized by the following general reactions[391–393]:

$$R_3SbX_2 + 2\,Na(ONCR'R'') \longrightarrow R_3Sb(ONCR'R'')_2 + 2\,NaX$$

$$R_3SbX_2 + 2\,HONCR'R'' \xrightarrow{Et_3N} R_3Sb(ONCR'R'')_2 + 2\,Et_3N \cdot HX$$

$$R_3Sb(OMe)_2 + 2\,HONCR'R'' \longrightarrow R_3Sb(ONCR'R'')_2 + 2\,MeOH$$

Triphenylantimony(V) oximates disproportionated into triphenylantimony and oximes when subjected to distillation under reduced pressure[392]. However, trialkyl derivatives are volatile under reduced pressure[393]. On the basis of IR and $^1$H NMR spectral studies, a trigonal bipyramidal geometry with planar $R_3Sb(V)$ group has been suggested [392, 393] for compounds of the type $R_3Sb(ONCR'R'')_2$. IR studies indicate the presence of hexacoordinated state of antimony in the solid state and pentacoordinated geometry in chloroform solution for $Me_3Sb(ONCNH_2Ph)_2$[393].

Pentamethylantimony reacts with oximes to yield methane and tetramethylantimony(V) oximates[294].

## 10 Other Compounds

Monomeric tetramethylantimony(V) diacetamido complex has been prepared by the reaction of pentamethylantimony and diacetamide[394].

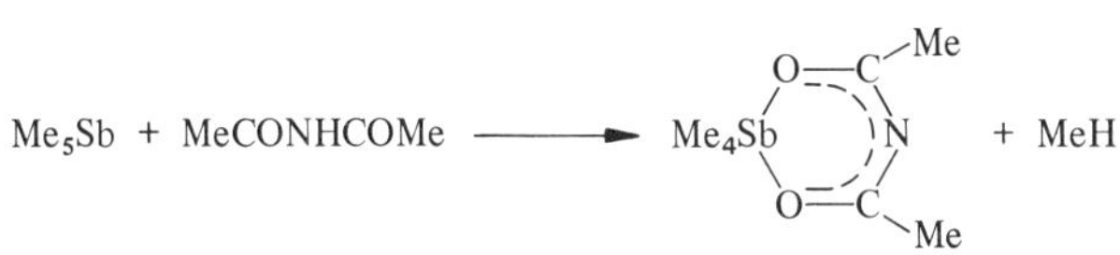


The first example of a molecule containing both penta- and hexa-coordinate antimony atoms has been reported in $\mu$-carbonato bis-tetraphenylantimony[395], the crystals of which belong to a triclinic system with space group $P\bar{1}$. One antimony atom has a distorted octahedral environment while the other has a slightly distorted trigonal bipyramidal geometry with the carbonato oxygen in an apical position:

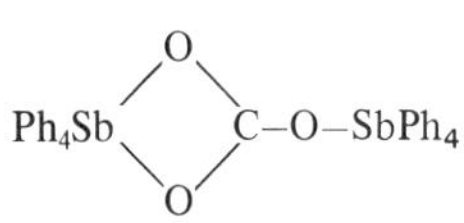


The reaction of antimonypentachloride with sodium salt of dimethyl formamide yields complexes of the type, $SbCl_4(CONMe_2)DMF$, $SbCl_3(CONMe_2)$ and $Cl_2Sb(CONMe_2)$[397]. These complexes have been characterized by IR spectra, conductivity measurements and thermal analysis.

With N,N'-dimethyloxamide (MeNHCOCONHMe), antimony pentachloride forms addition compounds of the type $SbCl_5 \cdot L$ and $(SbCl_5)_2 \cdot L$, which on pyrolysis lose HCl and yield product I and II[397]. Product I can also be obtained on heating $SbCl_5$ and ligand in chloroform or 1,2-dichloroethane[398]. Product I can further react with $SbCl_5$ yielding

Crystals of tetramethylantimony(dimethylthiophosphinate) belong to triclinic system with space group $P\bar{1}$[407]. The antimony atom has a distorted trigonal bipyramidal geometry while phosphorus atom is surrounded by a distorted tetrahedral environment.

A perusal of the above account shows that compounds of the type $SbL_nX_{5-n}$ (N = 1–5, X = halogen, L = ligand), $RSbL_nX_{3-n}$ and $R_2SbL_nX_{2-n}$ have hexacoordinated environment around antimony which can be acquired either by chelation of the ligand or by molecular association. Contrary to the above, tri-, tetra- and penta-organoantimony(V) derivatives generally do not show hexacoordinated geometry. The triorganoantimony(V) compounds exhibit a preference for the trigonal bipyramidal geometry and even the compounds which are six-coordinated in the solid state tend to change their geometry in solution. The tetraorganoantimony(V) moiety appears to form generally cationic (tetrahedral geometry) or covalent (Trigonal bipyramidal geometry) compounds. The pentaorganoantimony compounds exist generally in trigonal bipyramid but sometimes in square pyramid geometries.

In view of a renewed interest in the structural chemistry of organic derivatives of antimony(V) during the last decade, it is anticipated that future work will entail a more penetrating study in understanding the structural aspects of these compounds with techniques of ever increasing sophistications.

*Acknowledgements.* One of the authors (V.K.J.) is grateful to the Council of Scientific and Industrial Research, New Delhi for the award of a Post-Doctoral Fellowship.

## 11 References

1. Cotton, F. A., Wilkenson, G.: Advanced inorganic chemistry, John Wiley and Sons, Inc., New York 1972
2. Smith, J. D.: Comprehensive inorganic chemistry, (Ex. ed. Tortman-Dickenson, A. F.), Oxford, Pergamon Press, *Vol. II* (1973)
3. Peterson, L. K.: Int. Rev. Sci. *4*, 319 (1975)
4. Morgan, G. T.: Organic compounds of arsenic and antimony, London, Longmans, Green and Co. 1918
5. Kolditz, L.: Adv. inorgan. radiochem., Academic Press, New York, *7*, 1 (1965)
6. Coates, G. E., Wade, K.: Organometallic compounds, The main group elements, 1, Methuen, London 1967
7. Barnes, J. M., Magos, L.: Organometal. Chem. Rev. *3*, 137 (1968)
8. Dub, M.: Organometallic Compounds, Vol. III, 2nd Ed., Springer Verlag, New York 1968
9. Doak, G. O., Freedman, L. D.: Organometallic compounds of arsenic, antimony and bismuth, Wiley Interscience, 1970
10. Crow, J. P., Cullen, W. R.: Int. Rev. Sci. (MTP) *4*, 355 (1972)
11. Zingaro, R. A.: Ann. N. Y. Acad. Sci. *192*, 72 (1972)
12. Thayer, J. S.: J. Organometal Chem. *76*, 265 (1974)
13. Wittig, G., Torsell, K.: Acta Cham. Scand. *7*, 1293 (1953)
14. Nesmeyanov, A. N., Borisov, A. E., Novikova N. V.: Izv. Akad. Nauk SSSR *1960*, 952
15. Henry, M. C., Wittig, G.: J. Am. Chem. Soc. *82*, 563 (1960)
16. Nesmeyanov, A. N., Borisov, A. E., Novikova, N. V.: Izv. Akad. Nauk SSSR, Otd. Khim. Nauk *1960*, 147
17. Nesmeyanov, A. N., Borisov, A. E., Novikova, N. V.: Tetrahedron Lett. *1960*, 23

18. Nesmeyanov, A. N., Borisov, A. E., Novikova, N. V.: Izv. Akad. Nauk, Otd. Khim. Nauk *1961*, 1578
19. Wittig, G., Clauss, K.: Justus Liebigs Ann. Chem. *577*, 26 (1952)
20. Nesmeyanov, A. N., Borisov, A. E., Novikova, N. V.: Izv. Akad. Nauk SSSR, Otd. Khim. Nauk *1961*, 730
21. Nesmeyanov, A. N., Borisov, A. E., Novikova, N. V.: ibid., 612
22. Wittig, G., Hellwinkel, D.: Ber. *97*, 789 (1964)
23. Nesmeyanov, A. N., Borisov, A. E., Novikova, N. V.: Izv. Akad. Nauk SSSR, Ser. Khim. *1964*, 1197
24. Nesmeyanov, A. N., Borisov, A. E., Novikova, N. V.: ibid., 1202
25. Nesmeyanov, A. N., Borisov, A. E., Novikova, N. V.: Dokl. Akad. Nauk SSSR *155*, 1364 (1964)
26. Takashi, Y.: J. Organometal. Chem. *8*, 225 (1967)
27. Hellwinkel, D., Bach, M.: ibid. *20*, 273 (1969)
28. Meinema, H. A., Noltes, J. G.: ibid. *22*, 653 (1970)
29. Hellwinkel, D., Bach, M.: ibid. *28*, 349 (1971)
30. Doleshall, G., Nesmeyanov, A. N., Reutov, O. A.: ibid. *30*, 369 (1971)
31. Nesmeyanov, A. N. et al.: Izv. Akad. Nauk SSSR, Ser. Khim. *1973*, 1833; C. A. *80*, 48115
32. Tempel, N., Schwarz, W., Weidlein, J.: J. Organometal. Chem. *154*, 21 (1978)
33. Hellwinkel, D., Lindner, W., Schmidt, W.: Chem. Ber. *112*, 281 (1979)
34. Downs, A. J., Schmutzler, R., Steer, I. A.: Chem. Commun. *1966*, 221
35. Holmes, Sr., R. R., Deiters, R. M., Golen, J. A.: Inorg. Chem. *8*, 2612 (1969)
36. Borisov, A. E., Novikova, N. V., Chumaevskii, N. A.: Dokl. Akad. Nauk SSSR *136*, 129 (1961)
37. Hellwinkel, D., Bach, M.: Naturwissenschaften *56*, 214 (1969)
38. Wheatley, P. J., Wittig, G.: Proc. Chem. Soc. *1962*, 251
39. Wheatley, P. J.: J. Chem. Soc. *1964*, 3718
40. Beauchamp, A. L., Bennett, M. J., Cotton, F. A.: J. Am. Chem. Soc. *90*, 6675 (1968)
41. Beattie, I. R. et al.: J. Chem. Soc. Dalt. Trans. *1972*, 784
42. Kok, L. G.: Spectrochim. Acta *30A*, 961 (1974)
43. Cowley, A. H. et al.: J. Am. Chem. Soc. *93*, 2150 (1971)
44. Brabant, C., Blanck, B., Beauchamp, A. L.: J. Organometal. Chem. *82*, 231 (1974)
45. Brabant, C., Hubert, J., Beauchamp, A. L.: Can. J. Chem. *51*, 2952 (1973)
46. Hellwinkel, D.: Angew. Chem. *78*, 749 (1966)
47. Hellwinkel, D.: ibid. Int. Ed. *5*, 725 (1966)
48. Wittig, G., Hellwinkel, D.: ibid. *74*, 76 (1962)
49. Kuykendall, G. L., Mills, J. L.: J. Organometal. Chem. *118*, 123 (1976)
50. Muetterties, E. L. et al.: Inorg. Chem. *3* 1298 (1964)
51. Hellwinkel, D., Wuensche, C., Bach, M.: Phosphorus *2*, 167 (1973)
52. Long, G. G. et al.: J. Am. Chem. Soc. *92*, 4230 (1970)
53. Razuvaev, G. A. et al.: Zh. Obsch. Khim. *30*, 3234 (1960); C. A. *55*, 21010
54. Mc Eween, W. E., Lin, C. T.: Phosphorus *4*, 91 (1974)
55. Shen, K. W., McEween, W. E., Wolf, A. P.: J. Am. Chem. Soc. *91*, 1283 (1969)
56. Brock, C. P., Ibers, J. A.: Acta Crystallogr. *sect. A*, *32A*, 38 (1976)
57. Brock, C. P.: ibid. *33A*, 193 (1977)
58. Schmidt, H.: Liebigs. Ann. Chem. *421*, 174 (1920)
59. Worrall, D. E.: J. Am. Chem. Soc. *52*, 2046 (1930)
60. Biswell, C. B., Hamilton, C. S.: ibid. 57, 913 (1935)
61. Nishii, N., Hashimoto, K., Okawara, R.: J. Organometal. Chem. *55*, 133 (1973)
62. Morgan, G. T., Davies, G. R.: Proc. Roy. Soc. Ser. A, *110*, 523 (1926)
63. Nesmeyanov, A. N., Borisov, A. E.: Izv. Akad. Nauk SSSR, Otd. Khim. Nauk *1945*, 251
64. Doak, G. O., Long, G. G.: Trans. N. Y. Acad. Sci. *28*, 402 (1966)
65. Bruker, A. B., Makhlis, E. S.: Zh. Obsch. Khim. *7*, 1880 (1937)
66. Nesmeyanov, A. N. Kocheshkov, K. A.: Izv. Akad. Nauk. SSSR, Otd. Khim. Nauk *1944*, 416
67. Campbell, I. G. M.: J. Chem. Soc. *1950*, 3109
68. Reutov, O. A., Ptitsyna, O. A.: Izv. Akad. Nauk SSSR, Otd. Khim. Nauk *1952*, 93
69. Doak, G. O., Freedman, L. D., Efland, S. M.: J. Am. Chem. Soc. *74*, 830 (1952)

70. Morgan, G. T., Davies, G. R.: Proc. Roy. Soc. Ser. A, *127*, 1 (1930)
71. Sergeev, P. G., Bruker, A. B.: Zh. Obsch. Khim. *27*, 2220 (1957)
72. Bruker, A. B.: Zh. Obsch. Khim. *27*, 2223 and 2593 (1957)
73. Nesmeyanov, A. N. et al.: Izv. Akad. Nauk SSSR, Otd. Khim. Nauk *1958*, 1435
74. Yagupol'skii, L. M. et al.: Zh. Org. Khim. *11*, 459 (1975)
75. Dale, J. W. et al.: J. Chem. Soc. *1957*, 3708
76. Severengiz, T., Breunig, H. J.: Chem. Ztg. *104*, 202 (1980)
77. Campbell, I. G. M.: J. Chem. Soc. *1952*, 4448
78. Campbell, I. G. M., Morrill, D. J.: J. Chem. Soc. *1955*, 1662
79. Kolditz, L., Gitter, M., Roesel, E.: Z. Anorg. Allg. Chem. *316*, 270 (1962)
80. Polynova, T. N., Porai-Koshits, M. A.: Zh. Strukt. Khim. *2*, 477 (1961)
81. Polynova, T. N., Porai-Koshits, M. A.: Zh. Strukt. Khim. *7*, 642 (1966)
82. Gukasyan, S. E. et al.: Zh. Strukt. Khim. *14*, 650 (1973)
83. Ruddick, J. N. R., Sams, J. R., Scott, J. C.: Inorg. Chem. *13*, 1503 (1974)
84. Bordner, J., Doak, G. O., Peters, Jr., J. R.: J. Am. Chem. Soc. *96*, 6763 (1974)
85. Bowen, L. H., Long, G. G.: Inorg. Chem. *15*, 1039 (1976)
86. Bowen, L. H., Hedges, S. W.: Inorg. Nucl. Chem. Lett. *13*, 621 (1977)
87. Bertazzi, N., Gibb, T. C., Greenwood, N. N.: J. Chem. Soc. Dalt. Trans. *1976*, 1153
88. Bone, S. P., Sowerby, D. B.: J. Chem. Soc. Dalton Trans. *1979*, 715
89. Bone, S. P., Sowerby, D. B.: ibid. *1979*, 718
90. Begley, M. J., Bone, S. P., Sowerby, D. B.: J. Organometal. Chem. *165*, C47 (1979)
91. Nishii, N., Matsumura, Y., Okawara, R.: ibid. *30*, 59 (1971)
92. Beattie, I. R., Stokes, F. C., Alexander, L. E.: J. Chem. Soc. Dalton Trans. *1973*, 465
93. Dehnicke, K., Nadler, H. G.: Chem. Ber. *109*, 3034 (1976)
94. Pebler, J. et al.: Z. Anorg. Allg. Chem. *427*, 166 (1976)
95. Schwarz, W., Guder, H. J.: Z. Naturforsch. B: Anorg. Chem., Org. Chem. *33 B*, 485 (1978)
96. Barbieri, R., Bertazzi, N., Gibb, T. C.: J. Chem. Soc. Dalton Trans. *1979*, 1925
97. Popov, V. I., Kondratenko, N. V.: Zh. Obsh. Khim. *46*, 2597 (1976); C. A. *86*, 121457
98. Morgan, G. T., Davies, G. R.: Proc. Roy. Soc. Ser., A, *143*, 38 (1933)
99. Ptitsyna, O. A., Reutov, O. A., Ertel, G.: Izv. Aka. Nauk SSSR, Otd. Khim. Nauk *1961*, 265
100. Razuvaev, G. A., Shubenko, M. A.: Dokl. Akad. Nauk SSSR *67*, 1049 (1949); C. A. *44*, 1435
101. Shubenko, M. A.: Sb. Statei Obsh. Khim. *2*, 1043 (1953); C. A. *49*, 6856
102. Holmes, R. R., Bertant, E. F.: J. Am. Chem. Soc. *80*, 2983 (1958)
103. Burg, A. B., Grant, L. R.: ibid., *81*, 1 (1959)
104. Laudolt, H.: J. Prakt. Chem. *84*, 328 (1961)
105. Borisov, A. E., Novkova, N. V., Nesmeyanov, A. N.: Izv. Akad. Nauk SSSR, Ser. Khim. *1963*, 1506; C. A. *59*, 14021
106. Banister, A. J., Moore, L. F.: J. Chem. Soc. A *1968*, 1137
107. Nesmeyanov, A. N., Borisov, A. E.: Izv. Akad. Nauk SSSR, Ser. Khim. *1969*, 939; C. A. *71*, 39108
108. Nesmeyanov, A. N. et al.: ibid. *1969*, 1977
109. Nesmeyanov, A. N., Borisov, A. E.: ibid. *1969*, 974; C. A. *71*, 39104
110. Nesmeyanov, A. N., Borisov, A. E., Novikova, N. V.: ibid. *1969*, 1978; C. A. *72*, 21755
111. Sato, S., Matsumura, Y., Okawara, R.: Inorg. Nucl. Chem. Lett. *8*, 837 (1972)
112. Harris, J. I., Bowden, S. T., Jones, W. J.: J. Chem. Soc. *1947*, 1568
113. Hartmann, H., Kühl, G.: Z. Anorg. Allg. Chem. *312*, 186 (1961)
114. Long, G. G., Doak, G. O., Freedman, L. D.: J. Am. Chem. Soc. *86*, 209 (1964)
115. Doak, G. O., Long, G. G., Freedman, L. D.: J. Organometal. Chem. *4*, 82 (1965)
116. Beveridge, A. D., Harris, G. S., Inglis, F.: J. Chem. Soc. A, *1966,* 520
117. Clark, H. C., Goel, R. G.: Inorg. Chem. *5*, 998 (1966)
118. Doak, G. O., long, G. G., Key, M. E.: Inorg. Syn. *9*, 92 (1967)
119. Otero, A., Poya, P.: J. Organometal. Chem. *154*, 13 (1978)
120. Wiberg, E., Modritzer, K.: Z. Naturforsch B, *11*, 747 (1956)
121. Dessy, R. E., Chivers, T., Kitching, W.: J. Am. Chem. Soc. *88*, 467 (1966)
122. Herbstman, S.: J. Org. Chem. *29*, 986 (1964)
123. Lowry, T. M., Simons, J. H.: Ber. *63*, 1595 (1930)
124. Doak, G. O. et al.: J. Am. Chem. Soc. *88*, 2342 (1966)

125. Hantzsch, A., Hibbert, H.: Dtsch. chem. Ges. Ber. *40*, 1508 (1907)
126. Nefedov, V. D. et al.: Zh. Obsch. Khim. *33*, 2407 (1963)
127. Lile, W. J., Menzies, R. C.: J. Chem. Soc. *1950*, 617
128. Jensen, K. A.: Z. Anorg. Allg. Chem. *250*, 257 (1943)
129. Kataeva, L. M., Rydvanskii, Yu. V., Trofimova, N. I.: Zh. Fiz. Khim. *50*, 814 (1976); C. A. *85*, 5029
130. Parab, N. K., Dasai, D. M.: J. Indian Chem. Soc. *35*, 569 (1958)
131. Jaffe, H. H.: J. Chem. Phys. *22*, 1430 (1954)
132. Rao, C. N. R., Ramachandran, J., Balasubramaniam, A.: Can. J. Chem. *39*, 171 (1961)
133. Jensen, K. A., Nielsen, P. H.: Acta Chem. Scand. *17*, 1875 (1963)
134. Shindo, M., Okawara, R.: J. Organometal. Chem. *5*, 537 (1966)
135. Borisov, A. E. et al.: Dokl. Akad. Nauk SSSR *173*, 855 (1967)
136. Mackay, K. M., Sowerby, D. B., Yound, W. C.: Spectrochim. Acta A *24*, 611 (1968)
137. Goel, R. G., Maslowsky, Jr., E., Senoff, C. V.: Inorg. Nucl. Chem. Lett. *6*, 833 (1970)
138. Redington, R. L., Aljibury, A. L. K.: J. Mol. Spectroscop. *37*, 494 (1971)
139. Goel, R. G., Maslowsky, Jr., E., Senoff, C. V.: Inorg. Chem. *10*, 2572 (1971)
140. Woods, C., Long, G. G.: J. Mol. Spectroscop. *38*, 387 (1971)
141. Nevett, B. A., Perry, A.: J. Organometal. Chem. *71*, 399 (1974)
142. Verdonck, L., Van der Kelen, G. P.: Spectrochim. Acta, Part A *31 A*, 1707 (1975)
143. Nevett, B. A., Perry, A.: ibid. *33 A*, 755 (1977)
144. Dehnest, P., Demuth, R., Grobe, J.: ibid. *34 A*, 857 (1978)
145. Namarivayan, R., Viswanathan, S.: Bull. Chem. Soc. France *87*, 733 (1978)
146. Otero, A., Royo, P.: J. Organometal. Chem. *171*, 333 (1979)
147. Long, G. G. et al.: Inorg. Chem. *5*, 1358 (1966)
148. McKenney, R. L., Sisler, H. H.: ibid. *6*, 1178 (1967)
149. Moreland, C. G. et al.: ibid. *7*, 834 (1968)
150. Moreland, C. G., Long, G. G.: Inorg. Nucl. Chem. Lett. *8*, 347 (1972)
151. Kustes, W. A., Moreland, C. G., Long, G. G.: ibid. *8*, 695 (1972)
152. Moreland, C. G. Beam, R. J.: Inorg. Chem. *11*, 3112 (1972)
153. Ouchi, A., Uehiro, T., Yoshino, Y.: J. Inorg. Nucl. Chem. *37*, 2347 (1975)
154. Chermette, H. et al.: ibid. *34*, 1627 (1972)
155. Brichall, T., Conner, J. A., Hillier, I. H.: J. Chem. Soc. Dalton Trans. *1975*, 2003
156. Svergun, V. I. et al.: Izv. Akad. Nauk SSSR, Ser. Khim. *2*, 484 (1970); C. A. *73*, 9242
157. Brill, T. B., Hugus, Jr., Z. Z.: J. Chem. Phys. *53*, 1291 (1970)
158. Brill, T. B., Long, G. G.: Inorg. Chem. *9*, 1980 (1970)
159. Gukasyan, S. E., Shpinel, V. S.: Phys. Status Solidi, *29*, 49 (1968); C. A. 69, 82081
160. Stevens, J. G., Ruby, S. L.: Phys. Lett. A *32*, 91 (1970)
161. Shenoy, G. K., Friedt, J. M.: Phys. Rev. Lett. *31*, 419 (1973)
162. Bancroft, G. M. et al.: J. Chem. Soc. Dalton Trans. *1976*, 643
163. Wells, A. F.: Z. Kristallogr. *99*, 367 (1938)
164. Schwarz, W., Guder, H. J.: Z. Anorg. Allg. Chem. *444*, 105 (1978)
165. Polynova, T. N., Porai-Koshits, M. A.: Zh. Strukt. Khim. *1*, 159 (1960)
166. Polynova, T. N., Porai-Koshits, M. A.: ibid. *7*, 742 (1966)
167. Struchkov, Yu. T., Khotsyanov, T. L.: Dokl. Akad. Nauk SSSR *91*, 565 (1953); C. A. *48*, 422
168. Struchkov, Yu. T., Kitaigorodskii, A. I., Khotsyanova, T. L.: Zhur. Fiz. Khim. *26*, 530 (1952); C. A. *49*, 6686 (*cf.* Nesmeyanov a. Borisov, C. A. *40*, 2123)
169. Landolt, H.: Liebigs Ann. Chem. *78*, 91 (1851)
170. Landolt, H.: Ibid. *84*, 44 (1852)
171. Friedlander, S.: J. Prakt. Chem. *70*, 449 (1857)
172. Doering, W. E., Hoffmann, A. K.: J. Am. Chem. Soc. *77*, 521 (1955)
173. Chatt, J., Mann, F. G.: J. Chem. Soc. *1940*, 1192
174. Willard, H. H., Perkins, L. R., Blicke, F. F.: J. Am. Chem. Soc. *70*, 737 (1948)
175. Pullman, B. J., West, B. O.: Austr. J. Chem. *17*, 30 (1964)
176. Gruttner, G., Wiernik, M.: Ber. dtsch. chem. Ges. *48*, 1749 (1915)
177. Affsprung, H. E., May, H. E.: Anal. Chem. *32*, 1164 (1960)
178. Affsprung, H. E., Gainer, A. B.: Anal. Chim. Acta *27*, 578 (1962)
179. Doak, G. O., Long, G. G., Freedman, L. D.: J. Organometal. Chem. *12*, 443 (1968)

180. Siebert, H.: Z. Anorg. Allg. Chem. *273*, 161 (1953)
181. Cullen, W. R., Deacon, G. B., Green, J. H. S.: Can. J. Chem. *43*, 3193 (1965)
182. Schmidbaur, H., Weidlein, J., Mitschke, K. H.: Chem. Ber. *102*, 4136 (1969)
183. Schmidbaur, H. et al.: ibid. *106*, 1226 (1973)
184. Schmidbaur, H.: Spectrochim. Acta *20*, 1143 (1964)
185. Orenberg, J. B., Morris, M. D., Long, J. V.: Inorg. Chem. *10*, 933 (1971)
186. Bordner, J., Andrews, B. C., Long, G. G.: Crystal. Struct. Commun. *5*, 801 (1976)
187. Kok, G. L., Morris, M. D., Sharp, R. R.: Inorg. Chem. *12*, 1709 (1973)
188. Goel, R. G., Ridley, D. R.: Inorg. Nucl. Chem. Lett. *7*, 21 (1971)
189. Thayer, J. S.: Organometal. Chem. Rev. *1*, 157 (1966)
190. Schmidt, A.: Chem. Ber. *101*, 3976 (1968)
191. Revitt, D. M., Sowerby, D. B.: Inorg. Nucl. Chem. Lett, *5*, 459 (1969)
192. Revitt, D. M., Sowerby, D. B.: J. Chem. Soc. Dalton Trans. *1972*, 847
193. Goel, R. G. Ridley, D. R.: Inorg. Chem. *13*, 1252 (1974)
194. Hantzsch, A., Hibbert, H.: Chem. Ber. *40*, 1512 (1907)
195. Challanger, F., Smith, A. S., Paton, F. J.: J. Chem. Soc. *123*, 1052 (1923)
196. Wizemann, R. et al: J. Organometal. Chem. *20*, 211 (1969)
197. Ferguson, G., Goel, R. G., Ridley, D. R.: J. Chem. Soc. Dalton Trans. *1975*, 1288
198. Ferguson, G., March, F. C., Ridley, D. R.: Acta Crystallogr. sect. B-, *B31*, 1260 (1975)
199. Ferguson, G. et al.: J. Chem. Soc. D *1971*, 1547
200. Ferguson, G., Ridley, D. R.: Acta Crystallogr. sect. B, *29*, 2221 (1973)
201. Schmidbaur, H. et al.: Z. Anorg. Allg. Chem. *386*, 139 (1971)
202. Goel, R. G.: Cand. J. Chem. *47*, 4607 (1969)
203. Goel, R. G., Prasad, H. S.: Inorg. Chem. *11*, 2141 (1972)
204. Goel, R. G., Prasad, H. S.: J. Organometal. Chem. *59*, 253 (1973)
205. Downs, A. J., Steer, I. A.: ibid. *8*, P21 (1967)
206. Tranter, G. C., Addison, C. C., Sowerby, D. B.: ibid. *12*, 369 (1968)
207. Goel, R. G. et al.: Can. J. Chem. *47*, 1423 (1969)
208. Matsumura, Y., Okawara, R.: J. Organometal. Chem. *25*, 439 (1970)
209. Chremos, G. N., Zingaro, R. A.: ibid. *22*, 637 (1970)
210. Venezky, D. L. et al.: ibid. *35*, 131 (1972)
211. Briles, G. H., McEwen, W. E.: Tetrahed. Lett. *43*, 5299 (1966)
212. Chremos, G. N., Zingaro, R. A.: J. Organometal. Chem. *22*, 647 (1970)
213. Shindo, M., Matsumura, Y., Okawara, R. et al.: *11*, 299 (1968)
214. Zhdanov, G. S. et al.: Dokl. Akad. Nauk SSSR *92*, 983 (1953)
215. Glidwell, C.: J. Organometal. Chem. *116*, 199 (1976)
216. Beauchamp, A. L., Bennett, M. J., Cotton, F. A.: J. Am. Chem. Soc. *91*, 297 (1969)
217. Schmidbaur, H., Weidlein, J., Mitschke, K. H.: Chem. Ber. *102*, 4136 (1969)
218. Schulz, D. N.: Ph. D. Thesis, Univ. of Massachussetts, 1971
219. McEwen, W. E., Chupka, Jr., F. L.: Phosphorus *1*, 277 (1972)
220. Chupka, Jr., F. L., Knapezyk, J. W., McEwen, W. E.: J. Org. Chem. *42*, 1399 (1977)
221. Schmidt, H.: Ber. *55*, 697 (1922)
222. Fargher, R. G., Gray, W. H.: J. Pharmacol. Exp. Ther. *18*, 341 (1921)
223. Macallum, A. D.: J. Chem. Soc. Ind. *42*, 468 (1923)
224. Nakai, R., Toyoda, R., Tomono, H.: Repts. Inst. Chem. Res. Kyoto Univ. *18*, 22 (1949)
225. Gray, W. H., Lamb, I. D.: J. Chem. Soc. *1938*, 401
226. Doak, G. O.: J. Am. Chem. Soc. *68*, 1991 (1946)
227. Ida, M.: Yakugaku Zasshi *69*, 182 (1949)
228. Bowen, L. H., Long, G. G.: Inorg. Chem. *17*, 551 (1978)
229. Petrenko, L. P.: Zh. Obsch. Khim. *24*, 520 (1954)
230. Petrenko, L. P.: Voronezh. Gos. Univ. *49*, 19 (1958); C. A. *56*, 2470
231. Petrenko, L. P.: ibid. *49*, 25 (1958); C. A. *56*, 2471
232. Petrenko, L. P.: ibid. *57*, 145 (1959); C. A. *55*, 6425
233. Freeman, B. H., Lloyd, D., Singer, M. I. C.: Tetrahedron *28*, 343 (1972)
234. Shah, J. J.: J. Tenn. Acad. Sci. *51*, 130 (1976)
235. Radchenko, O. A., Nazaretyan, V. P., Yagupolskii, L. M.: Zh. Obsch. Khim. *46*, 565 (1976)
236. Pinchuk, A. M., Kuplennik, Z. I., Belaya, Zh. N.: ibid. *46*, 2242 (1976)

237. Kuplennik, Z. I., Pinchuk, A. M.: ibid. *49*, 155 (1979)
238. Bajpai, K., Srivastava, R. C.: Synth. React. Inorg. Met.-Org. Chem. *9*, 557 (1979)
239. Dahlmann, J., Winsel, K.: J. Prakt. Chem. *321*, 370 (1979)
240. Johnson, A. W.: Ylid Chemistry. Academic Press, New York 1966
241. Lloyd, D., Singer,M. I. C.: J. Chem. Soc. C *1971*, 2941
242. Lumbroso, H., Lloyd, D., Harris, G. S.: C. R. Hebd. Seances Acad. Sci. Ser. C *278*, 219 (1974)
243. Lloyd, D., Singer, M. I. C.: Chem. Ind. *1967*, 787
244. Lloyd, D., Singer, M. I. C.: ibid. *1968*, 1277
245. Harris, G. S. et al.: ibid. *1968*, 1483
246. Maillard, A. et al.: Bull. Soc. Chim. France *1962*, 843
247. Maillard, A., Deluzarche, A., Maire, J. C.: ibid. *1965*, 2962
248. Kijma, I., Takahashi, N.: Kogyokagaku Zasshi *70*, 796 (1967); C. A. *68*, 45654
249. Laber, R. A., Schmidt, A.: Z. Anorg. Allg. Chem. *409*, 129 (1974)
250. Arbuzov, B. A., Mareev, Yu. M., Vinogradova, V. S.: Izv. Akad. Nauk SSSR, Ser. Khim. *1979*, 170
251. Laber, R. A., Schmidt, A.: Z. Anorg. Allg. Chem. *405*, 71 (1974)
252. Meuwsen, A., Mogling, H.: Z. Anorg. Allg. Chem. *285*, 262 (1956)
253. Hass, D., Cech, D.: Z. Chem. *10*, 75 (1970)
254. Preiss, H.: Z. Anorg. Allg. Chem. *389*, 293 (1972)
255. Beger, F., Latscha, H. P.: Chem. Ztg. *97*, 272 (1973)
256. Paul, R. C., Bhasin, K. K., Sharma, R. D.: Indian J. Chem. *13*, 723 (1975)
257. Heimburger, R., Leroy, M. J. F.: Spectrochim. Acta, Part A *31 A*, 653 (1975)
258. Goetz-Grandmont, G. J., Leroy, M. J. F.: J. Chem. Res. Synop. *1979*, 158
259. Kolditz, L., Preiss, H.: Z. Anorg. Allg. Chem. *311*, 121 (1961)
260. Kolditz, L., Roensch, W.: ibid. *315*, 213 (1962)
261. Arbuzov, B. A., Mareev, Yu. M., Vinogradova, V. S.: Izv. Akad. Nauk SSSR, Ser. Khim. *1977*, 901
262. Arbuzov, B. A. et al.: Phosphorus Sulfur *4*, 53 (1978)
263. Arbuzov, B. A. et al.: Phosphorus Sulfur *5*, 201 (1978)
264. Preiss, H.: Z. Anorg. Allg. Chem. *380*, 65 (1971)
265. Preiss, H.: ibid. *362*, 24 (1968)
266. Paul, R. C., Madan, H., Chadha, S. L.: J. Inorg. Nucl. Chem. *36*, 737 (1974)
267. Paul, R. C., Madan, H., Chadha, S. L.: ibid. *37*, 447 (1975)
268. Paul, R. C., Madan, H., Chadha, S. L.: Indian J. Chem. *13*, 1188 (1975)
269. Chadha, S. L., Mohini, C., Gupta, P. K.: ibid *17 A*, 309 (1979)
270. Laber, R. A., Schmidt, A.: Z. Anorg. Allg. Chem. *416*, 41 (1975)
271. Laber, R. A., Schmidt, A.: ibid. *425*, 117 (1976)
272. Heimburger, R. et al.: Proc. Inst. Conf. Raman Spectroscopy *5 th*, 108 (1976); C. A. *87*, 175184
273. Paul, R. C. et al.: J. Inorg. Nucl. Chem. *38*, 169 (1976)
274. Paul, R. C. et al.: Indian J. Chem. *14 A*, 866 (1976)
275. Paul, R. C., Airon, D. K., Chadha, S. L.: ibid. *14 A*, 51 (1976)
276. Meinema, H. A., Noltes, J. G.: J. Organometal. Chem. *36*, 313 (1972)
277. Stevens, J. G., Trooster, J. M.: Inorg. Chim. Acta *40*, 263 (1980)
278. Schmidbaur, H., Arnold, H. S., Beinhofer, E.: Ber. *97*, 449 (1964)
279. Briles, G. H., McEwen, W. E.: Tetrahedron Lett. 5191 (1966)
280. Shen, K. W. et al.: J. Am. Chem. Soc. *90*, 1718 (1968)
281. McEwen, W. E., Briles, G. H., Giddings, B. E.: ibid. *91*, 7079 (1969)
282. S. Yoshida, S., Kitakawa, S.: Japan 7226183 (1969); C. A. *78*, 16795
283. Ouchi, A. et al.: Sci. Pap. Coll. Gen. Educ. Univ. Tokyo, *25*, 73 (1975); C. A. *86*, 5561
284. Jain, V. K.: Ph. D. Thesis, Rajasthan Univ. Jaipur (1980)
285. Dahlmann, J., Winsel, K.: East. Ger. Pat. 83134; C. A. *78*, 43710
286. Rieche, A., Dahlmann, J., List, D.: Ann. Chem. *678*, 167 (1964)
287. Dahlmann, J., Rieche, A., Austenat, L.: Monatsber. Deut. Akad. Wiss. Berlin *9*, 105 (1967); C. A. *68*, 49680
288. Yoshida, S., Kitakawa S.: Japan 7226384 (1972); C. A. *78*, 16794

289. Dahlmann, J., Austenat, L.: Just. Lieb. Ann. Chem. *729*, 1 (1969)
290. Moffett, K. D., Simmler, J. R., Potratz, H. A.: Anal. Chem. *28*, 1356 (1956)
291. Schmidbaur, H., Schmidt, M.: Angew. Chem. *73*, 655 (1961)
292. Razuvaev, G. A., Osanova, N. A.: J. Organometal. Chem. *38*, 77 (1972)
293. Razuvaev, G. A. et al.: ibid. *99*, 93 (1975)
294. Eberwein, B., Ott, R., Weidlein, J.: Z. Anorg. Allg. Chem. *431*, 95 (1977)
295. Matsumura, Y., Shindo, M., Okawara, R.: Inorg. Nucl. Chem. Lett. *3*, 219 (1967)
296. Schmidbaur, H., Mitschke, K. H.: Chem. Ber. *104*, 1842 (1971)
297. Chatterjee, S.: J. Inst. Chem. (India) *49*, 263 (1977)
298. Schmidbaur, H., Mitschke, K. H.: Chem. Ber. *104*, 1837 (1971)
299. Kravtsov, D. N. et al.: Izv. Akad. Nauk SSSR, Ser. Khim. *1974*, 927
300. Wardell, J. L., Grant, D. W.: J. Organometal. Chem. *140*, C 13 (1978)
301. Kravtsov, D. N. et al.: ibid. *86*, 303 (1975)
302. Wardell, J. L., Grant, D. W.: ibid. *198*, 121 (1980)
303. Wieber, M. et al.: ibid. *133*, 183 (1977)
304. Shindo, M., Okawara, R.: Inorg. Nucl. Chem. Lett. *5*, 77 (1969)
305. Ouchi, A. et al.: Bull. Chem. Soc. Japan *51*, 2427 (1978)
306. Baumstark, A. L., Landies, M. E., Brooks, P. J.: J. Org. Chem. *44*, 4251 (1979)
307. Rieche, A., Dahlmann, J., List, D.: Angew. Chem. *73*, 494 (1961)
308. Dahlmann, J., Rieche, A.: Chem. Ber. *100*, 1544 (1967)
309. Razuvaev, G. A., Zinov'eva, T. I., Brilkina, T. G.: Izv. Akad. Nauk SSSR, Ser. Khim. *1969*, 2007
310. Razuvaev, G. A. et al.: Dokl. Akad. Nauk SSSR, *193*, 355 (1970)
311. Razuvaev, G. A., Zinov'eva, T. I., Brilkina, T. G.: ibid. *188*, 830 (1969)
312. Razuvaev, G. A. et al.: J. Organometal. Chem. *40*, 151 (1972)
313. Dodonov, V. A. et al.: Khim. Elementorg. Soedin *4*, 69 (1976); C. A. *88*, 23077
314. Dahlmann, J., Winsel, K.: J. Prakt. chem. *318*, 390 (1976)
315. Dahlmann, J., Winsel, K.: ibid. *319*, 201 (1977)
316. Starikova, Z. A. et al.: Kristallografiya, *24*, 1211 (1979)
317. Starikova, Z. A. et al.: ibid. *23*, 969 (1978)
318. Laber, R. A., Schmidt, A.: Z. Anorg. Allg. Chem. *407*, 237 (1974)
319. Laber, R. A., Schmidt, A.: Chem. Ber. *108*, 1125 (1975)
320. Laber, R. A., Schmidt, A.: Z. Anorg. Allg. Chem. *414*, 261 (1975)
321. Laber, R. A., Schmidt, A.: Spectrochim. Acta *31 A*, 1589 (1975)
322. Marks, B. S., Schoepfle, B. O.: U. S. 3080406 (1963); C. A. *59*, 8598
323. Bullivant, D. P., Dove, M. F. A., Haley, M. J.: J. Chem. Soc. Chem. Commun. *1977*, 584
324. Schmidt, H.: Liebigs Ann. Chem. *429*, 142 (1922)
325. Okada, T., Okawara, R.: J. Organometal. Chem. *54*, 149 (1973)
326. Chang, M. M. Y., Su, K., Musher, J. I.: Isr. J. Chem. *12*, 967 (1974)
327. Ouchi, A., Honda, H., Kitazima, S.: J. Inorg. Nucl. Chem. *37*, 2559 (1975)
328. Lodochnikova, V. I., Panov, E. M., Kocheshkov, K. A.: Zh. Obsch. Khim. *34*, 946 (1964)
329. Thepe, T. C. et al.: Ohio J. Sci. *77*, 134 (1977); C. A. *87*, 61879
330. Havranek, J., Mleziva, J., Lycka, A.: J. Organometal. Chem. *157*, 163 (1978)
331. Ouchi, A.: Kagakuto Ryoiki *32*, 578 (1978); C. A. *91*, 74664
332. Bajpai, K., Singhal, R., Srivastava, R. C.: Indian J. Chem. *18 A*, 73 (1979)
333. Ouchi, A. et al.: Sci. Pap. Coll. Gen. Educ. Univ. Tokyo *28*, 73 (1978); C. A. *90*, 38371
334. Otera, J., Okawara, R.: J. Organometal. Chem. *17*, 353 (1969)
335. Musher, J., Su, K.: U. S. 3939190 (1976); C. A. *84*, 181136
336. Goel, R. G., Ridley, D. R.: J. Organometal. Chem. *38*, 83 (1972)
337. Okada, T., Okawara, R.: ibid. *42*, 117 (1972)
338. Goel, R. G., Ruddick, J. N. R., Sams, J. R.: J. Chem. Soc. Dalton Trans. *1975*, 67
339. Sowerby, D. B.: J. Chem. Res. Synop. *1979*, 80
340. Matsumura, Y., Shindo, M., Okawara,R.: J. Organometal. Chem. *27*, 357 (1971)
341. Schmidbaur, H., Mitschke, K. H., Weidlein, J.: Z. Anorg. Allg. Chem. *386*, 147 (1971)
342. Bone, S. P., Sowerby, D. B.: J. Chem. Res. Synop. *1979* 82
343. Schmidbaur, H., Mitschke, K. H.: Angew. Chem. *10*, 136 (1971)
344. Rosenheim, A., Lolwenstamm, W., Singer, L.: Ber. *36*, 1833 (1903)

345. Dilthey, W.: Ber. *36*, 923 (1903); *ibid* 1595 (1903); *ibid* 3207 (1903)
346. Kawasaki, Y., Tanaka, T., Okawara, R.: Spectrochim. Acta *22*, 1571 (1966)
347. Kawasaki, Y., Okawara, R.: Bull. Chem. Soc. Japan *40*, 428 (1967)
348. Kawasaki, Y., Tanaka, T., Okawara, R.: ibid, *40*, 1562 (1967)
349. Meinema, H. A., Noltes, J. G.: J. Organometal. Chem. *16*, 257 (1969)
350. Hammel, J. C., Smith, J. A. S., Wilkins, E. J.: J. Chem. Soc. A *1969*, 1461
351. Meinema, H. A., Mackor, A., Noltes, J. G.: J. Organometal. Chem. *37*, 285 (1972)
352. Okawara, R., Matsumura, Y., Nishii, N.: Proc. IV Int. Conf. Organometal. Chem., Bristol *1969*, Z 8
353. Nishii, N. et al.: Inorg. Nucl. Chem. Lett. *5*, 529 (1969)
354. Kawasaki, Y., Ito, T., Okawara, R.: Decomps. Organometal. Compds. Refract Cream, Metals, Metal Alloys Proc. Inst. Symp. *1968*, 47; C. A. *72*, 62366
355. Kanehisa, N., Kai, Y., Kasai, N.: Inorg. Nucl. Chem. Lett. *8*, 375 (1972)
356. Kanehisa, N. et al.: Bull. Chem. Soc. Japan *51*, 2222 (1978)
357. Nishii, N., Matsumura, Y., Okawara, R.: Inorg. Nucl. Chem. Lett. *5*, 703 (1969)
358. Nishii, N., Okawara, R.: J. Organometal. Chem. *38*, 335 (1972)
359. Meinema, H. A., Noltes, J. G.: ibid. *37*, C 31 (1972)
360. Uda, S. et al.: Cryst. Struct. Commun. *3*, 257 (1974)
361. Onuma, K., Kai, Y., Kasai, N.: Inorg. Nucl. Chem. Lett. *8*, 143 (1972)
362. Kroon, J., Hulscher, J. B., Peerdman, A. F.: J. Organometal. Chem. *37*, 297 (1972)
363. Ruddick, J. N. R., Sams, J. R.: Inorg. Nucl. Chem. Lett. *11*, 229 (1975)
364. Meinema, H. A., Noltes, J. G.: J. Organometal. Chem. *160*, 435 (1978)
365. Mackor, A., Meinema, H. A.: Recl. Trav. Chim. Pays-Bas, *91*, 911 (1972); C. A. *77*, 125478
366. Meinema, H. A., Mackor, A., Noltes, J. G.: J. Organometal. Chem. *70*, 79 (1974)
367. Meinema, H. A., Noltes, J. G.: ibid. *55*, C 77 (1973)
368. Jain, V. K., Bohra, R., Mehrotra, R. C.: Unpublished results
369. Goel, R. G., Ridley, D. R.: J. Organometal. Chem. *182*, 207 (1979)
370. Jain, V. K., Bohra, R., Mehrotra, R. C.: ibid. *184*, 57 (1980)
371. Matsumura, Y., Okawara, R.: Inorg. Nucl. Chem. Lett. *4*, 521 (1968)
372. Ebina, F. et al.: J. Chem. Soc. Chem. Commun. *1976*, 245
373. Ebina, F. et al.: Acta Crystallogr. sect. B *B 33*, 3252 (1977)
374. Ebina, F. et al.: ibid. *34 B*, 1512 (1978)
375. Ebina, F. et al.: ibid. *34 B*, 2134 (1978)
376. Di Bianca, F. et al.: Atti. Accad. Sci. Lett. Arti. Palermo, Part 1 *33*, 173 (1973); C. A. *83*, 114572
377. Meinema, H. A. et al.: J. Organometal. Chem. *107*, 249 (1976)
378. Bertazzi, N. et al.: J. Chem. Soc. Dalton Trans. *1977*, 957
379. Di Bianca, F., Rivarola, E.: Atti. Accad. Sci. Lett. Arti. Palermo, Part 1 *1972*, 167; C. A. *79*, 142431
380. Di Bianca, F. et al.: J. Organometal. Chem. *63*, 293 (1974)
381. Jain, V. K., Bohra, R., Mehrotra, R. C.: Aust. J. Chem. *33*, 2749 (1980)
382. Jain, V. K., Bohra, R., Mehrotra, R. C.: Unpublished Results
383. Ruddick, J. N. R., Sams, J. R.: J. Organometal. Chem. *128*, C 41 (1977)
384. Gopinathan, S., Gopinathan, C.: Indian J. Chem. 15 A, 660 (1977)
385. Meinema, H. A., Rivarola, E., Noltes, J. G.: J. Organometal. Chem. *17*, 71 (1969)
386. Kawasaki, Y.: Inorg. Nucl. Chem. Lett. *5*, 805 (1969)
387. Kawasaki, Y., Hashimoto, K.: J. Organometal. Chem. *99*, 107 (1975)
388. Kawasaki, Y.: Bull. Chem. Soc. Japan *49*, 817 (1976)
389. Kawasaki, Y.: ibid. *49*, 2319 (1976)
390. Kawasaki, Y., Takahashi, T., Fujioka, T.: J. Organometal. Chem. *131*, 239 (1977)
391. Harrison, P. G., Zuckerman, J. J.: Inorg. Nucl. Chem. Lett. *6*, 5 (1970)
392. Jain, V. K., Bohra, R., Mehrotra, R. C.: J. Indian Chem. Soc. *57*, 408 (1980)
393. Jain, V. K., Bohra, R., Mehrotra, R. C.: Inorg. Chim. Acta, *51* (1981)
394. Eberwein, B., Sille, F., Weidlein, J.: Z. Naturforsch., B *31 B*, 689 (1976)
395. Ferguson, G., Hawley, D. M.: Acta Crystallogr. sect. B *30*, 103 (1974)
396. Paul, R. C. et al.: Indian J. Chem. *18 A*, 77 (1979)
397. Laber, R. A., Schmidt, A.: Z. Anorg. Allg. Chem. *416*, 32 (1975)

398. Klein, W., Krauss, D., Latscha, H. P.: ibid. *401*, 85 (1973)
399. Kruss, B., Ziegler, M. L.: ibid. *401*, 89 (1973)
400. Ouchi, A. et al.: Bull. Chem. Soc. Japan *51*, 3511 (1978)
401. Stevens, J. G., Trooster, J. M.: J. Chem. Soc. Dalton Trans. *1979*, 740
402. Cras, J. A., Willemse, J.: Recl. Trav. Chim. Pays-Bas *97*, 28 (1978); C. A. *88*, 136742
403. Laber, R. A., Schmidt, A.: Z. Anorg. Allg. Chem. *428*, 209 (1977)
404. Graves, G. E., Van Wazer, J. R.: J. Organometal. Chem. *131*, 31 (1977)
405. Eberwein, B., Weidlein, J.: Z. Anorg. Allg. Chem. *420*, 229 (1976)
406. Graves, G. E., Van Wazer, J. R.: J. Organometal. Chem. *150*, 233 (1978)
407. Schwarz, W., Hausen, H. D.: Z. Anorg. Allg. Chem. *441*, 175 (1978)

Received September 25, 1981
R. J. P. Williams (editor)

## Author-Index Volumes 1–52

III which on heating yields II. All these products have been characterized by IR, Raman and $^{1}H$ NMR studies[397]. Mass spectrum of I has also been elucidated[398]. Structure I, (N,N′-dimethyloxamido)antimony tetrachloride, has been determined by single-crystal X-ray diffraction data[399]. Crystals of the compound belong to monoclinic system with space group $C^{5}_{2h}$-$P2_{1/n}$, with the antimony in an octahedral environment and the N,N′-dimethyloxamido ligand being N,O bonded.

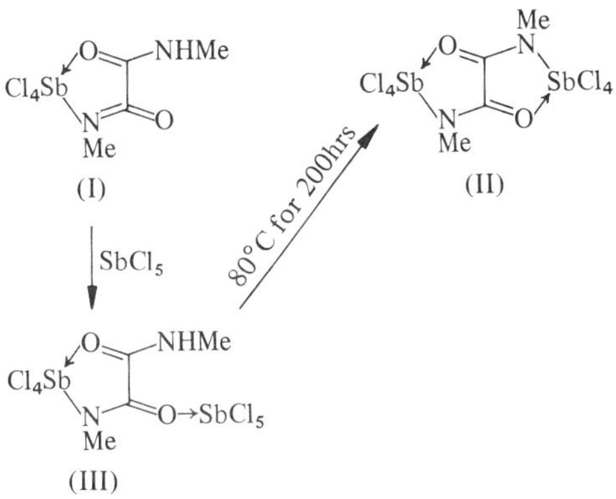


Some trimethylantimony(V) bis dithiocarbamates have been prepared by the reactions of $Me_3SbBr_2$ with ML (M = Na or K and L ≙ $RCS_2$, R = piperidino, pyrrolidino, PrO, $Me_2CHO$, $Me_2CHCH_2O$, $PhCH_2O$)[400]. A trigonal bipyramidal geometry has been proposed for these complexes. Mössbauer spectra of $Me_nSb(S_2CNR_2)_{5-n}$ (n = 3 or 4 and R = Me or Et) have recently been determined[401] and the observed isomer shifts, quadrupole coupling constants and asymmetry parameters have been interpreted with structure and bonding of the ligand to antimony atom. A slightly deformed trigonal bipyramidal geometry with three equatorial methyls has further been confirmed for trimethylantimony(V) bis N,N′-dimethyldithiocarbamate by X-ray analysis[402].

On heating, the 1:1 addition compound, $Cl_5Sb \cdot OPCl_n(OR)_{3-n}$ (R = Me or Et, n = 0–2) yields RCl and the dimers, $[Cl_4SbO_2PCl_2]_2$ and $[Cl_4SbO_2PCl(OR)]_2$ respectively, when n = 2 and 1. Both dimeric and tetrameric derivatives $[Cl_4SbO_2P(OR)_2]_2$ and $[Cl_3SbO_3P(OR)]_4$ are obtained, when n = O. The reaction of $K[O_2P(OR)_2]$ with $SbCl_5$ yields $[Cl_4SbO_2P(OR)_2]_2$ in methylene dichloride. These compounds have been investigated by IR, $^{1}H$ and $^{31}P$ NMR spectral data[403].

Graves and Van Wazer[404] have prepared triorganoantimony derivatives of phosphinic acids of the type, $R_3Sb[OP(O)(H)R']_2$ (R = Me or Ph and R′ = Ph or PhCH=CH). These compounds have been characterized by $^{1}H$ and $^{31}P$ NMR data, indicating the presence of Sb-O-P rather than Sb-P linkages.

Reactions of pentamethylantimony with phosphinic and phosphonic acids have been investigated[405, 406]. On the basis of IR, Raman, $^{31}P$ and $^{19}F$ NMR studies, an ionic structure in which $Me_4Sb^+$ cations have partly distorted tetrahedral geometry has been proposed for $Me_4Sb(O_2PF_2)$, $Me_4Sb(O_2PCl_2)$ and $Me_4Sb(O_2PH_2)$[405] $MeEt_3Sb(O_2PF_2)$[32]. However, a non-ionic structure has been proposed for $Me_4Sb(O_2PPh_2)$[406], $Me_4Sb(O_2PMe_2)$[405] and $Me_4Sb(OP(S)Me_2)$[405] $Et_3MeSb(O_2PMe_2)$[32]. Compounds of the type $Me_4Sb(O_2P(OH)R)$ (R = Me or Ph) are associated probably through hydrogen bonds[406].

# NMR

## Basic Principles and Progress
## Grundlagen und Fortschritte

Editors:
**P. Diehl, E. Fluck, R. Kosfeld**

Volume 12
**B. Lindman, S. Forsén**
### Chlorine, Bromine and Iodine NMR
Physico-Chemical and Biological Applications
1976. 74 figures, 29 tables. XIV, 368 pages
ISBN 3-540-07725-1

Volume 13
### Introductory Essays
Editor: **M. M. Pintar**
1976. 48 figures. XI, 154 pages
ISBN 3-540-07754-5

Volume 14
**H. Nöth, B. Wrackmeyer**
### Nuclear Magnetic Resonance Spectroscopy of Boron Compounds
1978. 1 figure, 96 tables. XII, 461 pages
ISBN 3-540-08456-8

Volume 15
**H. W. Spiess, A. Steigel**
### Dynamic NMR Spectroscopy
Corrected Reprint. 1982.
Approx. 76 figures. Approx. 220 pages
ISBN 3-540-08784-2

Volume 16
**P. S. Pregosin, R. W. Kunz**
### $^{31}$P and $^{13}$C NMR Transition Metal Phosphine Complexes
1979. 26 figures, 37 tables. IX, 156 pages
ISBN 3-540-09163-7

Volume 17
**J.-P. Kintzinger, H. Marsmann**
### Oxygen–17 and Silicon–29
1981. 31 figures. V, 235 pages
ISBN 3-540-10414-3

Volume 18
**G. J. Martin, M. L. Martin, J.-P. Gouesnard**
### $^{15}$N-NMR Spectroscopy
1981. 11 figures, 142 tables. VII, 382 pages
ISBN 3-540-10459-3

Volume 19
### NMR in Medicine
Editor: **R. Damadian**
1981. 77 figures. V, 174 pages
ISBN 3-540-10460-7

Volume 20
**G. Govil, R. V. Hosur**
### Conformation of Biological Molecules
New Results from NMR
1982. 92 figures. VIII, 216 pages
ISBN 3-540-10769-X

Springer-Verlag
Berlin
Heidelberg
New York